Lecture Notes in Physics

The Editorial Policy for Proceedings

The series Lecture Notes in Physics reports new developments in physical research and teaching – quickly, informally, and at a high level. The proceedings to be considered for publication in this series should be limited to only a few areas of research, and these should be closely related to each other. The contributions should be of a high standard and should avoid lengthy redraftings of papers already published or about to be published elsewhere. As a whole, the proceedings should aim for a balanced presentation of the theme of the conference including a description of the techniques used and enough motivation for a broad readership. It should not be assumed that the published proceedings must reflect the conference in its entirety. (A listing or abstracts of papers presented at the meeting but not included in the proceedings could be added as an appendix.)

When applying for publication in the series Lecture Notes in Physics the volume's editor(s) should submit sufficient material to enable the series editors and their referees to make a fairly accurate evaluation (e.g. a complete list of speakers and titles of papers to be presented and abstracts). If, based on this information, the proceedings are (tentatively) accepted, the volume's editor(s), whose name(s) will appear on the title pages, should select the papers suitable for publication and have them refereed (as for a journal) when appropriate. As a rule discussions will not be accepted. The series editors and Springer-Verlag will normally not interfere with the detailed editing except in fairly obvious cases or on technical matters.

Final acceptance is expressed by the series editor in charge, in consultation with Springer-Verlag only after receiving the complete manuscript. It might help to send a copy of the authors' manuscripts in advance to the editor in charge to discuss possible revisions with him. As a general rule, the series editor will confirm his tentative acceptance if the final manuscript corresponds to the original concept discussed, if the quality of the contribution meets the requirements of the series, and if the final size of the manuscript does not greatly exceed the number of pages originally agreed upon.

The manuscript should be forwarded to Springer-Verlag shortly after the meeting. In cases of extreme delay (more than six months after the conference) the series editors will check once more the timeliness of the papers. Therefore, the volume's editor(s) should establish strict deadlines, or collect the articles during the conference and have them revised on the spot. If a delay is unavoidable, one should encourage the authors to update their contributions if appropriate. The editors of proceedings are strongly advised to inform contributors about these points at an early stage.

The final manuscript should contain a table of contents and an informative introduction accessible also to readers not particularly familiar with the topic of the conference. The contributions should be in English. The volume's editor(s) should check the contributions for the correct use of language. At Springer-Verlag only the prefaces will be checked by a copy-editor for language and style. Grave linguistic or technical shortcomings may lead to the rejection of contributions by the series editors.

A conference report should not exceed a total of 500 pages. Keeping the size within this bound should be achieved by a stricter selection of articles and not by imposing an upper limit to the length of the individual papers.

Editors receive jointly 30 complimentary copies of their book. They are entitled to purchase further copies of their book at a reduced rate. As a rule no reprints of individual contributions can be supplied. No royalty is paid on Lecture Notes in Physics volumes. Commitment to publish is made by letter of interest rather than by signing a formal contract. Springer-Verlag secures the copyright for each volume.

The Production Process

The books are hardbound, and quality paper appropriate to the needs of the authors is used. Publication time is about ten weeks. More than twenty years of experience guarantee authors the best possible service. To reach the goal of rapid publication at a low price the technique of photographic reproduction from a camera-ready manuscript was chosen. This process shifts the main responsibility for the technical quality considerably from the publisher to the authors. We therefore urge all authors and editors of proceedings to observe very carefully the essentials for the preparation of camera-ready manuscripts, which we will supply on request. This applies especially to the quality of figures and halftones submitted for publication. In addition, it might be useful to look at some of the volumes already published.

As a special service, we offer free of charge LATEX and TEX macro packages to format the text according to Springer-Verlag's quality requirements. We strongly recommend that you make use of this offer, since the result will be a book of considerably improved technical quality.

To avoid mistakes and time-consuming correspondence during the production period the conference editors should request special instructions from the publisher well before the beginning of the conference. Manuscripts not meeting the technical standard of the series will have to be returned for improvement.

For further information please contact Springer-Verlag, Physics Editorial Department V, Tiergartenstrasse 17, W-6900 Heidelberg, FRG

R. Schmidt H. O. Lutz R. Dreizler (Eds.)

Nuclear Physics Concepts in the Study of Atomic Cluster Physics

Proceedings of the 88th WE-Heraeus-Seminar
Held at Bad Honnef, FRG, 26-29 November 1991

Springer-Verlag
Berlin Heidelberg GmbH

Editors

Rüdiger Schmidt
Institut für Theoretische Physik, Technische Universität Dresden
Mommsenstraße 13, O-8027 Dresden, Fed. Rep. of Germany

Hans O. Lutz
Fakultät für Physik, Universität Bielefeld
Universitätsstraße 25, W-4800 Bielefeld, Fed. Rep. of Germany

Reiner Dreizler
Institut für Theoretische Physik, Universität Frankfurt
Robert-Mayer-Straße 8-10, W-6000 Frankfurt a. M. 1, Fed. Rep. of Germany

ISBN 978-3-662-13912-7 ISBN 978-3-540-47264-3 (eBook)
DOI 10.1007/978-3-540-47264-3

Originally published by Springer-Verlag Berlin Heidelberg New York in 1992
Softcover reprint of the hardcover 1st edition 1992

Typesetting: Camera ready by author/editor

58/3140-543210 - Printed on acid-free paper

Preface

This volume contains papers presented at the 88th WE-Heraeus-Seminar "Nuclear Physics Concepts in Atomic Cluster Physics" in Bad Honnef, November 26–29, 1991.

Cluster physics, a field with an inherently interdisciplinary nature, is of rapidly growing scientific interest. Aspects of atomic and molecular physics, as well as condensed matter and chemical physics, are important for understanding the rich variety of atomic cluster phenomena. Moreover, it has become apparent that phenomena in atomic nuclei often have corresponding analogies in atomic clusters. Impressive examples are shell closing effects, collective excitations, fission, and scattering processes. It is thus not surprising that concepts originally developed in nuclear physics provide very powerful tools for the investigation and interpretation of atomic cluster properties. A discussion of nuclear and cluster phenomena, aiming at increasing the interaction between the two communities, could therefore cross-fertilize both cluster and nuclear physics, and thus also act as an important stimulus for both fields. To this end, approximately 65 scientists from the two fields met in Bad Honnef to discuss such problems of common interest. In order to provide the scientific program with a clear structure, it was organized in four topical sections:

- *Electronic and Nuclear Shells:*

  Shell closing effects and the corresponding special stabilities ("magic numbers") provide the best-known analogy between nuclear and cluster physics. Recently, shell effects have been found in atomic clusters with up to several thousand particles, giving evidence of "super shells".

- *Fission of Clusters and Nuclei:*

  Coulomb explosion of clusters is still a poorly understood phenomenon. It is supposedly quite similar to nuclear fission, although electronic shell structures are expected to play a dominant role.

- *Cluster Collisions and Nuclear Reactions:*

  The physics of cluster collisions is rather new territory. Here a very lively and promising development is taking place. Many fascinating analogous phenomena have been found, for example the dominant reaction channels in collisions between clusters, or the break-up processes in cluster-atom collisions.

- *Structures and Excited States of Clusters and Nuclei:*

  An intense discussion centered around the traditional areas of atomic and electronic cluster structure, which still provide some of the most challenging problems to theory as well as experiment. Good progress has been made in the interpretation of collective excitation in clusters ("giant resonances"), strongly aided by nuclear theory, which is also providing interesting new concepts and viewpoints.

The discussions were of high quality, always lively and sometimes quite controversial, partly due to the different "languages" spoken in the different scientific areas; particularly this latter fact once more confirmed the importance of such a meeeting as a forum for clarifying interaction.

The Physikzentrum provided an excellent frame for a meeting in a very friendly and stimulating atmosphere. Our special thanks are due to the WE-Heraeus-Stiftung for providing the organizational frame. We gratefully acknowledge financial support from the Ministerium für Wissenschaft und Forschung (MWF) of the State of Nordrhein-Westfalen.

Bielefeld
March 1992

R. Schmidt
H.O. Lutz
R. Dreizler

Contents

1. Shells

2. Fission

3. Reactions

4. Structure and Excitation

List of Participants

J.A. Alonso	Departamento de Fisica Teórica Universidad de Valladolid 47011 Valladolid Spain
G.S. Anagnostatos	NCSR Demokritos Institute of Nuclear Physics Aghia Pardskevi 15310 Greece
W. Andreoni	IBM Research Division Zürich Research Laboratory Säumerstr. 4 8803 Rüschlikon Switzerland
K.H. Bennemann	Fachbereich Physik (WE 5) Freie Universität Berlin Arnimallee 14 W–1000 Berlin 33 Germany
I. Bergström	Manne Siegbahninstitutet for Fysik Frescativägen 24 104 05 Stockholm Sweden
S. Bjørnholm	Niels Bohr Institutet Blegdamsvej 17 2100 Copenhagen Ø Denmark
V. Bonačić-Koutecký	Fachbereich Chemie Freie Universität Berlin Takustr. 3 W–1000 Berlin 33 Germany
G. Borstel	Fachbereich Physik Universität Osnabrück 4500 Osnabrück Germany
K.H. Bowen	Chemistry Department The Johns Hopkins University Baltimore, MD 21218 USA
M. Brack	Institut für Theoretische Physik Universität Regensburg Universitätsstr. 31 8400 Regensburg Germany

C. Bréchignac	Centre National de la Recherche Scientifique Laboratoire Aimé Cotton Campus d'Orsay — Bâtiment 505 91405 Orsay Cedex France
U. Buck	Max-Planck-Institut für Strömungsforschung Bunsenstr. 10 3400 Göttingen Germany
E.E.B. Campbell	Fakultät für Physik Albert-Ludwigs Universität Hermann-Herder Straße 3 7800 Freiburg Germany
R. Dreizler	Institut für Theoretische Physik Universität Frankfurt Robert-Mayer-Str. 8–10 6000 Frankfurt a. M. 11 Germany
W. Ekardt	Fritz-Haber-Institut der Max-Planck-Gesellschaft Faradayweg 4–6 W-1000 Berlin 33 Germany
F. Garcias	Depart. de Fisica Universidad Illes Balears 07071 Palma de Mallorca Spain
O. Genzken	Institut für Theoretische Physik Universität Regensburg Universitätsstr. 31 8400 Regensburg Germany
D. Gerlich	Fakultät für Physik Albert-Ludwigs Universität Hermann-Herder-Straße 3 7800 Freiburg Germany
W. Greiner	Institut für Theoretische Physik Universität Frankfurt Robert-Mayer-Str. 8–10 6000 Frankfurt a. M. 11 Germany

D.H.E. Groß	Hahn-Meitner-Institut für Kernforschung Glienicker Str. 100 W-1000 Berlin 39 Germany
J. Gspann	Institut für Kernverfahrenstechnik Postfach 3640 7500 Karlsruhe 1 Germany
C. Guet	DRFMC/SPHAT Centre d'Etudes Nucléaires 85X 38041 Grenoble Cedex France
H. Haberland	Fakultät für Physik Universität Freiburg Hermann-Herderstr. 3 7800 Freiburg Germany
E. Hammaren	Department of Physics University of Jyväskylä P. O. Box 35 40351 Jyväskylä Finland
R. Hasse	GSI Darmstadt Postfach 110552 6100 Darmstadt 11 Germany
M. Irion	Physikalische Chemie Technische Hochschule Darmstadt 6100 Darmstadt Germany
J. Jellinek	Chemistry Division Argonne National Laboratory Argonne, Illinois 60439 USA
R.O. Jones	IFF — Forschungszentrum Jülich GmbH Postfach 1913 5170 Jülich Germany
M. Kappes	Department of Chemistry Northwestern University 2145 Sheridan Rd. Evanston, Illinois 60208 USA

I. Katakuse	Department of Physics Faculty of Science Osaka University, Toyonaka-shi 560 Osaka Japan
H.-J. Kluge	Institut für Physik Universität Mainz Staudingerweg 7 6500 Mainz 1 Germany
O. Knospe	Institut für Theoretische Physik Technische Universität Dresden Mommsenstr. 13 O–8027 Dresden Germany
T. Kondow	Department of Chemistry Faculty of Sciences University of Tokyo Tokyo 113 Japan
U. Kreibig	I. Physikalisches Institut Technische Hochschule Aachen Sommerfeldstraße, Turm 28 5100 Aachen 1 Germany
V.V. Kresin	Department of Physics University of California Berkeley, CA 94720 USA
U. Lammers	Fachbereich Physik Universität Osnabrück Postfach 4469 4500 Osnabrück Germany
U. Landman	School of Physics Georgia Institute of Technology Atlanta, GA 30332 USA
E. Lipparini	Dipartimento di Fisica Università degli Studi di Trento 38050 Povo Italy
H.O. Lutz	Fakultät für Physik Universität Bielefeld Universitätsstraße 25 4800 Bielefeld 1 Germany

K. Lützenkirchen
Institut für Physik
Universität Mainz
Staudingerweg 7
6500 Mainz 1
Germany

T.P. Martin
Max-Planck-Institut für Festkörperforschung
Heisenbergstr. 1
7000 Stuttgart 80
Germany

T.D. Märk
Institut für Ionenphysik
Universität Innsbruck
Technikerstr. 25
6020 Innsbruck
Austria

K.H. Meiwes-Broer
Fakultät für Physik
Universität Bielefeld
Universitätsstr. 25
4800 Bielefeld 1
Germany

K. Möhring
Hahn-Meitner-Institut für Kernforschung Berlin
Glienicker Str. 100
W-1000 Berlin 39
Germany

B. Mühlschlegel
Institut für Theoretische Physik
Universität Köln
Zülpicher Str. 77
5000 Köln 41
Germany

S. Ohnishi
NEC Corporation
34 Miyukigaoka, Tsukuba
Ibaraki 305
Japan

J.M. Pacheco
Fritz-Haber-Institut der Max-Planck-Gesellschaft
Faradayweg 4–6
W-1000 Berlin 33
Germany

E. Recknagel
Fachbereich Physik
Universität Konstanz
Bücklestr. 13
7750 Konstanz
Germany

P.-G. Reinhard
Institut für Theoretische Physik
Universität Erlangen
Staudtstr. 7
8520 Erlangen
Germany

V. A. Rubchenya
V.G. Khlopin Radium Institute
Prospekt Shvernika 28
Leningrad
SU–194021
USSR

A. Sandulescu
Institut für Theoretische Physik
Universität Frankfurt
Robert-Mayer-Str. 8-10
6000 Frankfurt a. M.
Germany

W.A. Saunders
California Institute of Technology
Applied Physics
Mail Code 128–95
Pasadena, CA 91125
USA

R. Schmidt
Institut für Theoretische Physik
Technische Universität Dresden
Mommsenstr. 13
O–8027 Dresden
Germany

M. Schöne
Fakultät für Physik
Universität Bielefeld
Universitätsstraße 25
4800 Bielefeld 1
Germany

G. Seifert
Institut für Theoretische Physik
Technische Universität Dresden
Mommsenstr. 13
O–8027 Dresden
Germany

N. Takahashi
College of General Education
Osaka University
Toyonaka — Osaka, 560
Japan

M. Vollmer
Fachbereich Physik
Universität Kassel
Postfach 101380
3500 Kassel
Germany

1. Shells

Clusters in Nuclear Physics: From Nuclear Molecules to Cluster Radioactivities

Walter Greiner
Institut für Theoretische Physik der Universität
Frankfurt am Main, Germany

1 Introductory Remarks

Nuclear molecules have been first experimentally discovered by Bromley et al. [1] in the scattering of ^{12}C on ^{12}C, as resonant phenomena. Other systems where they could be observed since then include $^{17}O + ^{12}C$, $^{16}O + ^{16}O$, $^{28}Si + ^{28}Si$, $^{24}Mg + ^{24}Mg$, etc [2]. Resonances in the scattering excitation functions typically exhibit gross structure widths ($\Gamma_{CM} \sim 2$ - 4 MeV) and superimposed intermediate structure widths ($\Gamma_{CM} \sim 0.5$ MeV). The correlation of structures between different exit channels have been successfully explained by the double resonance mechanism [3] based on two centers shell modell (TCSM) [4] [5].

The principle behind an atomic molecule is well known. There are usually two or more centers (the nuclei) around which electrons are orbiting and thus binding them. An atomic quasimolecule is a short-living intermediate system, formed in nuclear scattering, in which electrons feel - during the collision - the binding of both atomic nuclei. Similarly, in a nuclear molecule the outermost bound nucleons are orbiting around both nuclear cores, binding them together for times τ(mol) longer than the collision time τ(coll). Depending on the ratio of these times, a whole variety of more or less pronounced intermediate phenomena do occur : from molecules of virtual and quasibound type to longer living intermediate, nearly compound systems.

The analogy to the atomic case should be stressed. There the potential minimum between the atoms comes from certain two-center orbitals with lower energy, which expresses the binding effect due to valence electrons. Similarly, in the nuclear case the two-center shell structure of the various valence nucleons causes the additional binding. The nuclear molecules are, however, much more complex than the atomic ones. Because of the strong force many more nucleons participate in the molecular interaction, depending on the overlap and fragmentation of the two subsystems. This leads to various types of nuclear molecules. Also there are many more closed and open channels and collective degrees of freedom, which damp the molecular states more than in atomic physics. At higher energies compression effects become important.

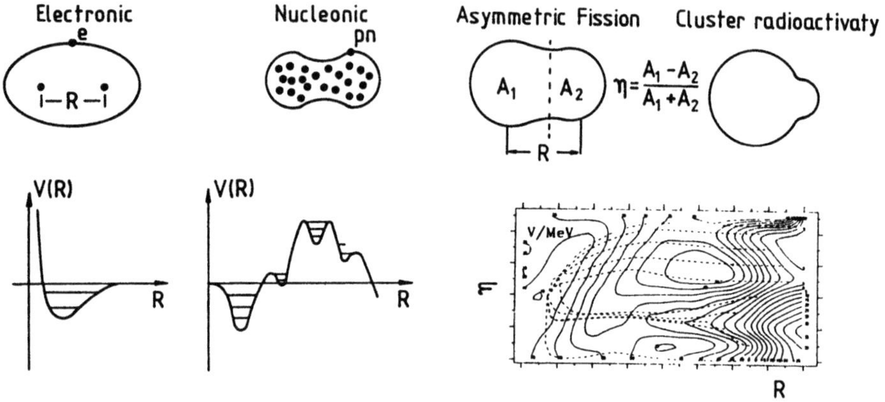

Figure 1: Electronic and nucleonic molecules. Shape isomers are also nuclear molecules. Shapes and potentail energy surfaces for asymmetric fission and cluster radioactivity indicate intermediate molecular configurations. The radial R and mass asymmetry η coordinates in the fragmentation theory are illustrated. The last diagram indicates the focussing collective flow leading to asymmetric or superasymmetric fission or even to cluster decay. The valley at $\eta = 0.8$ corresponds to cluster radioactivity and at $\eta = 0.2$ to fission. Only the half of the figure (for $\eta > 0$) was plotted; the other half is symmetric.

Even the fission isomers or the recently discovered shape isomers (superdeformed or hyperdeformed shapes produced at very high spins) as well as the nuclear systems exhibiting cluster radioactivities, might be considered to be nuclear molecules. At least there are quasimolecular stages in a fission and cluster decay process. For such objects it is essential that the potential energy has a certain "pocket", i. e. one or more minima, as a function of the two-center distance R, or as a function of the mass asymmetry coordinate $\eta = (A_1 - A_2)/(A_1 + A_2)$. In these minima the system is captured for some time during their relative motion (see Fig. 1), or the collective flow is channeled (focussed) during the separation process.

In order to explain the ground state deformations of many nuclei, the asymmetric distribution of fission fragments, fission isomers, intermediate structure resonances in fission cross-sections, etc, it is important to take into account both the collective and single-particle aspects of nucleonic motion, or , in other words, to add a shell and pairing correction to the liquid-drop energy [6] when the potential barrier is calculated. Otherwise, within the liquid-drop model, the nuclear shapes are always highly symmetric : spherical in the ground state, axially and reflection symmetric at the saddle point of the interaction potential. By using the TCSM for describing the single-particle states, one can follow the shell structure all the way from the original nucleus, over the potential barriers, up to the final stage of individual well separated fragments. Within fragmentation theory [7] [8] all kinds of fission fragment mass distributions experimentally observed (symmetric, asymmetric, superasymmetric) have been explained. Typical shapes are shown in Fig. 1. Superasymmetric fission and bimodal fission (experimentally discovered by Hulet et al.) [18] [19] are particular forms of the influence of shell structure on the dynamics of low energy collective motion. The basic information about these phenomena can be found in Figs. 2 and 5 - 8. The cold fission phenomenon is intimately connected with the high energy mode in bimodal fission and also with cluster radioactivity dicussed further below. In fact, extrapolating from asymmetric to superasymmetric fission and further to extreme asymmetric break-up of a nucleus we are lead to cluster radioactivity; in particular to the inside that there is no difference between fission and radioactivity.

Cluster radioactivity has been predicted in a classical paper by Sandulescu, Poenaru and Greiner [9] . Among other examples given in this paper, it was shown that ^{14}C should be the most probable cluster to be emitted from ^{222}Ra and ^{224}Ra. Four years later, Rose and Jones [10] detected ^{14}C radioactivity of ^{223}Ra. Since then, other types of emissions (O, Ne, Mg, Si) have been identified. Theories and experiments have been recently reviewed [11] [12] [13]. Cluster radioactivities to excited states of the daughter and of the cluster, predicted by Martin Greiner and Werner Scheid [14] have also been observed as a fine structure [15].

Other experimental facts prove the presence of strong shell effects in cold rearrangement processes. Cold fusion with one of the reaction partner ^{208}Pb was of practical importance for the synthesis of the heaviest elements [16]. This has been suggested rather early [17]. ^{132}Sn and its neighbours plays a key role in cold and bimodal fission phenomena [18] [19].

A pocket due to shell effects is present in the potential barrier of superheavy nuclei for which liquid drop barrier vanishes. It was estimated that magic numbers for protons and neutrons higher than 82; 126, are 114; 184, respectively, hence ^{298}114 is supposed to be double magic. Unfortunately no experimental evidence was obtained up to now in spite of a long search in nature and many attempts to produce superheavies by nuclear reactions.

From the rich variety of interconnected phenomena we have mentioned only few; some others will be presented in the following sections.

2 The Nuclear Two-Center Shell Model

A single-particle shell model asymptotically correct for fission and heavy-ion reactions was developed [4] [5] on the basis of the two-center oscillator, previously used in molecular

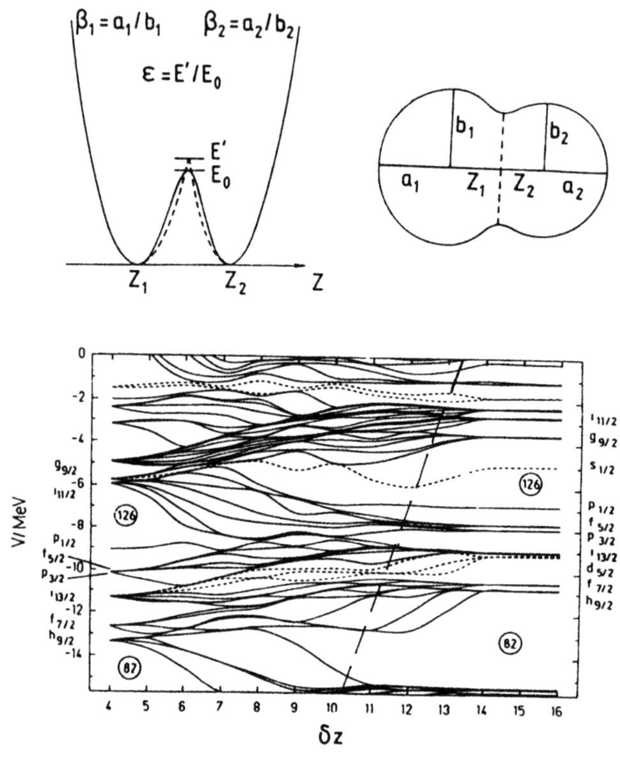

Figure 2: Two-center shell model potential along the z-axis, nuclear shape and level diagram for superasymmetric fission of ^{232}U with the light fragment ^{24}Ne. The levels leading to final states of the light fragment are plotted with dashed-lines. The asymptotic shells are reaching for inwards, practically close or up to the the saddle point. The approximate position of the saddle as a function of the two-center distance $\delta z = z_1 - z_2$ is indicated by a straight line.

physics. Besides the usual kinetic energy [20] $p^2/2m$, a shape dependent potential V_{shp} , a spin-orbit coupling term V_{ls} and a l^2-correction term V_{l^2}

are present in the Hamiltonian, $\hat{H}$ of the TCSM. It is convenient to use a cylindrical system of coordinates (z, ρ, φ) due to the assumed axial symmetry of the problem. One has :

$$\hat{H} = \mathbf{p}^2/2m + V_{shp}(\rho, z; R, \eta, \varepsilon, \beta_1, \beta_2) + V_{ls}(\mathbf{r}, \mathbf{p}, \mathbf{s}; \kappa_i) + V_{l^2}(\mathbf{r}, \mathbf{p}; \kappa_i, \mu_i) \quad (1)$$

where $\mathbf{r}, \mathbf{p}, \mathbf{s}$ are the coordinate, momentum and spin of a single nucleon of mass m, $R = z_2 - z_1$ is the separation distance between centers, $\eta = (V_1 - V_2)/(V_1 + V_2)$ is the mass asymmetry parameter, V_i $(i = 1, 2)$ are the volumes of the fragments, $\varepsilon = E'/E_o$ is the neck parameter, $\beta_i = a_i/b_i$ are the deformations, κ_i and μ_i are mass-dependent quantities defined in the Nilsson model (which is obtained for $R = 0$ within TCSM). E_o is the actual barrier height and $E' = m\omega_{z_i}^2 z_i^2/2$ is the barrier of the two-center oscillator. For $R = 0$ and $R \rightarrow \infty$ one nucleus and two fragments, respectively, are described.

The Hamiltonian is diagonalized in a basis of eigenfunctions analytically determined. As a result, the proton and neutron energy level diagrams for a given set of deformation parameters are obtained. An example of neutron levels for the split of ^{232}U in ^{24}Ne and ^{208}Pb is presented in Fig. 2. In this case a folded-Yukawa potential has been used for a nuclear shape parametrization obtained by smoothly joining two spheres with a third surface generated by rotating a circle of radius $R_3 = 1/c_3$ around the symmetry axis. A small radius $R_3 = 1$ fm has been chosen in the above diagram.

3 Nuclear Molecules and Nuclear Landau–Zener Effect

In peripheral heavy ion reactions, only the nucleons located near the surface of each participant are moving on molecular orbits. The lower shells nuclei are forming the nuclear "cores" of the molecule. Molecular orbits are formed and can be observed when the ratio of the collision time to nuclear period (orbiting time) of the valence nucleon, roughly given by $\sqrt{\epsilon_F A_{pr}/E_{lb}}$, is high enough. Here ϵ_F is Fermi energy, and E_{lb}/A_{pr} is the bombarding energy per nucleon. For example in a collision of ^{17}O with ^{12}C at energies near the Coulomb barrier, this ratio is about 3.5, and consequently the effect may be viewed. Resonances are observed in a narrow window of the excitation energy versus angular momentum plane - the *molecular window*.

Signatures of the formation of molecular orbits are observable in reaction channels involving enhanced nuclear transitions of nucleons between molecular levels (elastic and inelastic transfer, or inelastic excitation of the valence nucleons), due to avoided crossing of molecular single nucleon levels - the *nuclear Landau - Zener effect* . According to a theorem of Neumann and Wigner, levels with the same symmetry cannot cross. This avoided level crossing is the point of nearest approach of two levels with the same projection of angular momentum Ω, on the internuclear axis. The nuclear Landau - Zener effect [21] consists in a large enhancement of transition probability between the two levels at pseudocrossing (avoided crossing). For a constant relative velocity v of two nuclei, the transition probability P from the lower adiabatic state to the higher one (see Fig. 3) is

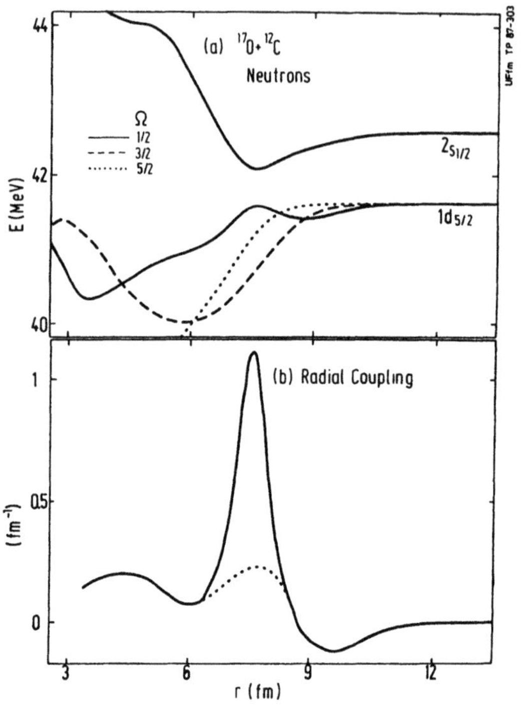

Figure 3: (a) TCSM neutron level diagram for $^{17}O + ^{12}C$ system. Only the levels used in calculations are shown. The full, dashed, and dotted lines corresponds to $\Omega = 1/2$, $3/2$, and $5/2$, respectively. (b) The radial coupling matrix element for the transition from the ground state to the first excited state of ^{17}O by promoting the valence neutron from the $1d$ to the $2s$ level. Dotted curve was obtained by supressing the peak at the avoided level crossing.

$$P_{1\to2} = 2P_{LZ}(1 - P_{LZ});\, P_{LZ} = exp(-2\pi G)$$

$$G = \mid H_{12}\mid^2/[\hbar v \mid \frac{d}{dR}(\varepsilon_1 - \varepsilon_2) \mid] \tag{2}$$

where $2H_{12}$ is the closest distance between the adiabatic eigenvalues E_1, E_2 at $R = R_C$, and ε_i are the corresponding diabatic (crossing) levels given by :

$$E_{1,2} = \frac{1}{2}(\varepsilon_1 + \varepsilon_2)^{1/2} \mp [\frac{1}{4}(\varepsilon_1 - \varepsilon_2)^2 + \mid H_{12} \mid^2]^{1/2} \tag{3}$$

As a function of incident energy, $P_{1\to2}$ steeply rises to 0.5 and then drops to zero.

An example of avoided level crossing is shown in Fig. 3 at $R = R_C$, where two $\mid \Omega \mid = 1/2$ levels, are approaching the single particle states $1d_{5/2}$ and $2s_{1/2}$ of ^{17}O for $R \to \infty$. Assuming the valence neutron on $1d_{5/2}$ level in the ground state of ^{17}O and the $2s_{1/2}$ level in its first excited state, signatures of nuclear molecular resoances have been predicted [21] in the inelastic excitation function of $^{17}O + {}^{12}C$ reaction. The valence neutron of ^{17}O is strongly excited from ground state to the $2s_{1/2}$ excited state ($1/2^+$) at $E^* = 0.871$ MeV.

Two excitation mechanisms between molecular single-particle levels are effective radial and rotational couplings. The former acts between states with the same Ω, and is largest (see the lower part of Fig. 3) at the point of an avoided level crossing (Landau - Zener transition). The later (Coriolis coupling) arises from the rotation of the body fixed coordinate system. It causes transitions between states with $\Omega \pm 1$, and is the largest at the crossing point of levels. The TCSM level diagrams (as in Fig. 3) are used to predict enhanced transitions and to determine the most important couplings in specific reactions.

The oscillatory behaviour of the differential cross sections for elastic scattering of ^{17}O on ^{12}C and inelastic excitation of the first $1/2^+$ state of ^{17}O, at incident energies of 50 and 62 MeV, experimentally determined by the Strasbourg group [22] (see Fig. 4) is an evidence of nuclear molecular phenomena. Complete quantum - mechanical molecular reaction calculations for this system have been performed within *molecular particle-core model.* In this model, first introduced in 1972 [23] and improved successively [24] [25] [26] [27] the outermost bound nucleons are eigenstates of the TCSM. The mean field is generated by all nucleons and contains adiabatic polarisation effects between nuclei. By considering two unequal cores and one valence neutron, the Hamiltonian of the model contains two operators of kinetic energy (for valence neutron and two cores) and an optical potential $(U + iW)$ function of the separation distance between the cores.

The wave function ψ_{IM} of the scattering problem , written in the laboratory system for a given angular momentum I, has the same structure as in the strong coupling (Nilsson) model, but here $\Phi_{\alpha j \Omega}$ is eigenfunction of the asymmetric TCSM Hamiltonian. The coupled channel equations for the radial wavefunctions $R_{\alpha l I}$ are obtained by projection. In the coupled channel calculations, the elastic scattering and the excitation of the $1/2^+$ state at 0.871 MeV have been included. The radial coupling shown in Fig. 3, has a strong narrow peak due to Landau- Zener effect at $R = 7.6$ fm. This peak is located just behind the barrier of the real optical potential U. Rotational couplings also induce transitions between elastic and inelastic channels. As it is shown in Fig. 4, the experiment is in good agreement with the theory. The agreement of the theory with the backward rise of the neutron–transfer crossection demonstrates the existence of the Landau–Zehner–effect and hence the level–crossing. This in turn is proof that we are dealing with nuclear molecules, because level crossings as a function of the two–center distance are characteristic molecular effects.

We believe that fusion reaction calculations including channels with enhanced single-particle transitions due to the nuclear Landau-Zener effect, could clarify the role of molecular states in the fusion process. A possible connection of nuclear molecules with cluster radioactivity has been emphasized [28].

A pocket in the potential barrier of giant systems (U-U for example) is a good indication about the possibility to produce experimentally giant atomic or nuclear molecules. These are of extremely high importance for testing the electrodynamics of strong fields, predicting spontaneous production of electron-positron pairs.

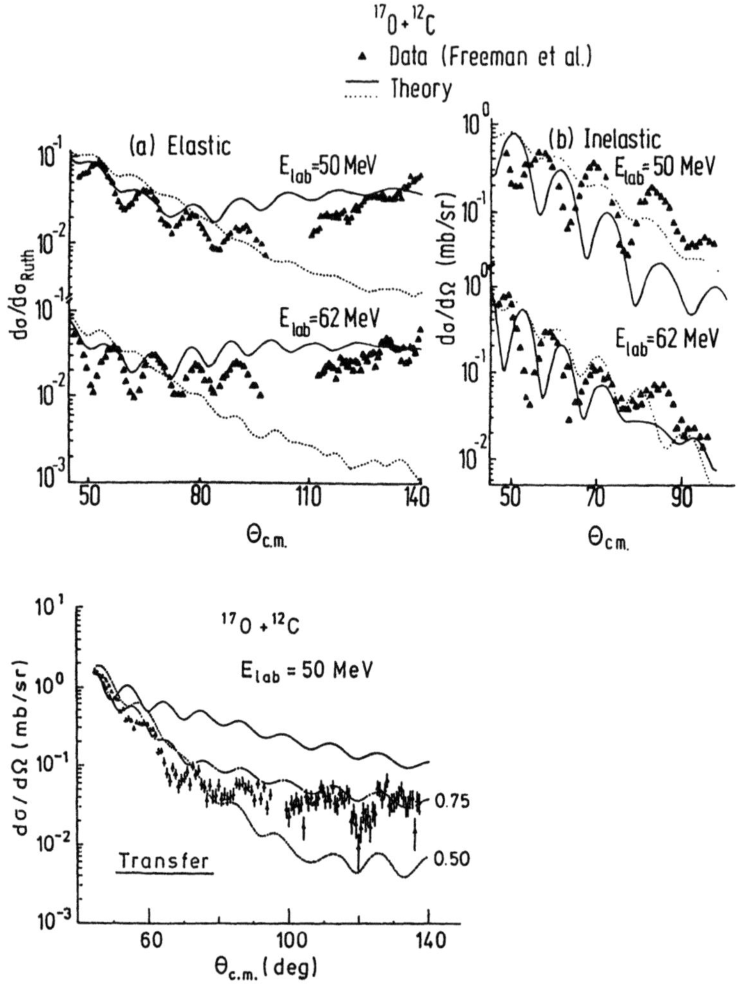

Figure 4: Differential cross sections for: (a) the elastic scattering of ^{17}O on ^{12}C; (b) inelastic excitation of the first $1/2^+$ state of ^{17}O at the incident energies 50 and 62 MeV. The solid and dotted lines are calculations in correspondence with similar curves from Fig. 3b. (c) The calculated transfer yield is found to be very sensitive to the energy splitting of the avoided crossing. The figure demonstrates the influence of a variable intensity of the radial coupling.

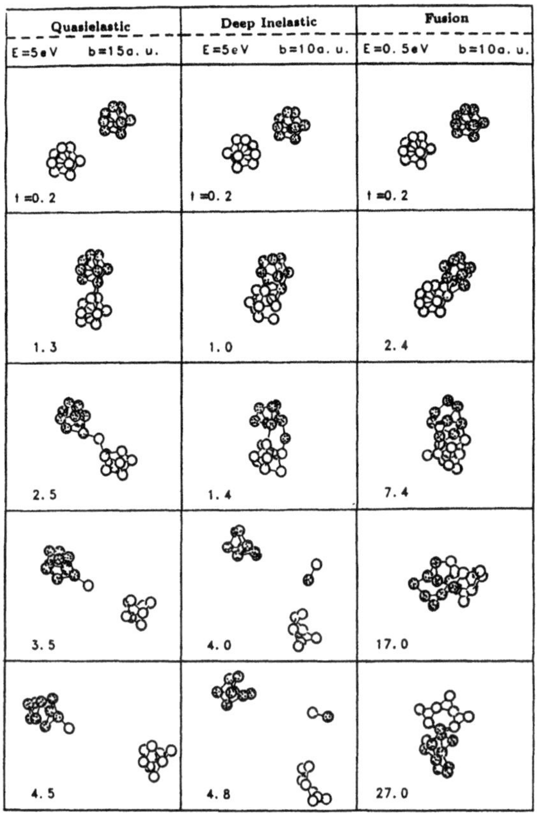

Figure 5: The formation of a $(Na)_9$–$(Na)_9$ atomic cluster molecule within a molecular dynamic simulation.

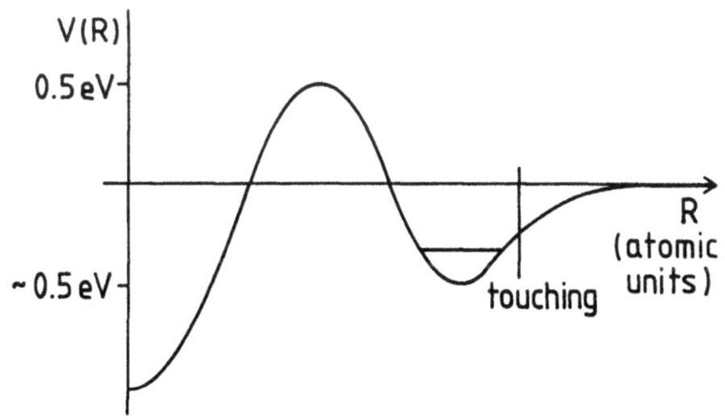

Figure 6: Typical form of the inter–cluster–potential for $(Na)_9$–$(Na)_9$ according to R. Schmidt.

Let us also mention the analogy of these nuclear molecules with atomic cluster molecules, as first discussed by R. Schmidt (these proceedings). For these do also exist internuclear potentials with boundary pockets. Figure 5 illustrates a molecular dynamics–calculation for a $(Na)_9$–$(Na)_9$–cluster collision clearly visualizing the formation of a $(Na)_9$–$(Na)_9$–molecule. In figure 6 the inter–cluster potential is sketched.

4 Fission Processes within Fragmentation

The theory of fragmentation is a consistent method allowing to treat two-body and many-body breakup channels in fission, fusion, and heavy ion scattering. Within this theory we get a unified point of view on fission, superasymmetric fission, cold- and bimodal- fission and cluster decay.

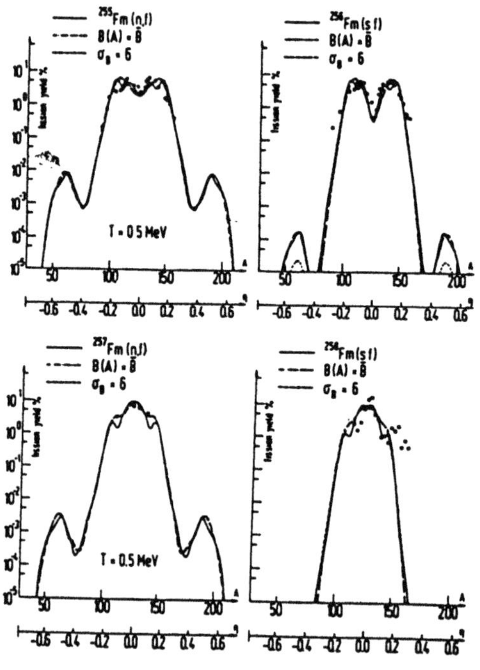

Figure 7: Mass yields for various Fm isotopes calculated by using different inertia : $B_{\eta\eta}$ (full curve); average B (chain curve), and smoothed B (dashed curve). Experimental data are denoted by points. The observation of superasymmetric fission with one of the clusters being ^{78}Ni is a formidable task.

The main collective coordinates have been already defined in section 2. By assuming the adiabatic approximation, for each pair of (R, η) coordinates, the liquid drop energy $V_{LDM}(R, \eta, \varepsilon, \beta_1, , \beta_2)$ can be minimised with respect to $\varepsilon, \beta_1, \beta_2$, leading to $E_{LDM}(R, \eta)$. Then the shell and pairing corrections $\delta U(R, \eta)$ are added. In this way the time consuming procedure of calculating single-particle levels is applied only once for a given combination (R, η). Also the collective mass tensor $B_{ik}(R, \eta)$ can be calculated by using the states of the TCSM supplemented with pairing forces.

At high excitations the shell effects could be completely washed up. To account for this physical behavior, a Gaussian factor is introduced :

$$V(R, \eta, \Theta) = E_{LDM}(R, \eta) + \delta U(R, \eta) \exp(-\Theta^2/\Theta_o^2) \tag{4}$$

where at a given excitation energy, E^*, the nuclear temperature equals

$$\Theta = (10E^*/A)^{1/2}.$$

One can choose $\Theta_o = 1.5$ MeV in order to obtain a vanishing contribution of shell effects at excitation energies of the order of 60 MeV. The wave function $\Phi_k(R, \eta, \Theta)$ is a solution of a Schroedinger equation in η:

$$\left(-\frac{\hbar^2}{2\sqrt{B_{\eta\eta}}}\frac{\partial}{\partial\eta}\frac{1}{\sqrt{B_{\eta\eta}}}\frac{\partial}{\partial\eta} + V(R, \eta, \Theta)\right)\Phi_k(R, \eta, \Theta) = E_k \Phi_k(R, \eta, \Theta) \tag{5}$$

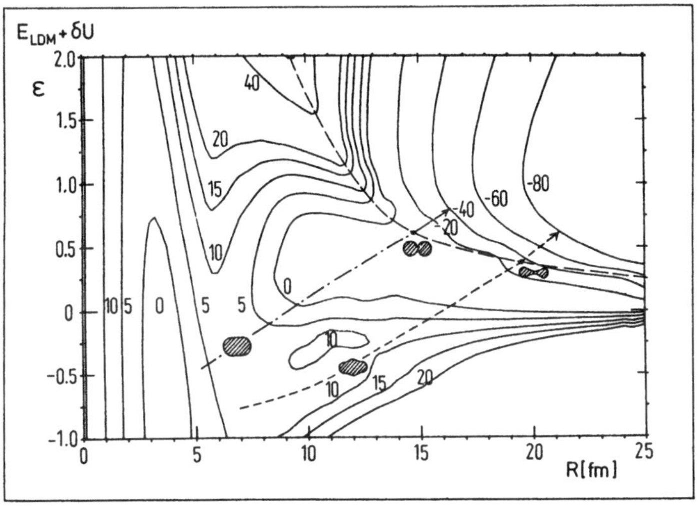

Figure 8: Potential energy surface for symmetric fission of ^{256}Fm The dashed curve with arrow is the elongated shapes fission path. The dot-and-dashed curve is the compact shapes (cold) fission path.

Here R is kept fixed and only the η-degree of freedom is treated in order to simplify the procedure and the understanding. This can be justified as long as fission is an adiabatic process.

The fission fragment mass yield normalized to 200% is given by :

$$Y(A_2) = \mid \Psi(R,\eta,\Theta)\mid^2 \sqrt{B_{\eta\eta}}(400/A) \tag{6}$$

where $B_{\eta\eta}$ is a diagonal component of the inertia (effective mass) tensor, and:

$$\mid \Psi(R,\eta,\Theta)\mid^2 = \sum_k \mid \Phi(R,\eta,\Theta)\mid^2 \exp(-E_k/\Theta) \tag{7}$$

The example of transition from asymmetric to symmetric fission in Fm isotopes (see Fig. 7) illustrates again the dominant influence of shell effects [29] on fission fragment mass asymmetry. This transition was confirmed by experiment [30]. Moreover, bimodal fission has been observed in this region of nuclei [19]. A typical potential energy surface for a nucleus undergoing bimodal fission is shown in Fig. 8, containing the two paths for normal and cold fission [45]. Rather quantitative investigations of this phenomenon have been given by Sobichewski te al. [33].

The side peaks in Fig. 7 have also been obtained in other calculations based on fragmentation theory, as for example in U and No. First experimental indications for superasymmetric fission can be deduced from Gönnenwein's experiments [18]. Clearly the mass distributions in Eq. (6) yield also contributions for extreme break-ups of the fissioning nucleus, i. e. for $\eta \sim 0.8 - 1.0$. This means that nuclei emit small clusters, even though with small probability. We are thus lead to the idea of cluster radioactivity and, furthermore, to the inside that there is no difference between (cold) fission and a radioactive process. To further illustrate this idea, let us look at the groundstate wavefunction and its mass distribution for a two-center potential , as shown in Fig. 7. Obviously

$$P(\eta) = \int \Psi_o^*(R,\eta)\Psi_o(R,\eta)d^3R \tag{8}$$

can be interpreted as the cluster preformation probability in a nucleus of mass A within the interval $\eta, \eta + d\eta$ of cluster formation. In order to describe cluster radioactivity one has to extend the theory to very asymmetric shapes.

The ATCSM contains odd multipole moments due to the use of the mass asymmetry coordinate η. The octupole moment is given by :

$$Q_{30}(R,\eta) = \sqrt{\frac{7}{4\pi}} \int Y_{30}^*(\theta,0) r^3 \rho_e(\mathbf{r},R,\eta) d^3r \tag{9}$$

where ρ_e is the charge density resulting for this isolated fragmentation.

Another interesting feature of the ground state wave function follows from the quantity $P(\eta) = \int \mid \Psi_o(R,\eta)\mid^2 dR$. As mentioned above, this is the preformation probability for the cluster configuration characterized by mass asymmetry η. With this interpretation we can see that the observable octupole

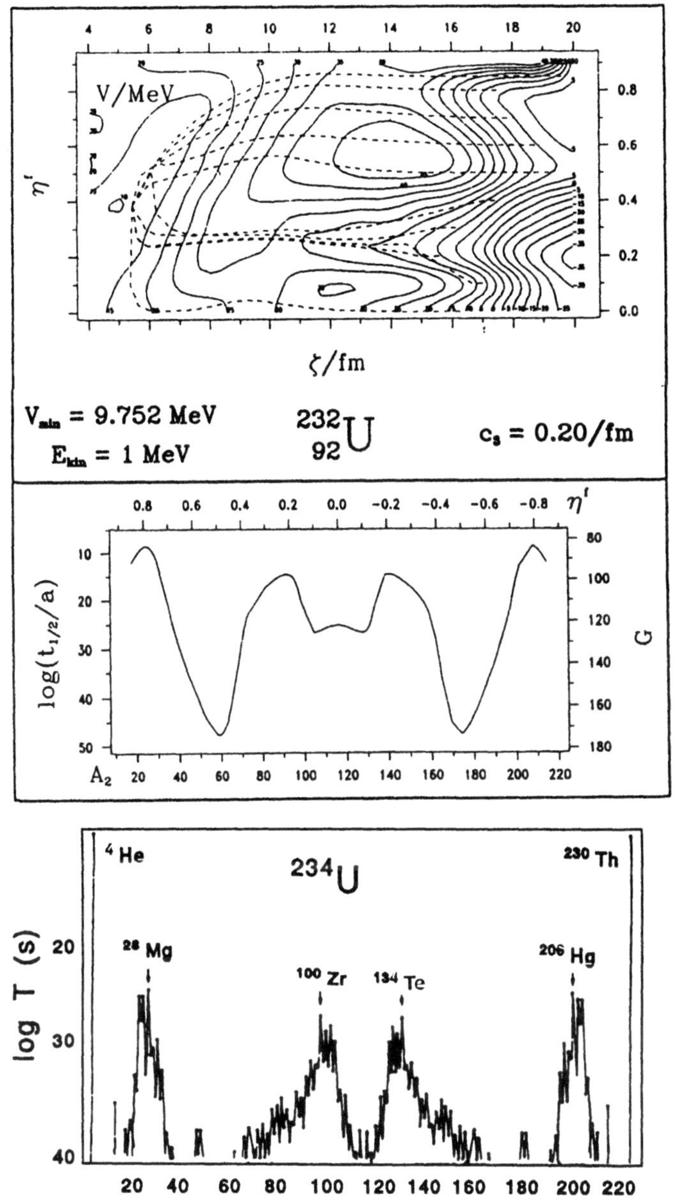

Figure 9: Potential energy surfaces versus mass asymmetry and separation distance between centers for ^{232}U at a fixed neck coordinate $c_3 = 0.2/\text{fm}$. The dashed lines are fission paths leading to a given final mass asymmetry. The corresponding half-life (in years) along these paths are plotted in the middle part of the figure. It is comparable with the lifetime (in seconds) spectrum of ^{234}U calculated within ASAFM, given in the lower part of the figure. The upper two pictures have been calculated by D. Schnabel and H. Klein.

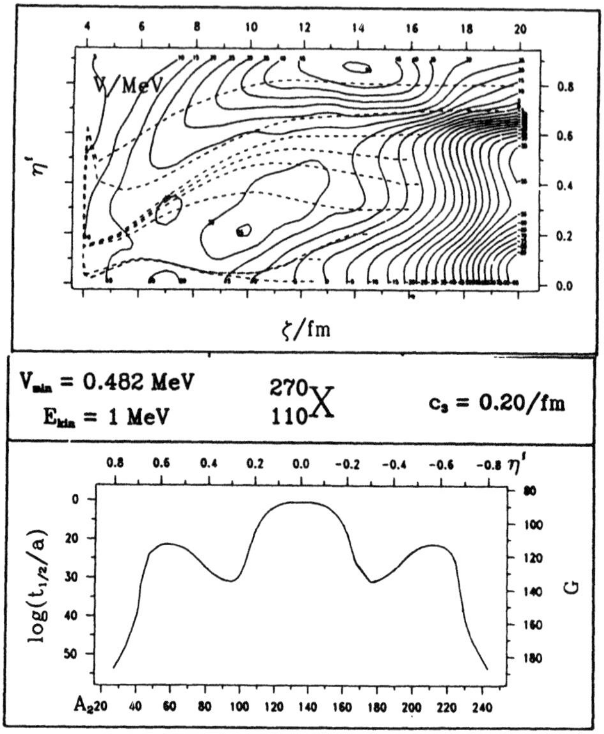

Figure 10: The same as the preceding figure for the nucleus $^{270}110$. Both diagrams have been obtained by D. Schnabel and H. Klein.

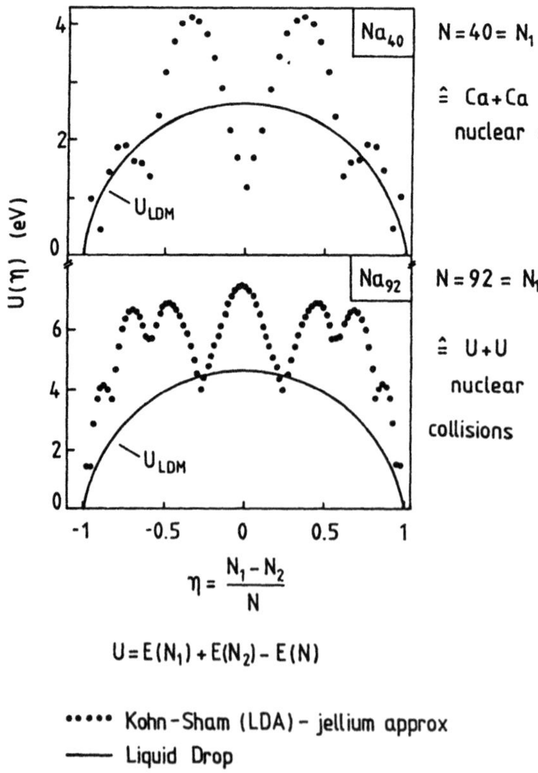

Figure 11: Fragmentation potentials for atomic clusters according to R. Schmidt. They must be seen in analogy to the fragmentation potentials of u (asymmetric fission) and Ca (symmetric fission) respectivly.

moment is not a result of one particular configuration of the type shown in Fig. 1, but a superposition of all clusters, weighted by the corresponding preformation probability.

On the potential energy surface [32] given in Figs.9 and 10, the fission paths leading to various final asymmetries are shown with dashed lines. They are determined by minimizing [31] the WKB-integral. Shell corrections based on the energy levels of the kind shown in Fig. 2 have been added to the Yukawa-plus-exponential potential energy extended for fragments with different charge densities [34]. The inertia tensor has been computed by using Werner-Wheeler approximation within hydrodynamical model. A fixed neck radius has been assumed. The corresponding halflife obtained along each path in the upper part of Fig. 9 is plotted in the middle part, showing that according to these calculations, Ne radioactivity of ^{232}U has a larger probability than the spontaneous fission of the same nucleus. It is interesting to note the similarity with the time spectrum of the ^{234}U nucleus calculated within ASAFM. Finally we mention that fragmentation potentials can also be calculated for metallic clusters, as done by Rüdiger Schmidt (see contribution to this conference) which must be seen in analogy to the nuclear ones for U fission and Ca fission respectivly.

5 Cluster Decay Modes

Spontaneous cluster emission is allowed if the released energy $Q = M - (M_e + M_d)$ is a positive quantity. In this equation M, M_e, M_d are the atomic masses of the parent, the emitted cluster and the daughter nuclei, respectively, expressed in units of energy.

The most important observable is the parent nucleus life-time, T relative to this disintegration mode, and the corresponding branching ratio with respect to α-decay, $b = T_\alpha/T$. If T is low enough and b is sufficiently high, the phenomenon may be detected experimentally, and the kinetic energy of the emitted cluster $E_k = QA_d/A$ is also measured. Up to now the longest measured lifetime is of the order of 10^{26} seconds and the lowest branching ratio is almost 10^{-17} !

The difficult task of the theory is to study the dynamics of the process. Three fission models and one cluster preformation model have been used in our papers of 1980 to predict the new decay modes (see [11], [12] and the references therein). All are using the shape parametrization of TCSM. In order to be able to take into consideration the large number of combinations parent - emitted cluster (at least of the order of 10^5), we developed since 1980, the analytical superasymmetric fission model (ASAFM) with which we made the first predictions of nuclear lifetimes. In 1984 , before any other model was developed, we published the first estimates of the half-lives and branching ratios relative to α decay for more than 150 decay modes, including all cases experimentally confirmed up to now on ^{14}C, ^{22}O, $^{24-26}Ne$, $^{28,30}Mg$, and $^{32,34}Si$ radioactivities. A comprehensive table was produced by performing calculations within that model. Subsequently, the numerical predictions of ASAFM have been improved by taking better account of the pairing effect in the correction energy [35]. Cold fission fragments [36] [37] were also considered in a new version of the tables [38]. The above mentioned systematics was further extended in the region of heavier clusters with mass numbers $A_e > 24$ [39]. Recently, the half-life estimations within ASAFM have been updated and the region of parent nuclei expanded far from stability and toward superheavy elements using the 1988 mass tables as input data for the Q-value [40] calculation.

The region of cluster emitters (see Fig. 12) extends well beyond that of α-emitters. Light clusters are preferentially emitted from neutron-deficient nuclei and the heavy ones from neutron-rich parents. Nevertheless, from practical point of view, the above mentioned conditions to observe cluster radioactivities are fulfilled in a few smaller areas.

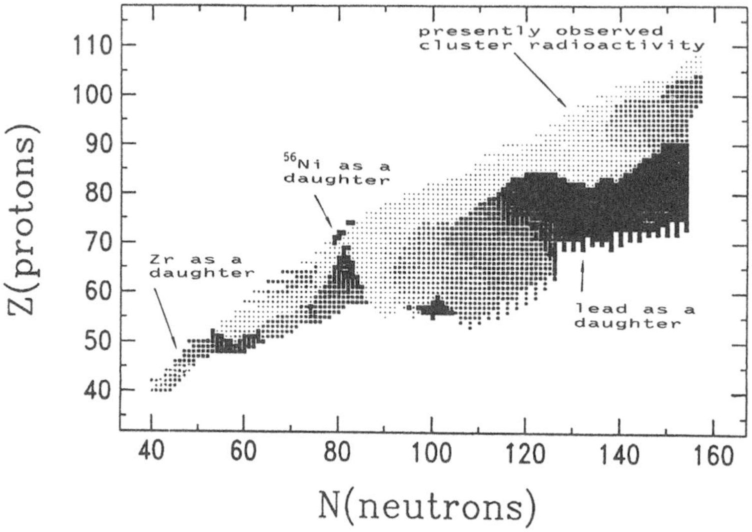

Figure 12: Chart of nuclides with the regions of the most probable cluster emissions ($Z_e = 4 - 28$). Light points are clusters with small number of protons; dark points - with large Z.

The half-life of a nucleus (A, Z) against the split into a cluster (A_e, Z_e) and a daughter (A_d, Z_d), is calculated with analytical relationships derived from

$$T = [(h \ln 2)/(2E_v)]exp(\frac{2}{\hbar}\int_{R_a}^{R_b}\{2\mu[(E(R) - E_{cor}) - Q]\}^{(1/2)}dR \qquad (10)$$

where $\mu = mA_eA_d/A$ is the reduced mass, m is the nucleon mass, and $E(R)$ is the interaction energy of the two fragments separated by the distance R

between centers. It is known that the fission barrier heights are too large within the liquid drop model. E_{cor} is a correction energy allowing to get a more realistic, lower and thiner barrier, and to introduce shell and pairing effects. R_a and R_b are the turning points of the WKB integral, h is the Planck constant, and E_v is the zero point vibration energy. For practical reasons (to reduce the number of fitting parameters) we took $E_v = E_{cor}$ though it is evident that, owing to the exponential dependence, any small variation of E_{cor} induces a large change of T, and thus plays a more important role compared to the preexponential factor variation due to E_v. By confusion, this correction was misquoted [41] as violating the energy conservation. In fact one can see easily that the Q-value is not changed; it only affects the height and the width of fission barrier in a way similar with the shell correction method.

The unified approach of three groups of decay modes (cold fission, cluster radioactivities and α-decay) within ASAFM is best illustrated on the example of ^{234}U nucleus (see the lower part of Fig. 9), for which all these processes have been measured. For α-decay, Ne- and Mg- radioactivities the experimental half-lives are in good agreement with our calculations (in fact the predicted lifetimes for cluster radioactivities have been used as a guide to perform the experiment). Spontaneous cold fission of this nucleus was not measured, but the experiments on induced cold fission are showing that ^{100}Zr is indeed the most probable light fragment.

a
b
c

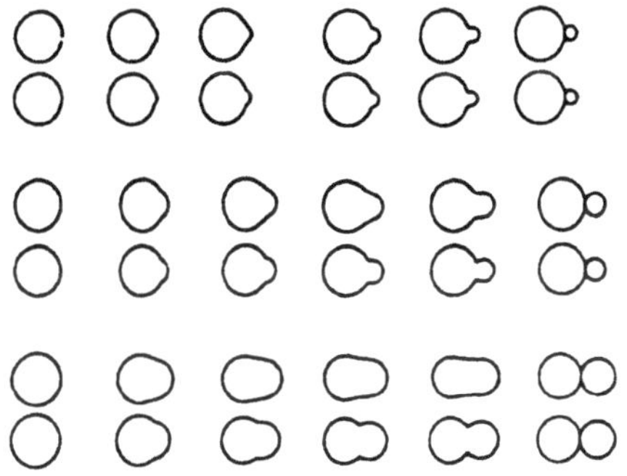

Figure 13: Neck influence on fission dynamics of ^{234}U in a wide range of mass asymmetry. The nuclear shape along optimum fission path are compared with two intersected spheres with the same separation distance between centers for α-decay (a), ^{28}Mg radioactivity (b), and cold fission with light fragment ^{100}Zr.

An extensive study of the fission dynamics over a wide range of mass asymmetry has been performed [42] [43] [44], by replacing the reduced mass with the Werner-Wheeler inertia tensor and the liquid-drop model energy with Yukawa-plus-exponential potential extended to fragments with different charge densities [34].

In order to calculate the inertia tensor for very asymmetric fission, we had demonstrated the importance of taking into account a correction term, B^c_{ij}, due to the center of mass motion.

We got [42] analytical expressions of the inertia for two parametrizations of nuclear shapes, obtained by intersecting two spheres of radii R_1, R_2 and volumes V_1, V_2 : "cluster-like" (with R_2 = constant) and more compact (V_2 = constant). Pik-Pichak [41] claimed that the later is the best parametrization for studying cluster radioactivities as fission processes. In fact he made some errors. By ignoring the above mentioned contribution of the center of mass motion he obtained wrong values for the inertia. As we have shown [42] in case of α-decay , the ratio of wrong to correct value of inertia may be as high as 30/4. A comparison of action integral K along the fission paths just mentioned shows that cluster-like shapes are more suitable (action integral lower) when the mass number of the emitted cluster is less than 34, giving additional support for the two-center shape parametrization within ASAFM.

From the same dynamical calculations [42] we concluded that the choice of the shape coordinate has no consequence on the value of action integral, determining the observable halflife. In conclusion the separation distance of geometrical centers R is as good as the distance between mass centers of the fragments, again in contrast to what was claimed in Ref. [41].

A three-dimensional parametrization [45] has been used [43] to study the influence of a smooth neck on the fission dynamics. We performed calculations for α-decay, ^{28}Mg radioactivity and cold fission (with the light fragment ^{100}Zr) of ^{234}U . For every mass asymmetry, the optimum fission path in the plane of two variables: separation distance R, and the neck radius R_3 was determined by minimization of action integral. The corresponding shapes, along the fission path and in the absence of the smooth neck, are compared in Fig.13. One can see that the neck influence is stronger when the mass asymmetry parameter is small; for a very large mass asymmetry the parametrization of two intersected spheres is a good approximation of the optimum fission path. As we have shown recently, fission trajectories in the space of the deformation coordinates $(R, S \equiv R_3)$ at a given mass asymmetry η can be found by solving the following nonlinear equation of second degree:

$$DS'' + D_3 S'^3 + D_2 S'^2 + D_1 S' + D_0 = 0 \tag{11}$$

where the coefficients D are expressions containing partial derivatives with respect to S and R of the inertia tensor components and of the energy.

$$D = 2E(AC - B^2) \tag{12}$$

$$D_3 = -2EC\frac{\partial B}{\partial S} + C\left(C\frac{\partial E}{\partial R} - B\frac{\partial E}{\partial S}\right) + E\left(\frac{\partial C}{\partial R} + B\frac{\partial C}{\partial S}\right) \tag{13}$$

$$\begin{aligned} D_2 &= 3B\left(C\frac{\partial E}{\partial R} + E\frac{\partial C}{\partial R}\right) - (2B^2 + AC)\frac{\partial E}{\partial S} \\ &\quad -2E\left(C\frac{\partial A}{\partial S} + B\frac{\partial B}{\partial S}\right) + AE\frac{\partial C}{\partial R} \end{aligned} \tag{14}$$

$$D_1 = -3B\left(A\frac{\partial E}{\partial S} + E\frac{\partial A}{\partial S}\right) + (2B^2 + AC)\frac{\partial E}{\partial R} + 2E\left(A\frac{\partial C}{\partial R} + B\frac{\partial B}{\partial R}\right) - EC\frac{\partial A}{\partial R} \tag{15}$$

$$D_0 = 2EA\frac{\partial B}{\partial R} + A\left(B\frac{\partial E}{\partial R} - A\frac{\partial E}{\partial S}\right) - E\left(A\frac{\partial A}{\partial S} + B\frac{\partial A}{\partial R}\right) \tag{16}$$

and E is the deformation energy (Q-value subtracted out). The other quantities:

$$A = B_{RR} + 2B_{RR_2}\frac{\partial R_2}{\partial R} + B_{R_2R_2}\left(\frac{\partial R_2}{\partial R}\right)^2 \tag{17}$$

$$B = B_{RS} + B_{R_2S}\frac{\partial R_2}{\partial R} \tag{18}$$

$$C = B_{SS} \tag{19}$$

are dependenton the components of the nuclear inertia tensor.

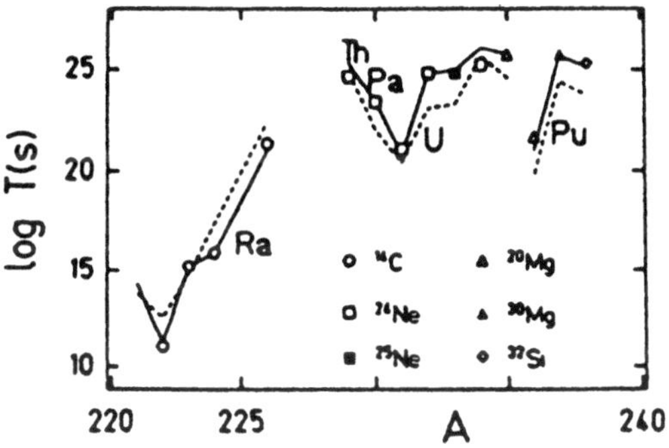

Figure 14: Comparison of our early life-time predictions (dashed line) and the calculations of 1986 including an even-odd effect (full lines) with experimental points for cluster radioactivities confirmed up to now.

Some authors are trying to show [46] that fission models are in conflicting relations with preformation cluster models. They have invoked even naive arguments, as we have mentioned [47]. We have compared the two kinds of models and our predictions with experimental data [48] pointing out that there was no method available which would allow preformation probabilites or quantum penetrabilities to be measured and thereby check the validity of the two kinds of theories. Recently [47] we gave a new interpretation of the cluster preformation probability within a fission model, as the penetrability of the prescission part of the barrier. On this basis we have shown that preformation cluster models are equivalent with fission models and we introduced one universal curve for each kind of cluster radioactivity. A number of successful experiments have been performed since 1984 (see the review [13], and [52]). They are compared with our early predictions and with those of

1986, made after including an even-odd effect in Fig.14. One can see the good agreement between the calculated and experimentally determined half-lives in a range of 15 orders of magnitude. The strong shell effect predicted by the theory has been confirmed.

6 Summary

A unifying point of view of various nuclear phenomena may be best achieved on the basis of two center shell model, allowing to describe di-nuclear systems and the dynamical evolution from one initial nucleus to two final nuclei, or viceversa.

Nuclear molecules are very general phenomena. Light and intermediate nuclear molecule signatures have been already detected as resonances in the scattering cross sections, as various kinds of shape isomers (fission isomers, superdeformed and hyperdeformed nuclei at very high spin). In a sens even the nuclear systems exhibiting cluster radioactivities may be viewed as nuclear molecules.

Many important physical phenomena are expected to be seen in very heavy dinuclear systems (U-U, U-Cm, etc). They could be produced by using the existing heavy ion accelerators. Potential energy pockets supporting giant nuclear molecules have been predicted in various theoretical approaches [49] [50]. Such giant nuclear molecules are of fundamental importance for the observation of the decay of the vacuum in supercritical fields [51]. Moreover, these giant molecules seem to be the most superdeformed nuclei one can imagine. They deserve to be studied in their own right and may lead to exciting new nuclear structure in the years ahead.

Cluster radioactivity is a well established phenomenon. Conceptually it broadened our horizon by telling us that radioactivity is not limited to the three classical modes namely, α-decay, β-decay, and γ-decay. Furthermore, fission and especially cold fission and cluster radioactivity are the same phenomena.

References

[1] D. A. Bromley, J. A. Kuehner, and E. Almquist, Phys. Rev. Lett. **4** (1960) 65.

[2] N. Cindro, W. Greiner, and R. Caplar, eds., *Frontiers of Heavy Ion Physics* (World Scientific, Singapore), 1987.

[3] W. Scheid, W. Greiner,and R. Lemmer, Phys. Rev. Lett. **25** (1970) 176.

[4] P. Holzer, U. Mosel, and W. Greiner, Nucl. Phys. **A 138** (1969) 241.

[5] J. Maruhn, and W. Greiner, Z. Phys. **251** (1972) 431.

[6] M. Brack, J. Damgaard, A. Jensen, H. C. Pauli, V. M. Strutinsky, and C. Y. Wong, Rev. Mod. Phys. **44** (1972) 320.

[7] H. J. Fink, J. Maruhn, W. Scheid, and W. Greiner, Z. Phys. **268** (1974) 321.

[8] J. A. Maruhn, W. Greiner and W. Scheid, *in Heavy Ion Collisions* edited by R. Bock (North Holland, Amsterdam), Vol. 2, 1980, p. 399.

[9] A. Sandulescu, D. N. Poenaru, and W. Greiner, Sov. J. Part. Nucl. **11** (1980) 528.

[10] H. J. Rose and G. A. Jones, Nature **307** (1984) 247.

[11] W. Greiner, M. Ivaşcu, D. N. Poenaru, and A. Sandulescu, *in Treatise on Heavy Ion Science* edited by D. A. Bromley (Plenum, New York), Vol. 8, 1989, p. 641.

[12] D. N. Poenaru, M. Ivaşcu, and W. Greiner, *in Particle Emission from Nuclei* edited by D. N. Poenaru and M. Ivaşcu (CRC, Boca Raton, Florida), Vol. III, 1989, p. 203.

[13] P. B. Price, Annual Rev. Nucl. Part. Sci. **39** (1989) 19.

[14] M. Greiner and W.Scheid, J. Phys. G. **12** (1986) L 285.

[15] L. Brillard, A.G. Elayi, E. Hourani, M. Hussonnois, J.F. Le Du, L.H. Rosier, and L. Stab, C.R. Acad. Sci. **309** (1989) 1105.

[16] G. Münzenberg, Rep. Prog. Phys. **51** (1988) 57.

[17] A. Sandulescu, R. K. Gupta, W. Scheid, and W. Greiner, Phys. Lett. **60 B** (1976) 225.

[18] F. Gönnenwein, B. Börsig and H. Löffler, in *Proc. Internat. Symp. on Collective Dynamics*, Bad Honnef, edited by P. David (World Sci., Singapore), 1986, p. 29.

[19] E. K. Hulet, J. Wild, R. Dougan, R. Lougheed, J. Landrum, A. Dougan, M. Schädel, R. hahn, P. Baisden, C. Henderson, R. Dupzyk, K. Sümmerer, and G. R. Bethune, Phys. Rev. Lett. **56** (1986) 313.

[20] J. M. Eisenberg and W. Greiner, *Nuclear Theory*, Vol. 1, (North Holland, Amsterdam), 3d edition, 1987.

[21] J. Y. Park, W. Greiner, and W. Scheid, Phys. Rev. **C 21** (1980) 958.

[22] R. M. Freeman, C. Beck, F. Haas, A. Morsad, and N. Cindro, Phys. Rev. **C 33** (1986) 1275.

[23] J. Y. Park, W. Scheid, and W. Greiner, Phys. Rev. **C 6** (1972) 1565.

[24] J. Y. Park, W. Scheid, and W. Greiner, Phys. Rev. **C 20** (1979) 1988.

[25] G. Terlecki, W. Scheid, H. J. Fink, and W. Greiner, Phys. Rev. C **18** (1978) 265.

[26] R. Könnecke, W. Greiner, and W. Scheid, Phys. Rev. Lett. **51** (1983) 366.

[27] A. Thiel, W. Greiner, J. Y. Park, and W. Scheid, Phys. Rev. C **36** (1987) 647.

[28] N. Cindro and M. Bozin, Phys. Rev. C **39** (1989) 1665.

[29] H. J. Lustig, J. A. Maruhn, and W. Greiner, J. Phys. G. **6** (1980) L 25.

[30] D. C. Hoffman and L. P. Somerville *in Particle Emission from Nuclei* edited by D. N. Poenaru and M. Ivaşcu (CRC, Boca Raton, Florida), Vol. III, 1989, p. 1.

[31] R. Herrman, J.A.Maruhn, and W.Greiner, J. Phys. G. **12** (1986) L 285.

[32] D. Schnabel and H. Klein, private communication.

[33] S. Cwiok, P. Rozmej, A. Sobichewski, and Z. Patyk, Nucl. Phys. A **491** (1989) 281.

[34] D. N. Poenaru, M. Ivaşcu, and D. Mazilu, Comp. Phys. Communications, **19** (1980) 205.

[35] D. N. Poenaru, W. Greiner, M. Ivaşcu, D. Mazilu, and I. H. Plonski, Z. Phys. A **325** (1986) 435.

[36] D. N. Poenaru, M. Ivaşcu, and W. Greiner, Nucl. Tracks, **12** (1986) 313.

[37] D. N. Poenaru, J. A. Maruhn, W. Greiner, M. Ivaşcu, D. Mazilu, and R. Gherghescu, Z. Phys., A **328** (1987) 309.

[38] D. N. Poenaru, M. Ivaşcu, D. Mazilu, R. Gherghescu, K. Depta, and W. Greiner, Central Institute of Physics, Bucharest, Report NP-54-86, 1986.

[39] D. N. Poenaru, M. Ivaşcu, D. Mazilu, I. Ivaşcu, E. Hourani, and W. Greiner, in *Developments in Nuclear Cluster Dynamics* edited by K. Akaishi et al. (World Scientific, Singapore, 1989), p.76.

[40] D. N. Poenaru, D. Schnabel, W. Greiner, D. Mazilu, R. Gherghescu, Atomic Data and Nuclear Data Tables, in print.

[41] G. A. Pik-Pichak, Sov. J. Nucl. Phys. **44** (1987) 923.

[42] D. N. Poenaru,J. A. Maruhn, W. Greiner, M. Ivaşcu, D. Mazilu, and I. Ivaşcu, Z. Phys. A **333** (1989) 291.

[43] D. N. Poenaru, M. Ivaşcu, I. Ivaşcu, M. Mirea, W. Greiner, K. Depta, and W. Renner, in *50 Years with Nuclear Fission* (American Nuclear Society, Lagrange Park) 1989, p 617.

[44] D. N. Poenaru, M. Mirea, W. Greiner, I. Căta, and D. Mazilu, Modern Physics Letters A **5** (1990) 2101.

[45] K. Depta, R. Herrmann, J. A. Maruhn, and W. Greiner, in *Dynamics of Collective Phenomena* edited by P. David (World Scientific, Singapore) 1987, p. 29.

[46] B. G. Novatsky and A.A.Ogloblin, Vestnik AN SSSR, 1988, p. 81.

[47] D. N. Poenaru and W. Greiner, J. Phys. G, in print.

[48] D. N. Poenaru, W. Greiner, and M. Ivaşcu, Nucl. Phys., **A 502** (1989) 59c.

[49] M. Seiwert, W. Greiner, and W. T. Pinkston, J. Phys. G., **11** (1985) L-21.

[50] J. F. Berger, J. D. Anderson, P. Bonche, and M. S. Weiss, Phys. Rev., **41** (1990) R2483.

[51] W. Greiner, B. Müller, and J. Rafelski, *Quantum Electrodynamics of Strong Fields* (Springer Verlag, Berlin), 1985.

[52] A. A. Ogloblin et al., Phys. Lett. B **235** (1990) 35.

Balian-Bloch Supershells for Pedestrians

S. Bjørnholm

The Niels Bohr Institute, University of Copenhagen
Blegdamsvej 17, DK-2100 Copenhagen Ø, Denmark

Abstract:

The occurrence of electronic shells and supershells in sodium clusters can be understood in a rather simple way by invoking Bohr's quantum postulate from 1913 and the Fermi gas expression for the relation between density and Fermi momentum, which in turn follows from the Pauli principle. An analysis of theoretical and experimental data on this basis lets the closed electronic orbits of triangular and square shape emerge as the main contributors to the supershells, in agreement with the idealized predictions by Balian and Bloch (1971).

1. Energy Eigenvalues in a Spherical Potential

Atoms and nuclei are limited in size to a few hundred constituent fermions. This is not so with metallic clusters. The shell structure encountered in these clusters is therefore potentially more than just a repeat of previously known physical effects. Metal clusters indeed add new dimensions to the phenomenon of shell structure. Here we will discuss a particularly striking aspect, the supershells. In essence, this aspect amounts to following a pure quantum phenomenon towards the limit of large quantum numbers where – according to Niels Bohr – a correspondence between classical and quantal motion will become apparent.

To begin with, we may look at pure quantum mechanical calculation – a very simple one. Approximating real clusters as systems of independent electrons in a common mean field of the Woods-Saxon type

$$V(r) = \frac{V_\circ}{1 + \exp\left((r - R)/a\right)} \ . \tag{1}$$

one can generate the energy eigenvalues ε_i by solving the one particle Schrödinger equation with this radial potential. The sum of eigenvalues will represent a measure of the total binding energy $E(N)$ of a spherical cluster of size N.

$$E(N) = \sum_{i=1}^{N} \epsilon_i \ . \tag{2}$$

Plotting this quantity as a function of N one discovers that it can be approximated rather accurately by

$$\widetilde{E}(N) = -aN + bN^{2/3} \ , \tag{3}$$

with a and b being positive numbers. This expression describes the binding of a classical liquid drop, a is the binding energy per particle inside the drop(volume energy), while $bN^{2/3}$ is a correction term, the surface energy, accounting for the reduced binding of the particles on the surface of the drop. The approximation is not perfect. The difference between $E(N)$ and $\widetilde{E}(N)$ is finite as it must be, since $E(N)$ oscillates as a result of the non-uniform distribution of eigenvalues associated with shell structure while $\widetilde{E}(N)$ is perfectly smooth. This difference is a measure of the modulation of the total binding energy due to shell structure, viz.

$$E_{shell} = E(N) - \widetilde{E}(N) \tag{4}$$

Figure 1 is a plot of this quantity, taken from ref. [1]. The downward cusps occur at shell closings and represent the magic numbers. As one sees, the shells appear to be regularly spaced. They lie almost equidistantly in the plot where the abcissa is a measure of the cluster radius, proportional to the cube root of the size N.

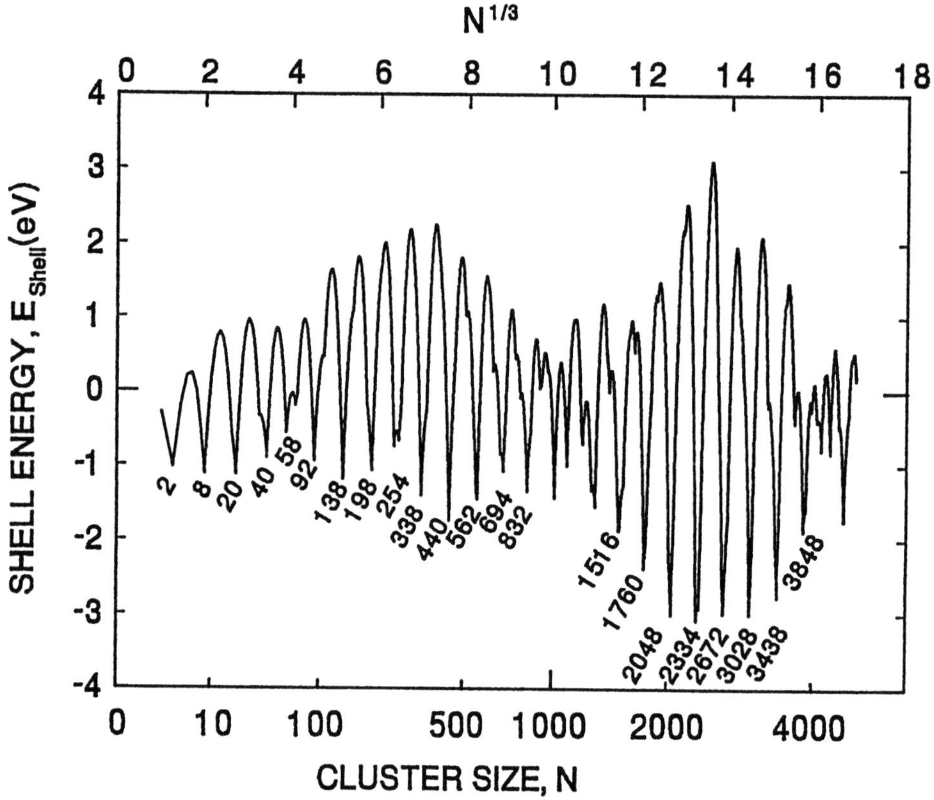

Fig. 1 - *The periodically varying contribution to the calculated binding energy of a spherical sodium cluster. The total binding energy is about one electron volt (eV) per atom, i.e.* $E_{tot} \approx N$ *eV. The periodic variations are due to the quantized motion of the conduction electrons in a spherical field of the type eq. (1). They are small compared to the total binding energy, but they are responsible for the intensity variations observed in experiment. From [1].*

In addition to the ordinary periodic shell structure, the calculation in Figure 1 also shows a higher-order periodicity. The shell oscillations reach a maximum in amplitude around the tenth period; they then tend to die out at period fifteen, only to rise again to a new maximum at the twentieth period, around size 2300. This higher-order beat mode actually keeps repeating as long as the calculation goes, i.e. to a size of about ten thousand. The calculation thus predicts a sequence, where the shell structure appears to terminate at about size one thousand. Here the model clusters exhibit bulk-like, structureless electronic properties. In fact, however, this is just a passing stage in the development. The shell structure reappears.

These two kinds of periodicity, the primary shell structure and the beat mode, or supershell, appear here as the result of the purely numerical exercise of solving the one-particle Schrödinger equation with the Woods-Saxon potential eq. (1). How can one understand the two periods? As always when one has to do with a beat mode, one can either think of it in terms of two close-lying frequences or wavenumbers, or in terms of their mean value and their difference. The ratio of the latter two is 10-15 in the present case, i.e. the two base frequencies or wavenumbers differ by some 7-10 pct. How does that come about?

2. Quantum Shells and Closed Classical Orbits

To answer this it is helpful to express the periodicities in terms of the relation between magic numbers $N_\circ$ and the shell number n and then compare with some more familiar cases. As one sees from the plot in Fig. 1, the shell closings, i.e. the magic numbers $N_\circ$ are equidistantly spaced. The abscissa is linear in $N^{1/3}$, so $N_\circ^{1/3}$ is linear in n, or $N_\circ$ is proportional to n^3 to leading order. The numerical coefficient can be estimated by inserting the calculated magic numbers between 100 and 1000 to obtain, for the Woods-Saxon potential:

$$N_\circ \approx 0.21\, n^3 + \ldots \,, \tag{5}$$

Let us compare this with the eigenvalue spectrum of the hydrogen atom, or more generally the Kepler problem of a particle in an r^{-1}-potential. Each shell is perfectly degenerate with all odd and even numbered ℓ-values represented in a given shell n up to $\ell_{max} = (n-1)$. The number of (Fermi) particles in a shell is equal to

$$\sum_{\ell=0}^{\ell_{max}} 2(2\ell+1) = 2(\ell_{max}+1)^2 = 2n^2 \ . \tag{6}$$

The magic numbers for the r^{-1} potential are:

$$N_\circ = \sum_{\nu=1}^{n} 2\nu^2 = \frac{2}{3}n\left(n+\frac{1}{2}\right)(n+1) \approx \frac{2}{3}n^3 + \ \dots \ , \tag{7}$$

with n = 1,2,3, ... Thus there are more levels per shell than with the Woods-Saxon potential.

In the spherical harmonic oscillator potential each (perfectly degenerate) shell holds alternating even (0,2,4, ...) or odd (1,3,5, ...) ℓ-values. The magic numbers (harmonic oscillator) are:

$$N_\circ = \frac{1}{3}(n+1)(n+2)(n+3) \approx \frac{1}{3}n^3 + \ \dots \tag{8}$$

with n = 0,1,2,3, ... Here the number of particles per shell is intermediate between the Kepler potential and the Woods Saxon potential. On the other hand, the third power dependence is common to all three spherical potentials; as opposed to the subshell ℓ-degeneracies that increase as ℓ^2 or, effectively, as n^2.

As a next step we may look at the three dimensional harmonic oscillator.

$$E_n = \hbar\omega_x \left(n_x + \frac{1}{2}\right) + \hbar\omega_y \left(n_y + \frac{1}{2}\right) + \hbar\omega_z \left(n_z + \frac{1}{2}\right) \tag{9}$$

It is equivalent to three independent one-dimensional harmonic oscillators. The motion is separable in the three coordinates, and the total energy is just the sum of three independent contributions. Expressing the change in total eigen-energy eq. (9) in terms of the changes in individual quanta, expressed in units of action $\hbar$, leads to:

$$\delta E_n = \frac{\partial E_n}{\partial(\hbar n_x)}\,\delta(\hbar n_x) + \frac{\partial E_n}{\partial(\hbar n_y)}\,\delta(\hbar n_y) + \frac{\partial E_n}{\partial(\hbar n_z)}\,\delta(\hbar n_z)$$

$$= \omega_x\delta(\hbar n_x) + \omega_y\delta(\hbar n_y) + \omega_z\delta(\hbar n_z)\ . \tag{10}$$

The particular way of writing this example shows how the derivative of the quantum energy E_n with respect to action quantum numbers $\hbar n_i$ is equal to the frequency of the motion of the *classical* harmonic oscillator. This observation is actually not as particular as one may suspect. It is valid for any quantum system, where the corresponding classical motion is periodic with characteristic frequencies ω_i. Another simple example is circular motion in a radial potential. The quantum energy is $E_\ell = \ell(\ell + 1)\hbar^2/2mR^2$. Thus, with the angular momentum $mvR = \hbar(\ell + 1/2)$,

$$\frac{\partial E_\ell}{\partial(\hbar\ell)} = \frac{\hbar(\ell + 1/2)}{mR^2} = \frac{mvR}{mR^2} = \frac{v}{R} = \omega_\ell\ . \tag{11}$$

Returning to the question of shell structure, a perfectly degenerate shell in a three-dimensional harmonic oscillator requires that δE_n in eq. (10) is equal to zero for several values of δn_x, δn_y, δn_z, i.e.

$$\omega_x\delta n_x + \omega_y\delta n_y + \omega_z\delta n_x = 0\ , \tag{12}$$

with δn_x, etc. being positive or negative integers. Clearly, this is equivalent to requiring $\omega_x : \omega_y : \omega_z = a^{-1} : b^{-1} : c^{-1}$ where a, b, and c are integers.

Integer ratios for the three independent periods implies that the classical motion is described by *closed orbits.* After a period divisible with the three individual periods, the motion will be back at the initial point. Thus, in this example, there is a direct connection between closed classical orbits and quantal shell structure. In addition, we have seen that the partial derivative of the total eigenenergy with respect to a quantized action variable is equal to the frequency of the corresponding classical motion.

The situation with the three dimensional harmonic oscillator is particularly transparent because not only is the motion separable in x,y, and z; the total energy is also a linear sum of three independent contributions, each governed by its specific quantum number. (A particle in a rectangular box is another slightly more complicated example).

The motion in the r^{-1} potential is also separable in the polar coordinates (r, θ, φ), but the total eigenvalue is not a sum of three independent contributions. (The radial wave equation depends on ℓ). Still, one can write the eigenenergy E_n for hydrogen:

$$E_n = Ry\, n^{-2} = Ry\,(n_r + \ell)^{-2} \; , \tag{13}$$

where $(n_r - 1)$ denotes the number of nodes in the radial wave function.(For a given ℓ, the m-states are always degenerate, and the radial wave function is independent of m). Using the generalized result that the derivative of the total eigenenergy with respect to an action quantum number equals the classical frequency and that $(n_r + \ell) =$ const. for any shell leads to

$$\omega_r = \omega_\ell \tag{14}$$

or that the angular period and the radial period of the classical motion are equal. The Kepler ellipses describing orbiting in the r^{-1}-potential with its center at one of the focal points have exactly this property, whatever their size or eccentricity. Thus the occurrence of highly and fully degenerate eigenstates reflects the existence of closed elliptical orbits as the only kind of motion for a bound state in a r^{-1}-potential.

The motion in the three-dimensional, spherical harmonic oscillator potential is also elliptic. Here, the center of the orbits coincide with the center of the potential. This means that the radial motion goes through two complete periods (in-out-in-out) for each angular turn. Thus, classically:

$$\omega_r = 2\omega_\ell \tag{15}$$

for the spherical, harmonic oscillator. This is again reflected in the eigenvalue equation. In polar coordinates it can be written

$$E_n = \hbar\omega_\circ \left(n + \frac{3}{2}\right) = \frac{1}{2}\hbar\omega_\circ \left(2n_r + \ell + \frac{3}{2}\right) \; , \tag{16}$$

where n_r describes the number of nodes in the radial wave function. With $\partial E_n/\partial n_r \equiv \hbar\omega_r$ and $\partial E_n/\partial \ell \equiv \hbar\omega_\ell$ we again obtain eq. (15); this time from the quantal description.

Figure 2 summarizes and illustrates the correspondence between closed classical orbits and quantal shell structure. To illustrate further one can think of plots of E_{shell} versus $N^{1/3}$ for the hydrogen atom and the spherical harmonic oscillator in analogy to Fig. 1. Compared to this figure such plots (not shown) differ in three respects. *First*, the shell spacings are different, eqs. (5, 7 and 8)

$$\text{Hydrogen} : N_\circ \approx \frac{2}{3}n^3 + \; ... \qquad ; \; \omega_\ell = \omega_r \tag{17a}$$

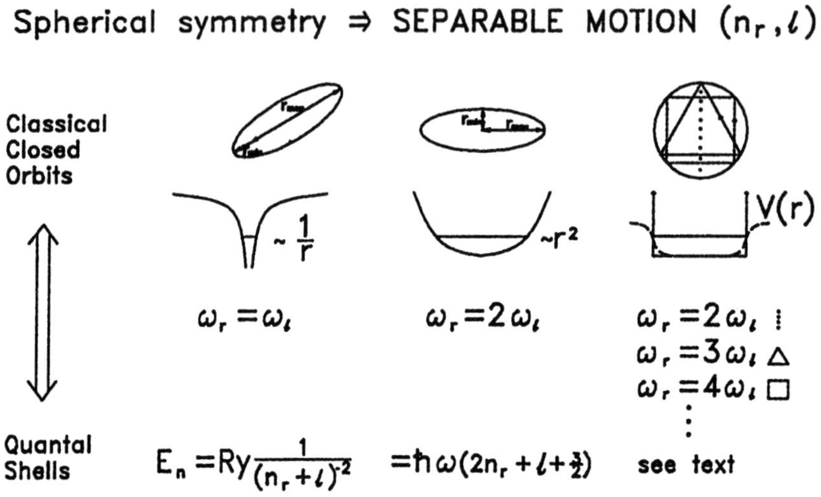

Fig. 2 - *For spherically symmetric potentials the motion is separable in the radial and angular coordinates. Classically one can therefore speak about separate frequencies in the radial and angular motion, respectively. If the frequencies stand in a (simple) integer ratio to each other, the orbits will be closed. Quantum mechanically, the energy eigenvalues can be labelled with both radial and an angular quantum numbers. Closed periodic orbits in the classical case, and degeneracies of levels with different radial and angular quantum numbers, (shell structure) in the quantal case are closely related to each other. In the two cases, left and center, there is only one type of closed orbit. In the third case, right, several different orbits are possible.*

$$\text{Harm. osc.}: N_\circ \approx \frac{1}{3}n^3 + \ldots \qquad ; \; \omega_\ell = \frac{1}{2}\omega_r \tag{17b}$$

$$\text{Woods} - \text{Saxon}: N_\circ \approx 0.21 \cdot n^3 + \ldots \quad ; \; \omega_\ell \approx \frac{1}{3}\omega_r \; ? \tag{17c}$$

Comparing these spacings with the classical frequency relations eqs. (14 and 15) seems to suggest that classical triangular or pear-shaped orbits are playing a role in the case of the Woods-Saxon potential, or equivalently that one out of three ℓ-values contribute to each shell. *Secondly*, the hydrogen spectrum and the harmonic oscillator exhibits no beat pattern, as opposed to Fig. 1. This reflects the fact that there is only one type of orbit in each case and only one frequency relation. A beat pattern requires two different orbits with two different frequency relations, i.e. periods. In other words, this is a hint that triangular orbits cannot alone account for the shell and supershell periodicities obtained with the Woods-Saxon potential. *Thirdly*, the amplitudes in plots analogous to Fig. 1 are larger for the hydrogen and harmonic oscillator cases and also more regular. This reflects the perfect degeneracies within a shell in these two models. The eigenenergy spectrum of the Woods-Saxon model is more spread out. There is bunching, but no perfect degeneracies.

3. Triangles and Squares

These qualitative considerations lead naturally to ask what can be said about classical orbits in a Woods-Saxon potential, eq. (1). Without much loss of generality one may instead think simply of a spherical cavity.

There is a large number of orbits possible. Some are closed, others never return to the starting point. The closed orbits may be simple, closing after a single angular turn, or they may form more complicated, starlike figures. The situation is not at all as simple as in the hydrogen or harmonic oscillator cases. Nevertheless one can order the orbits according to increasing length. The simplest are the pendulating orbits going through the origin. There is a two-dimensional multitude of those, corresponding to the two polar angles. Next comes the triangular and then the square orbits, representing a three-dimensional multitude each (corresponding to the three Euler angles), and so forth. What is needed is a theory that connects this system of classical orbits with the eigenvalue spectrum of stationary waves in the cavity. Such a theory actually exists. It is due to Balian and Bloch [2] and to Gutzwiller [3]. In ref. [2] the theory is applied to the cavity, while in ref. [1] it is extended to the spherical Woods-Saxon potential. In essence, the complete set of delta functions describing eigenvalues as a function of energy can be expressed as a sum over contributions from an infinite series of closed classical orbits of increasing length. The major periodicities, the ones seen in Fig. 1, on the other hand, emerge already from the sum truncated after the third i.e. square, orbit. The pendulating orbits contribute little, and thus the shell and supershell periodicities can be understood as the quantum signatures of the classical triangular and square orbits contributing with about equal strength to the eigenvalue spectrum.

The beat pattern can be qualitatively visualized by looking at a particle of definite speed, v_F, in a cavity of increasing radius $R \sim N^{1/3}$, where N is the N'^s eigenstate, degeneracies included. If $R = r_{WS} \cdot N^{1/3}$, where r_{WS} is the Wiegner-Seitz radius, it corresponds to looking at an electron in a metallic drop where the electron moves with Fermi velocity. The action J of such a particle in a closed orbit of length L is (2 · energy · time) $= mv^2\tau$, τ being the period of the orbit equal to L/v_F. This action is quantized in units of h. For a triangular orbit, $L_3 = 3\sqrt{3} \cdot R$. For a square orbit $L_4 = 4\sqrt{2} \cdot R$. Hence

$$J_3 = n_3 h = mv_F^2 \frac{L_3}{v_F} = 3\sqrt{3}\, mv_F R_3 \ , \tag{18a}$$

$$J_4 = n_4 h = mv_F^2 \frac{L_4}{v_F} = 4\sqrt{2}\, mv_F R_4 \ , \tag{18b}$$

or,

$$R_3 = \left(\frac{h}{mv_F}\right) \frac{n_3}{3\sqrt{3}} \ ; \qquad n_3 = 1,\ 2,\ 3,\ \ldots \tag{19a}$$

$$R_4 = \left(\frac{h}{mv_F}\right) \frac{n_4}{4\sqrt{2}} \ ; \qquad n_4 = 1,\ 2,\ 3,\ \ldots\ . \tag{19b}$$

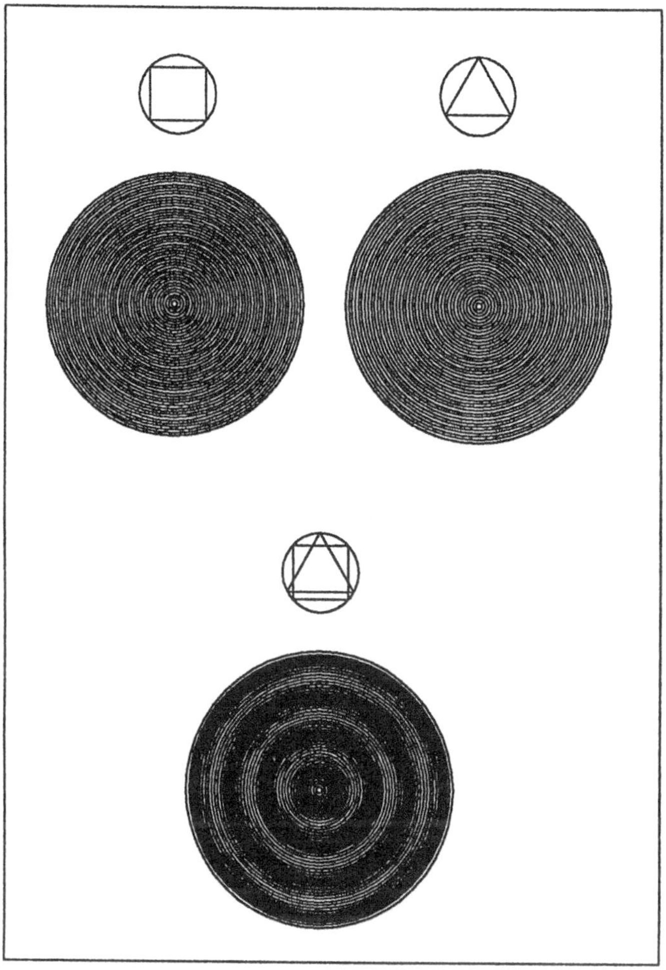

Fig. 3 - *If two classically closed orbits e.g. triangles and squares contribute equally to the quantal shell structure, it will give rise to a higher periodicity, supershells.*

If only triangular orbits were contributing to shell structure, closed shells would occur for drop sizes shown in the top right of Figure 3. Analogously, if square orbits were dominating, the sizes at the top left would represent closed shells. If both orbits contribute equally it results in the size pattern shown below in Fig. 3. The triangular and square orbits are alternatingly being in phase and out of phase, resulting in the beat pattern shown. Simple quantization of the action of classical triangular and square orbits having the same energy thus leads to the same results as solving the Schrödinger wave equation for a particle in the same potential. This is the correspondence principle.

The correspondence becomes more accurate as the action quantum number increases; and metal clusters offer special opportunities for observing systems with large quantum numbers [4,5].

The ultimate correspondence with classical orbital motion of a well-localized particle with continuously variable total energy requires that one goes beyond a description in terms of sharp energy eigenvalues. By combining wave functions with close-lying eigenvalues one can build wave packets moving like a classical particle. With increasing quantum numbers, the relative uncertainty in momentum and position of the wave packet decreases, converging in this way towards the classical limit.

References

1. H. Nishioka, K. Hansen, and B. R. Mottelson, *Phys. Rev.* **B42**, 9377 (1990).
2. R. Balian, and C. Bloch, *Ann. Phys.* **69**, 76 (1971).
3. M. C. Gutzwiller, *J. Math. Phys.* **8**, 1979, (1967), *ibd* **11**, 1791 (1969), and *ibd* **12**, 343 (1971).
4. J. Pedersen, S. Bjørnholm, J. Borggreen, K. Hansen, T.P. Martin, and H. D. Rasmussen, *Nature* **353**, 733 (1991)
5. T. P. Martin, present *Lecture Notes in Physics*, Springer (1991)

SELFCONSISTENT CALCULATION OF ELECTRONIC SUPERSHELLS IN METAL CLUSTERS

O. Genzken and M. Brack

Inst. f. Theor. Physik, Universität Regensburg, D-8400 Regensburg, Germany

Abstract: We report on selfconsistent microscopic calculations of the electronic shell and supershell structure of sodium clusters with up to $N \sim 3000$ atoms. The spherical jellium model in local density approximation is used and the Kohn-Sham equations are solved numerically. The finite temperature of the valence electrons is included by treating them as a canonical subsystem embedded in the heat bath of the ions. In particular, we evaluate the total free energy $F(N)$ and investigate its fluctuating part, the shell-correction energy $\delta F(N)$, as a function of temperature T and particle number N. We also discuss the second difference $\Delta_2 F(N) = F(N+1) + F(N-1) - 2F(N)$ and its relation to the recently measured cluster mass abundance spectra.

Metal clusters provide a unique example for the study of shell effects in finite fermion systems containing up to several thousand particles [1]. A selfconsistent microscopic description of such large systems is possible only in the so-called self-consistent jellium model [2], where the interacting valence electrons move in the field of a uniformly charged sphere ('jellium') representing the ions. We report here on recent Kohn-Sham calculations for sodium clusters with up to $N \sim 3000$ atoms at finite temperatures [3].

Nishioka *et al.* [4], using a phenomenological Woods-Saxon potential fitted to the microscopic potentials of Ekardt [2], have drawn attention to the 'supershell structure' in the level density and the oscillating part of the total binding energy of clusters with N up to 4000: a pronounced beating pattern in which the shell structure is enveloped by a slowly oscillating amplitude. This is, in fact, a very general feature of discrete eigenmodes in a cavity or in any steep potential confining many particles to a limited domain of space. Balian and Bloch have shown in their fundamental work [5] that the beating pattern of the level density in an infinite square well is explained by the superposition of amplitudes associated to closed classical trajectories; they reproduced the approximate shape of the exact level density by summing up only the contributions from triangular and squared classical orbits. Extending this method to smooth potentials, Strutinsky *et al.* [6] were able to explain the gross shell structure of atomic nuclei using realistic deformed shell model potentials. Nuclei, however, are not big enough to exhibit the 'supershell' beating of their shell structure.

The experimental observation of supershell structure in cluster expansion sources is inhibited by the fact that the clusters so produced have, at least initially, a finite temperature which tends to reduce the shell effects [7,8,9]. Nevertheless, in the newest sodium vapour expansion experiments of the Copenhagen - Orsay - Stuttgart collaboration [1], a supershell beating in the mass abundance spectrum of sodium clusters has been put into evidence. We shall show here that the experimental results can be explained semi-quantitatively in the selfconsistent jellium model, if the effect of finite temperature on the valence electrons is properly included.

We use the finite-temperature density functional theory in the Kohn-Sham (KS) approach [10], employing the spherical jellium model and the local-density functional for exchange and correlations by Gunnarsson and Lundqvist [11]. The Wigner-Seitz radius of bulk sodium, $r_s = 3.96$ a.u., is used; otherwise our calculations are completely parameter free. We treat the valence electrons as a *canonical* ensemble in the heat bath of the ions and minimize the Helmholtz free energy $F(N) = E(N) - TS(N)$ of a cluster with N atoms, where E is its total internal energy and S the entropy of the electrons at a given temperature T. (See Ref. [9] for details and, in particular, for a fast algorithm for the exact calculation of the canonical partition function.) A canonical treatment with exactly conserved particle number N is important here since we investigate quantities like $\Delta_2 F(N) = F(N+1) + F(N-1) - 2F(N)$ which are very sensitive to temperature effects through the entropy part $-TS$: the large degeneracies of the spherical magic shells lead to large entropies even at small temperatures.

The quantity $\Delta_2 E(N) = \Delta_2 F(N)(T = 0)$ has often been taken as a measure for the stability of the cluster: since it represents the curvature of the total binding energy as a function of N, it is particularly large for the 'magic' systems which have a strongly negative shell correction. It has furthermore been argued [7] that if the *evaporation* process, which takes place immediately after the adiabatic expansion, is responsible for enhancing the most stable clusters in the final mass yield, $\Delta_2 F(N)$

should be proportional to $-\Delta_1 \ln I_N$, where I_N is the fluctuating part of the observed mass yield. In Refs. [8,9] it has been shown that $\Delta_2 F(N)$ decreases very fast for $N \sim 100 - 400$ and becomes practically zero for $N \geq 500$ already at $T \gtrsim 400$ K. Therefore, the smoothing effect of a finite temperature on quantities like $\Delta_2 F(N)$ is very crucial for the observability of shell structure and, in particular, the supershells in large metal clusters.

The temperature dependence of shell structure has been well studied in nuclei, both schematically [12] and in selfconsistent Hartree-Fock calculations [13]. In a schematic harmonic oscillator approximation, the amplitude of the shell-correction δF — and thus also of the quantity $\Delta_2 F(N)$ in which the average energies practically cancel — is found to go like $\delta F(T) = \delta F(0)\ \tau/(\mathrm{Sinh}\tau)$ with $\tau = 2\pi^2 T/\hbar\omega$. Expanding for large temperatures and using $\hbar\omega \propto N^{-1/3}$, this gives a temperature suppression factor $\propto \exp(-N^{1/3})$. Pedersen *et al.* [1] therefore multiplied the logarithmic derivatives of the mass yields I_N by $\sqrt{N}\exp(cN^{1/3})$, where c is a constant containing an effective temperature, and the root factor compensates the decrease of the shell-correction at $T = 0$ with increasing N [12]. In the resulting plot, magic shell closures with N up to ~ 2720 and a beating of the shell oscillations can clearly be seen (see also Fig. 2 below).

As an example of our theoretical results, we show in Fig. 1 the free energy shell-correction $\delta F(N)$ versus $N^{1/3}$ at the three temperatures $T = 0$ K, 400 K and 600 K. [Hereby we simply used $\delta F(N) = F(N) - \overline{F}(N)$ with a liquid drop model type expansion for the *average* free energy, $\overline{F}(N) = e_b N + a_s N^{2/3} + a_c N^{1/3}$, determining a_s and a_c at each temperature by a simple eye fit such that $\delta F(N)$ is oscillating around zero. The bulk energy is fixed at its theoretical value e_b = -2.2567 eV.]

The salient feature of the curves in Fig. 1 is the supershell beating of the otherwise quite regular shell structure. The $T = 0$ curve is very similar to that obtained by Nishioka *et al.* [4] for a phenomenological Woods-Saxon potential; note that the present results are fully selfconsistent. The amplitude of the shell effects is clearly reduced with increasing temperature.

In Figure 2, we have reproduced the relevant figure from the experimental analysis of Pedersen *et al.* [1] and compare it to our theoretical results. Here the negative second difference $-\Delta_2 F(N)$ is shown, multiplied by the same enhancement factor (with the value of c readjusted by $\sim 10\%$). In spite of the simplyfing assumptions underlying the identification of $-\Delta_2 F(N)$ with $\Delta_1 \ln I_N$ [7,9], the agreement of the two curves is striking. This demonstrates that the ***finite temperature of the valence electrons*** which alone contribute to the quantities shown here — the ionic parts of the free energies practically cancel in the differences $\Delta_2 F(N)$ and $\Delta_1 \ln I_N$ — plays an essential role in the mass yields and can be correctly taken into account in selfconsistent KS calculations even in the simple jellium model.

Much remains, however, to be understood – in particular the value of the factor c which, using the above harmonic oscillator estimates, is too large for the estimated temperatures [1,7] of $\sim 400 - 500$ K. A more realistic study of the evaporation mechanism and, more generally, a non-equilibrium treatment of the ions' dynamics would be desirable to this aim.

We are grateful to S. Bjørnholm for many enlightening discussions and a continuing encouragement, and to K. Hansen for important contributions at the early stages of our investigations.

References

[1] J. Pedersen, S. Bjørnholm, J. Borggreen, K. Hansen, T. P. Martin and H. D. Rasmussen, Nature **353**, 733 (1991); T. P. Martin, S. Bjørnholm, J. Borggreen, C. Bréchignac, Ph. Cahuzac, K. Hansen and J. Pedersen, Chem. Phys. Lett. **186**, 53 (1991); see also the talks of S. Bjørnholm and T. P. Martin, these Proceedings.

[2] W. Ekardt, Phys. Rev. **B 29**, 1558 (1984).

[3] O. Genzken and M. Brack, Phys. Rev. Lett. **76**, 3286 (1991).

[4] H. Nishioka, K. Hansen, and B. R. Mottelson, Phys. Rev. **B 42**, 9377 (1990).

[5] R. Balian, C. Bloch, Ann. Phys. (N.Y.) **69**, 76 (1971).

[6] V. M. Strutinsky, A. G. Magner, S. R. Ofengenden, and T. Døssing, Z. Phys. **A 283**, 269 (1977).

[7] S. Bjørnholm, J. Borggreen, O. Echt, K. Hansen, J. Pedersen, H. D. Rasmussen, Z. Phys. **D 19**, 47 (1991).

[8] M. Brack, O. Genzken, and K. Hansen, Z. Phys. **D 19**, 51 (1991).

[9] M. Brack, O. Genzken, and K. Hansen, Z. Phys. **D 21**, 65 (1991).

[10] For a review on finite-temperature density functional theory, see U. Gupta and A. K. Rajagopal, Physics Reports **87**, 259 (1982); for its justification for canonical systems, see R. Evans, Adv. in Phys. **28**, 143 (1979).

[11] O. Gunnarsson and B. I. Lundqvist, Phys. Rev. **B 13**, 4274 (1976).

[12] A. Bohr, and B. M. Mottelson, *Nuclear Structure II* (Benjamin, 1975).

[13] See, e.g., M. Brack and P. Quentin, Nucl. Phys. **A 361**, 35 (1981), and earlier references quoted there.

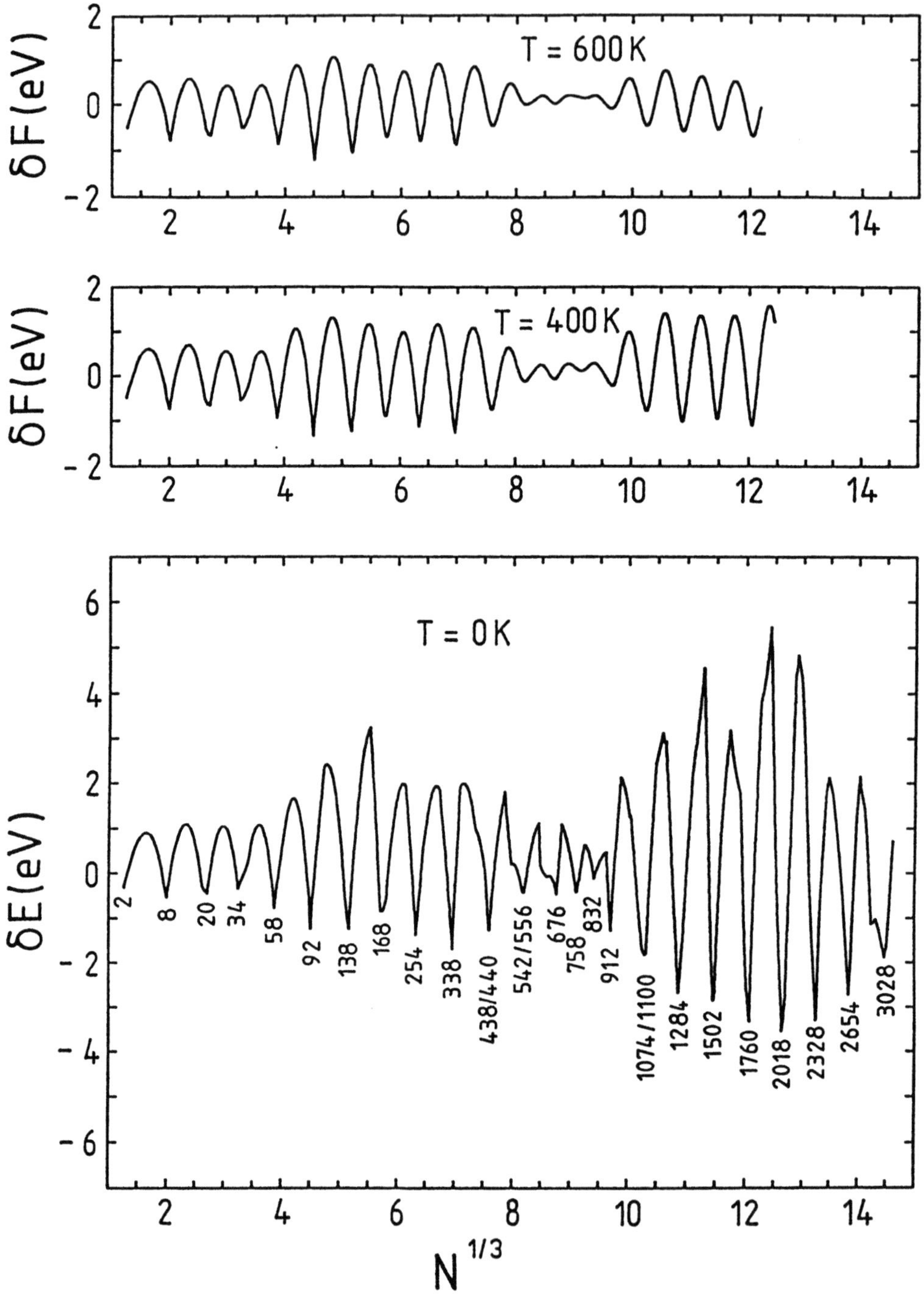

Figure 1: Free energy shell-correction $\delta F(N) = F(N) - \overline{F}(N)$ for spherical Na clusters versus $N^{1/3}$ for three different temperatures T, obtained in selfconsistent KS calculations [3]. LDM parameters used at $T = 0$ K: $a_s = 0.6259$, $a_c = 0.2041$; at $T = 400$ K: $a_s = 0.5918$, $a_c = 0.3796$; and at $T = 600$ K: $a_s = 0.5755$, $a_c = 0.4204$ (all in eV). Numbers near the bottom are the magic numbers of filled major spherical electronic shells.

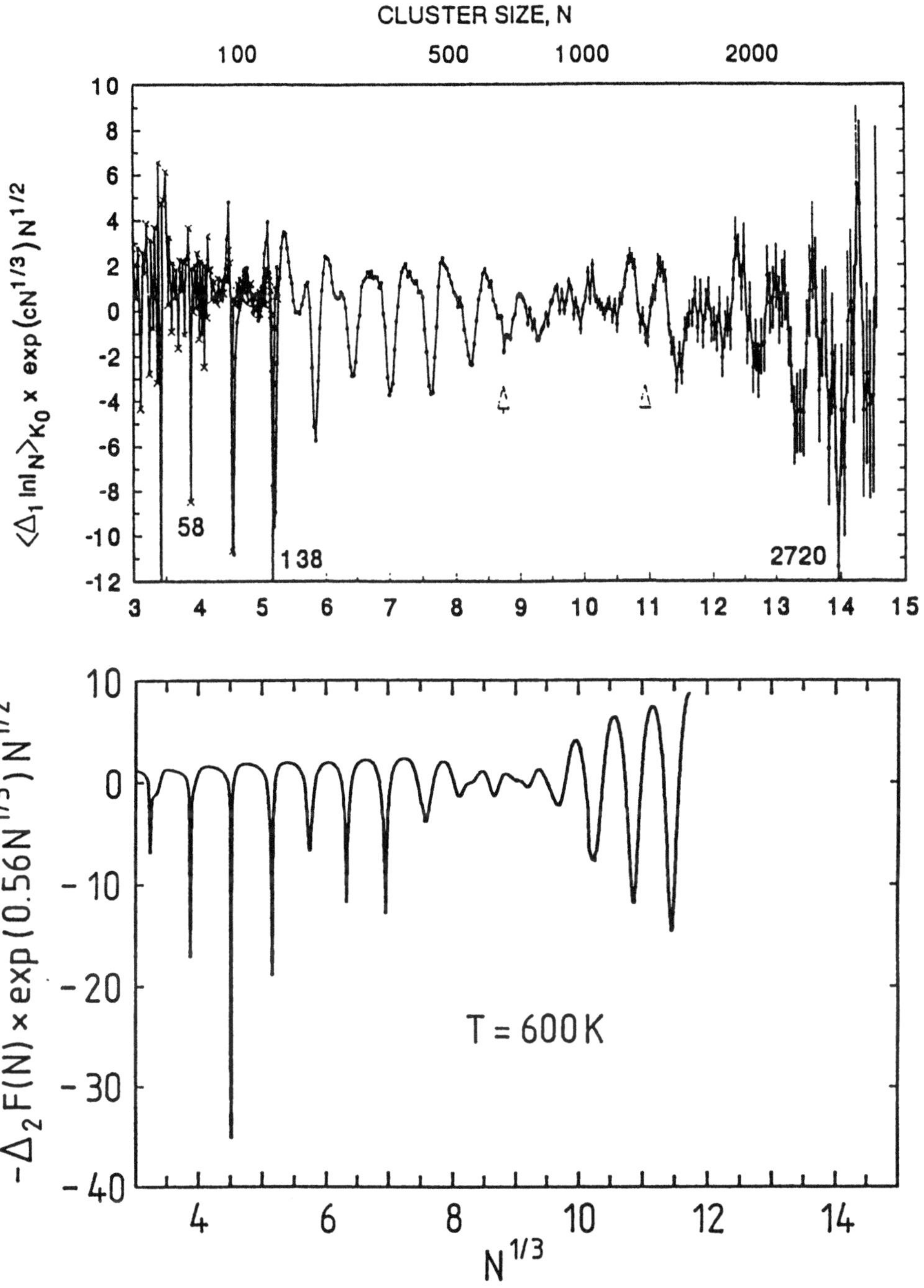

Figure 2: *Upper part:* Relative variation $< \Delta_1 \ln I_N >_{K_0}$ in experimental Na cluster abundance I_N versus $N^{1/3}$ (Pedersen *et al.* [1]; see this reference for the details). *Lower part:* Negative second difference $-\Delta_2 F(N)$ of free energy obtained in selfconsistent KS calculations [3] at $T = 600$ K. Both quantities are enhanced by a N-dependent factor to compensate for the temperature suppression (see text).

Supershells in Laser-Warmed Na-Clusters

T.P. Martin[1], *S. Bjørnholm*[2], *J. Borggreen*[2], *C. Bréchignac*[3], *Ph. Cahuzac*[3], *K. Hansen*[2] *and J. Pedersen*[2]

[1]Max-Planck-Institut für Festkörperforschung
Heisenbergstr. 1, 7000 Stuttgart 80, FRG
[2]The Niels Bohr Institute, University of Copenhagen,
4000 Roskilde, Denmark
[3]Laboratoire Aimé Cotton, CNRS II,
Bâtiment 505, 91405 Orsay Cedex, France

Abstract

Mass spectra are reported for large sodium clusters warmed by a continuous laser beam prior to ionization. During the 1ms warming period the clusters lose more than 10% of their mass by single atom evaporation. The resulting size distribution reveals what appears to be electronic supershell structure for clusters containing up to 2500 atoms.

Introduction

The one-particle eigenstates of fermions moving in a spherically symmetric potential are characterized by a well-defined angular momentum. In large systems the degeneracy of these so-called subshells is considerable, $2(2\ell+1)$. For certain forms of the radial dependence of the potential there can be a further condensation of subshells into highly degenerate shells. However, there exist only a few known cases for which this shell degeneracy is exact, e.g. the hydrogen atom and the spherical harmonic oscillator. More often, shells appear as an approximate grouping of subshells on an energy scale, as in atomic nuclei. In fact, the earliest model developed for nuclei [1,2] describes very nicely the elecronic structure of metal clusters [3-18] containing hundreds of electrons.

In this paper we present evidence for electronic shell formation in sodium clusters containing up to 2500 valence electrons.

Experimental

The technique we have used to study shell structure in metal clusters is photoioinization time-of-flight (TOF) mass spectromery. The mass spectrometer has a mass range of 600 000 amu and a mass resolution of up to 20 000. The cluster source is a low pressure, rare gas, condensation cell. Sodium vapor was quenced in cold He gas having a pressure of about 1 mbar. Clusters condensed out of the quenched vapor and were transported by the gas stream through a nozzle and through two chambers of intermediate pressure into a high vacuum chamber. The size distribution of the clusters could be controlled by varying the oven-to-nozzle distance, the He gas pressure, and the oven temperature. The clusters were photoionized with a 0.4 μJ, 2x1 mm, 15 ns, 308 nm (4.0 eV) excimer laser pulse.

The clusters were warmed prior to ionization with a continuous Ar ion laser beam running parallel to the neutral cluster beam. The laser light entered the ionization chamber through a heated window, passed through a 3.0 mm diameter nozzle, through the oven chamber and finally exited through a second window where the laser intensity was recorded. Short wavelength light was found to warm much more efficiently. Using the 458 nm (2.71 eV) laser line, 10 mW proved sufficient to appreciable alter the neutral size distribution.

Results

Electronic shell structure can be observed exerimentally in several ways. Perhaps the most easily understood method is to (i) measure the ionization potentials for each cluster size. Electrons in newly opened shells are less tightly bound, ie. have lower ionization energies. However, considerable experimental effort is required to measure the ionization energy of even a single cluster. A complete photoionization spectrum must be obtained and very often an appropriate source of tunable light is simply not available. It is much easier to observe shell closings in photoionization, TOF mass spectra. This can be done in either one of two ways. (ii) If mass spectra are recorded using light with energy near threshold ionization, a shell is announced as an abrupt increase in cluster ion intensity. (iii) If mass spectra are recorded with light well above threshold, the neutral size distribution can be sampled. Here it is generally found [3,15] that a shell is announced by a decreased ion intensity.

Figure 1 shows mass spectra of laser-warmed Na clusters obtained with two different wavelengths of the ionizing light, i.e., by using methods (ii) and (iii). Photons having an energy of 3.1 eV are very close to the size dependent ionization threshold of the sodium clusters investigated. The corresponding mass spectrum is characterized by steps. For example, the step at about $(Na)_{440}$ occurs because smaller cluster can hardly be ionized with 3.1 eV photons, but larger clusters, with a lower threshold energy, allow themselves to be ionized. The steps in the mass spectra obtained using threshold ionization reflect sudden changes in the ionization energy of clusters as a function of size. These features are relatively sharp and easily interpreted.

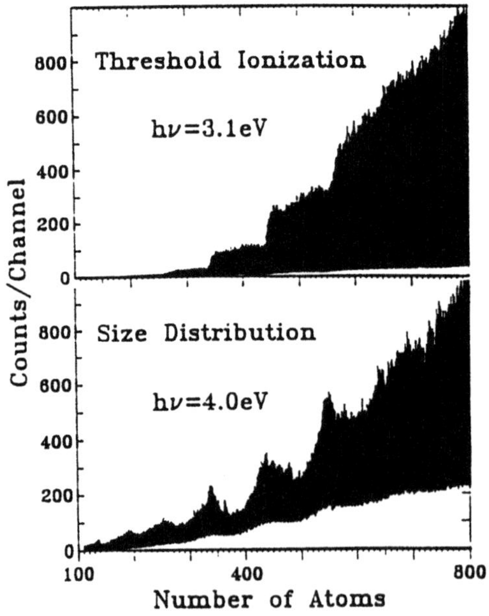

Fig. 1. Mass spectra of $(Na)_n$ clusters obtained using ionizing light near the ionization threshold (top) and well above the ionization threshold (bottom). In both cases the neutral cluster beam was heated with 2.54 and 3.41 eV laser light.

The mass spectra obtained with ionizing photons having energy well above threshold are quite different from the spectrum just discussed. Without the warming laser the mass spectra are without structure, i.e. the size distribution of the cold clusters emerging from our source is smooth. If the warming laser is turned on we obtain not steps but peaks as seen in the bottom of Fig. 1. We believe these peaks reflect the neutral size distribution of the laser-warmed clusters. It appears that it is usually possible to correlate a falling edge of the size distribution with a step in the threshold ionization spectrum. Because of this correlation, we will characterize mass spectra obtained using excimer light by the number of atoms at steep negative slopes. A more extended mass spectrum of laser-warmed sodium clusters obtained with 4.0 eV ionizing photons is shown at the top of Fig. 2.

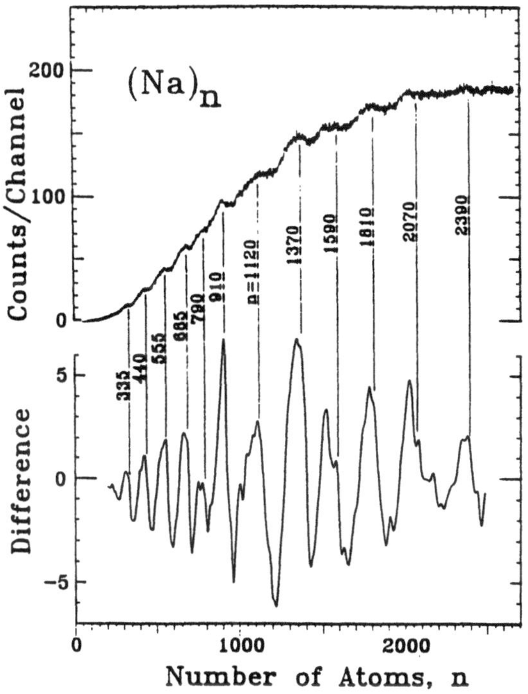

Fig. 2. Mass spectrum of $(Na)_n$ clusters using 4.0 eV ionizing light and (458 nm) 2.71 eV continuous axial warming light having an intensity of 500 mW/cm^{-2}. The spectrum has been smoothed over one hundred 16ns time channels (top). In order to emphasize the shell structure, an envelope function (obtained by smoothing over 20 000 time channels) is subtracted from a structured mass spectrum (smoothed over 1500 time channels). The difference is shown in the bottom spectrum.

Five independent measurements were made under the same experimental conditions. The positions, relative heights and widths of features in the mass spectra were well reproducible. This spectrum has been smoothed with a spline function extending over one hundered 16 ns time channels. Notice that the structure observed does not occur at equal intervals on a scale linear in mass. In order to present this structure in a form more convenient for analysis, the data have been processed in the following way. First, the raw data is averaged with a spline function extending over 20 000 time channels. The result is a smooth envelope curve containing no structure. Second, the raw data is averaged with a spline over 1500 channels. Finally, the two averages are subtracted. The result is shown in the bottom of Fig. 2.

Discussion

The clusters in this experiment have been warmed with a continuous laser beam running parallel to the neutral cluster beam. But what is implied by "warming"? Consider the fate of a typical 500-atom cluster as it moves from the nozzle to the detector.

It leaves the nozzle with the temperature of the He carrier gas (100K) traveling at a velocity of about 350 m/s. During its 1ms flight to the ionization volume it undergoes no further collisions but does begin to absorb photons. We don't really know the absorption cross-section of this cluster at the warming laser wavelength (458 nm). However, 1 $Å^2$/atom is a typical upper limit for smaller clusters. It can be expected that the cross-section will be cluster size dependent. This size dependency will be reflected in the final mass distribution. The cluster absorbs the first 25 photons without evaporating any atoms, gaining an excess energy of about 70 eV and reaching a temperture of about 500 K. This all takes place in the first 450 μs. The temperature of the cluster remains rather constant for the last half of its journey to the ionization volume. It continues to absorb photons of course but after each absorption it evaporates 2 or 3 atoms returning to its original temperature before absorbing the next phonon. It loses a total of 80 atoms, i.e. 16% of its original mass. It appears that this repeated heating and cooling through the "critical temperature for evaporation" on this time scale favors the evolution of a size distribution with relatively strong peak near sizes corresponding to closed electronic shells.

The photon energy (4.0 eV) of the ionizing laser has been chosen so that it is well above the ionization threshold (3.0 eV) of the sodium clusters investigated. The excess energy (1 eV) is sufficient to cause only one atom to evaporate. This is a negligible loss on the mass scale we will be considering. For this reason, we believe that the magic numbers obtained reflect variations in the size distribution of the neutral clusters induced by the warming laser.

The concept of shells can be associated with a characteristic length. Each time the radius of a cluster increases by one unit of this characteristic length, a new shell has been added. A good rough test of whether or not shell structure has been observed can be quickly carried out by plotting the shell index as a function of the radius or $n^{\frac{1}{3}}$. If the points fall on a straight line, the data is consistent with shell formation. That this is indeed the case here can be seen in Fig. 3.

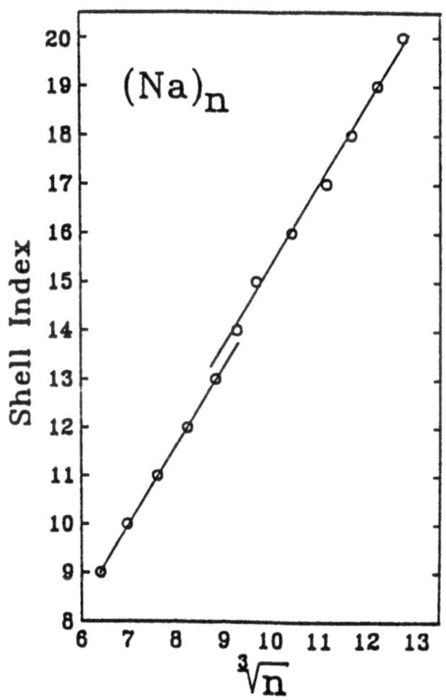

Fig. 3. The electronic shell closings fall approximately on a straight line if plotted on an $n^{\frac{1}{3}}$ scale. An even better fit is obtained using two straight lines with a break between shell 13 and 14. Such a break or "phase change" would be an indication of supershell structure.

However, an even better fit can be obtained using two straight lines with a break between shell 13 and 14. This too can be interpreted in an interesting way.

It has been suggested [20–22] that shell structure might periodically appear and disappear with increasing cluster size. Such a supershell structure can be understood as a beating pattern created by the interference of two nearly equal periodic contributions. Quantum mechanically the contributions can be described as arising from competing energy quantum numbers. Classically the contributions can be described as arising from two closed electron trajectories within a spherical cavity. One trajectory is triangular, the other square.

Concluding Remarks

The size distribution of sodium clusters produced in our source is completely smooth. However, if the clusters are heated with a continuous laser beam, atoms boil off, resulting in a structured mass spectrum, the features of which occur at equal intervals on an $n^{\frac{1}{3}}$ scale. For this reason we associate the features with electronic shell structure in clusters containing up to 2500 electrons. Our data [23] are consistent with a supershell minimum at about 800 electrons. More conclusive data on supershells has recently been obtained in the Niels Bohr Institute using a modified experimental approach [24] and in Orsay for Li clusters [25].

Acknowledgment

This work was supported in part by the Danish Natural Science Research Council.

References

1. M.G. Mayer, Phys. Rev. 75 (1949) 1969L.
2. O. Haxel, J.H.D. Jensen, H.E. Suess, Phys. Rev. 75 (1949) 1766L.
3. W.D. Knight, K. Clemenger, W.A. de Heer, W.A. Saunders, M.Y. Chou, M.L. Cohen, Phys. Rev. Lett. 52 (1984) 2141.
4. M.M. Kappes, R.W. Kunz, E. Schumacher, Chem. Phys. Lett. 91 (1982) 413.

5. I. Katakuse, I. Ichihara, Y. Fujita, T. Matsuo, T. Sakurai, T. Matsuda, Int. J. Mass. Spectrom. Ion Processes 67 (1985) 229.
6. C. Bréchignac, Ph. Cahuzac, J.-Ph. Roux, Chem. Phys. Lett. 127 (1986) 445.
7. W. Begemann, S. Dreihofer, K.H. Meiwes-Broer, H.O. Lutz, Z. Phys. D3 (1986) 183.
8. W.A. Saunders, K. Clemenger, W.A. de Heer, W.D. Knight, Phys. Rev. B 32 (1986) 1366.
9. T. Bergmann, H. Limberger, T.P. Martin, Phys. Rev. Lett. 60 (1988) 1767.
10. J.L. Martins, R. Car, J. Buttet, Surf. Sci. 106 (1981) 265.
11. W. Eckardt, Ber. Bunsenges, Phys. Chem. 88 (1985) 289.
12. K. Clemenger, Phys. Rev. B 32 (1985) 1359.
13. Y. Ishii S. Ohnishi, S. Sugano, Phys. Rev. B 33 (1986) 5271.
14. T. Bergmann, T.P. Martin, J. Chem. Phys. 90 (1989) 2848.
15. S. Bjørnholm, J. Borggreen, O. Echt, K. Hansen, J. Pedersen, H.D. Rasmussen, Phys. Rev. Lett. 65 (1990) 1627.
16. H. Göhlich, T. Lange, T. Bergmann, T.P. Martin, Phys. Rev. Lett. 65 (1990) 748.
17. T.P. Martin, T. Bergmann, H. Göhlich, T. Lange, Chem. Phys. Lett. 172 (1990) 209.
18. J.L. Persson, R.L. Whetten, Hai-Peng Cheng, R.S. Berry, to be published.
19. E.C. Honea, M.L. Horner, J.L. Persson and R.L. Whetten, Chem. Phys. Lett. 171 (1990) 147.
20. R. Balian, C. Bloch, Ann. Phys. 69 (1971) 76.
21. A. Bohr, B.R. Mottleson, Nuclear Structure (Benjamin, London) (1975).
22. H. Nishioka, K. Hansen, B.R. Mottelson, Phys. Rev. B 42 (1990) 9377.
23. T.P. Martin, S. Bjørnhlm, J. Borggreen, C. Bréchignac, Ph. Cahuzac, K. Hansen and J. Pedersen, Chem. Phys. Lett. 186 53 (1991).
24. J. Pedersen, S. Bjørnholm, J. Borggreen, K. Hansen, T.P. Hansen, T.P. Martin and H.D. Rasmussen, Nature 353 (1991) 733.
25. C. Bréchignac, et al., to be published.

Coexistence of Electronic Shells and Shells of Atoms in Microclusters

G.S. Anagnostatos
Institute of Nuclear Physics
National Center for Scientific Research "Demokritos"
GR-153 10 Aghia Paraskevi Attiki, Greece

Microclusters exhibit electronic shells and shells of atoms, which in some cases coexist in one and the same spectrum and in some other cases shells of atoms exist alone. That is, electronic shells alone never exist. Specifically, if the constituents of a cluster are the neutral atoms themselves (i.e., a delocalization of the valence electrons does not exist), then only shells of atoms appear (proper for either atomic fermions--odd number of electrons--or atomic bosons--even number of electrons). However, if the constituents of a cluster are the ion cores of atoms plus their delocalized valence electrons (i.e., a delocalization of the valence electrons does exist), then a coexistence of electronic shells (due to delocalized valence electrons) and shells of atoms (due to ion cores) appears in one and the same spectrum.

1. Introduction

In microclusters both electronic shells and shells of atoms have been observed. The question is if these two factors of stabilizing processes compete with each other or if they are two independent processes, if they are segregated or coexist on the same mass spectrum. A contribution to the answer of these questions is the purpose of the present paper.

The starting point of the present work is the differences of magic numbers observed in the mass spectra between born neutral and born ionized alkali clusters and their comparison with the predictions of the jellium models [1-5]. Finally, a new model is introduced based on the nature (fermionic or bosonic) of particles (electrons, atoms, ion cores) constituting a cluster [6]. According to this model atoms may form magic numbers by themselves, while electrons and ion cores may form independent magic numbers which coexist on the same mass spectrum.

2. Existing Experimental and Theoretical Situation

In Fig.1(a) the mass spectrum of born neutral sodium clusters [2] is shown, while in Fig.1(b)-(d) those of born ionized sodium [8], rubidium [9], and lithium [10] clusters are shown. In Fig.2(a) the theoretical predictions of the spherical jellium model [2] are shown, while in Fig.2(b) and (c) those of the spheroidal [4] and of the Nilsson-type [3] jellium models are shown.

It is interesting to make the following observations on Figs.1 and 2. Specifically, the magic numbers in Fig.1(a) are 2,8,20,40, and 58. Indeed, these numbers are the prominent predictions of Fig.2(a), but there some additional magic numbers at 18,34,... appear which are not present in Fig.1(a). However, while these additional numbers are absent from Fig.1(a), they are present in the spectra of Fig.1(b(-(c). Specifically, Fig.1(b) includes the magic number 18 (=19-1) and Fig.1(c) includes the magic numbers 18 and 34 (=35-1). At the same time, however, Fig.1(b)-(d), which (despite the fact that they correspond to small size clusters) appear to be closer to the predictions of the spherical jellium model [2], show magic numbers at 13, 19 and 25 (Fig.1(b)), at 13 and 19 (Fig.1(c)), and at 7, 13 and 19 (Fig.1(d)). These numbers are well known as magic numbers of rare gas clusters [11-13] and are completely irrelevant to the predictions of the spherical [2] and the spheroidal [3,4] jellium models shown in Fig.2.

Efforts to explain the celebrated mass spectrum of Fig. 1(a) more precisely than the spherical jellium model of Fig. 2(a) have led to the introduction of the spheroidal jellium model of Fig.2(b) and of the Nilsson-type deformation jellium model of Fig.2(c). It is apparent that Fig.2(b) presents a richer spectrum than that of Fig.2(a) and some peaks of the former explain some secondary experimental peaks of the latter. However, the questionable peaks at 18,34,... still exist as prominent peaks in Fig.2(b). These peaks, of course, do not show any significance in the theoretical spectrum of Fig.2(c). Thus, one could consider that Nilsson-type deformation of the cluster makes theory and experiment of born neutral alkali clusters consistent to each other. However, the problem rises again when we try to interpret Fig.1(b)-(d) of born ionized alkali clusters. In these figures, as mentioned earlier, the numbers 18 and 34, which were practically absent in the mass spectrum of Fig.1(a), are prominent peaks in the mass spectra of

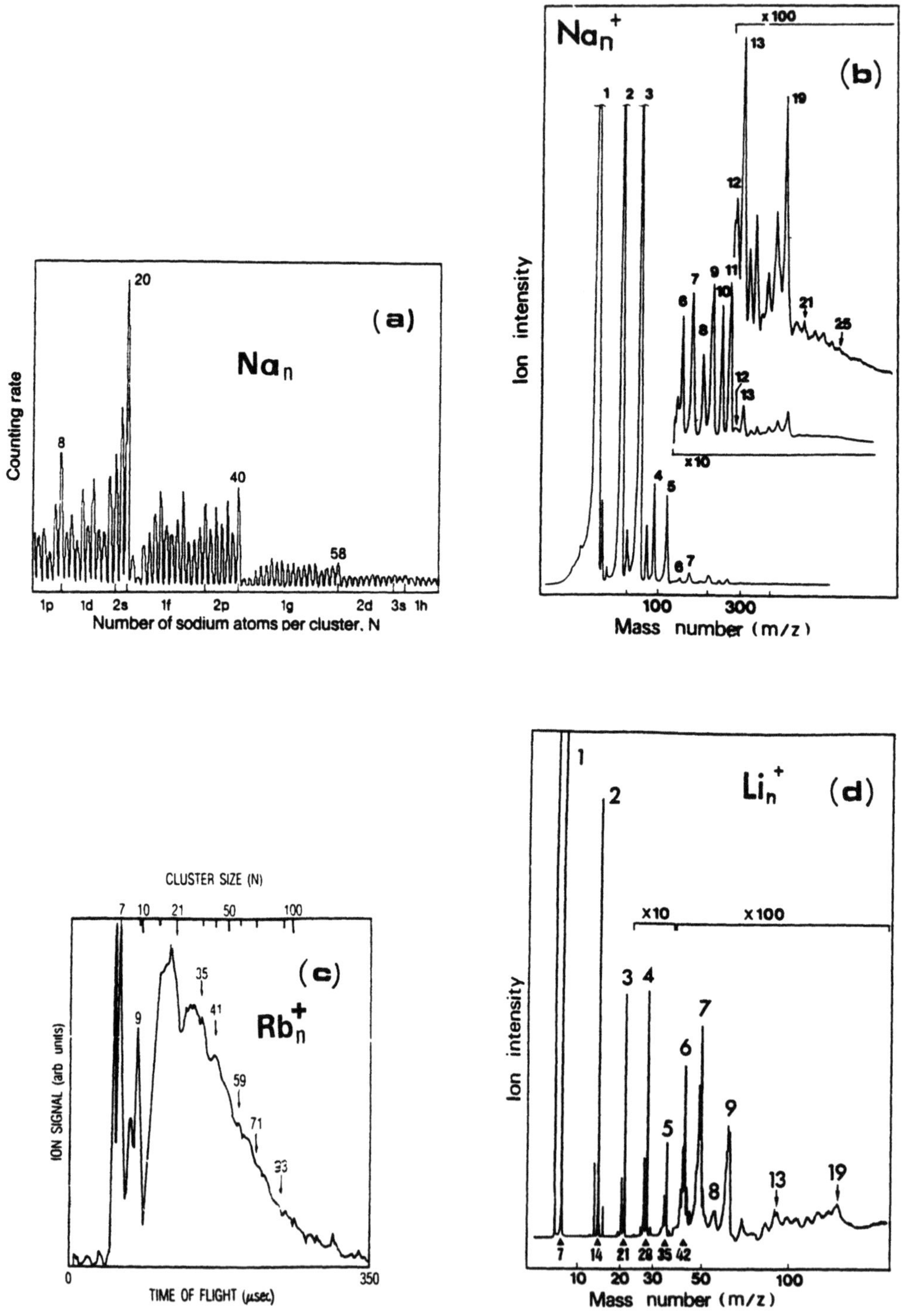

Fig. 1 Experimental mass spectra (a) of born neutral clusters for sodium, (b) of born ionized clusers for sodium, (c) of born ionized clusters for rubidium, and (d) of born ionized clusters for lithium.

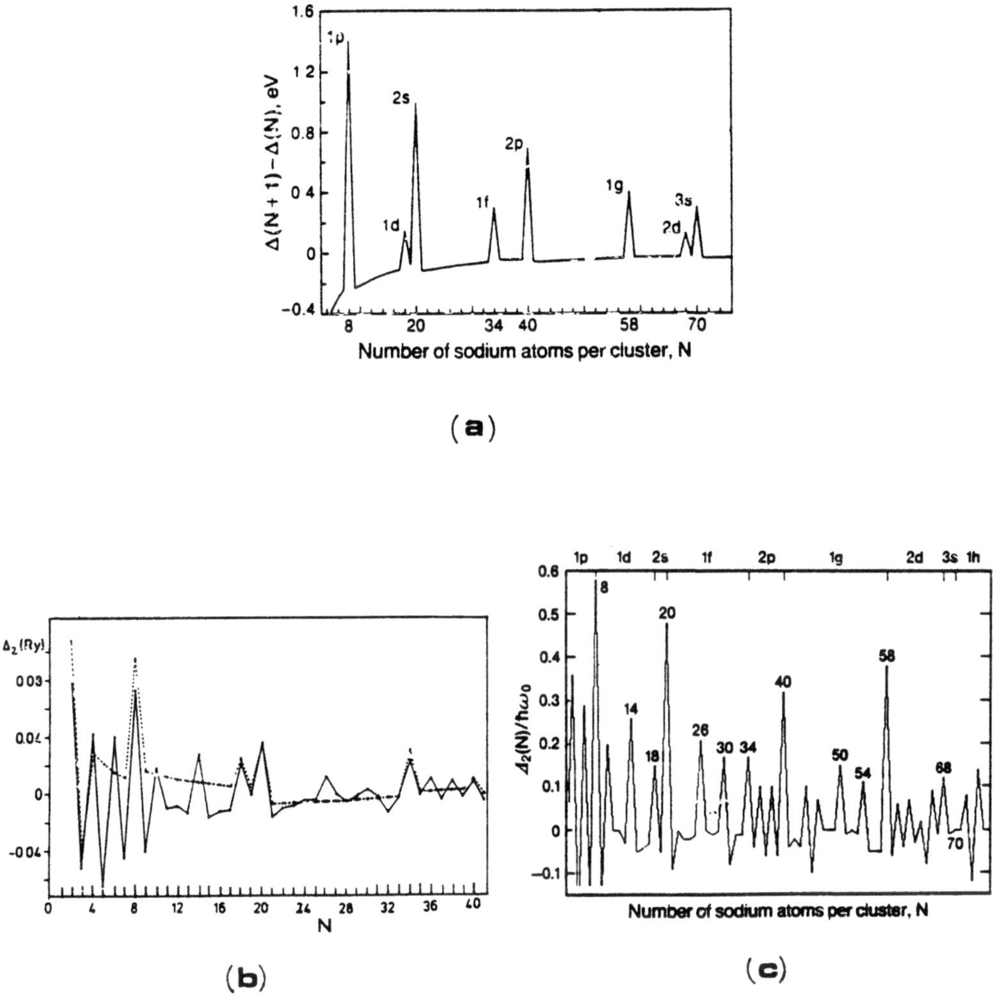

Fig. 2 Theoretical electron structure according to (a) spherical jellium model, (b) spheroidal jellium model, and (c) Nilsson-type deformation jellium model.

Fig.1(b)-(d). Thus, the spectra of Fig. 1(b)-(d) more closely resemble the predictions of the spherical jellium model than those of the spheroidal jellium models. Also, if born neutral alkali clusters are deformed between magic numbers [3,4], born ionized alkali-clusters should be at least equally deformed between magic numbers. This last remark raises the question if deformation of the jellium background really plays a role in the born neutral and the born ionized alkali mass spectra of Fig.1, since this assumption seems to explain only the first but not the latter spectra. On the other hand, it is known from previous works [7,14] that ion cores form close packing structures and thus a deformation of such a structure (due to the incomplete filling of the electronic shells) seems rather difficult. Hence, one could think that the explanation of alkali mass spectra is far from being complete and convincing, despite the impressive progress of the theories.

Through the present work and previous [6,7,15] publications a different explanation of the alkali mass spectra is provided, based on the remark that the nature of particles involved in the structure of the clusters should be considered. Specifically, besides the electron the neutral atom and its ion core are considered as composite particles in the framework of the present model. Thus, magic numbers due to all three categories of particles are expected. More details about the model are given below. What should be emphasized, however, is that the present model is not in contradiction with the jellium models. The difference is that the present model considers that there are experimental conditions (e.g., born neutral, low temperature, small size clusters) for which the valence electrons are not delocalized. Under these conditions the jellium models are not applicable and the magic numbers are due to neutral atoms alone. In this situation the valence electrons are localized in their own atoms and play no role in the formation of the magic numbers. In contrast, under different experimental conditions (e.g., born ionized, high temperature, large size clusters) there is delocalization of the valence electrons and thus the jellium models are applicable. Then, there are magic numbers due to electrons according to the jellium model, but at the same time there are magic numbers due to ion cores according to the present model.

3. Quantitative Treatment in the Model

It is specifically assumed that all atoms (or ion cores) in a shell of the cluster taken together create an average central potential (assumed harmonic) common for all atoms (or ion cores) in this shell and that in this potential each atom (or ion core) performs an independent particle motion obeying the Schrödinger equation. In other words, we consider a multi-harmonic potential description of the cluster, as follows

$$H\psi = E\psi, \quad H = T + V \qquad (3.1)$$

$$H = H_{1s} + H_{1p} + H_{1d2s} + \ldots \qquad (3.2)$$

where

$$H_i = V_i + T_i = -V + 1/2m(\omega_i)^2 r_i^2 + T_i \qquad (3.3)$$

That is, we consider a state-dependent Hamiltonian, where each partial harmonic oscillator potential has its own state-dependent frequency ω_i. All these ω_i's are determined from the harmonic oscillator relation (3.4).

$$\hbar\omega_i = (\hbar^2/m \langle r^2{}_i\rangle)(n+3/2), \qquad (3.4)$$

where n is the harmonic oscillator quantum number and $\langle r_i{}^2\rangle^{1/2})$ is the average radius of the relevant high fluximal shell made of either bosonic [12] or fermionic [12] atoms. Before applying (3.4), to each of the shells a value (0,1,2,3,...) of the harmonic quantum number n is assigned and a value of $\langle r_i{}^2\rangle^{1/2}$ is derived from the geometry of the shell taking the finite size of the atom sphere into account. Thus, $\hbar\omega_i$ changes value each time either n or $\langle r_i{}^2\rangle^{1/2}$ (or both) change value.

In the case of bosonic atoms there is no restriction for the number of atoms constituting the shell, since any number of such atoms is accepted for the same quantum state (symmetric total wave function). In the case of fermionic atoms, however, the atoms on each shell are restricted by the Pauli principle (antisymmetric total wave function). It is satisfying that all relevant shells for fermionic atoms [3,10,13] fulfil this fundamental requirement, as explained in detail in [12].

According to the Hamiltonian of (3.2), the binding energy of a cluster of N atoms is given by (3.5)

$$BE = 1/2\ (VN) - 3/4\ [\sum_{i=1}^{N} \hbar\omega_i(n+3/2)], \qquad (3.5)$$

where V is the average potential depth given [12] by (3.6)

$$V = -aN+b+c/N, \qquad (3.6)$$

where the coefficient c expresses the sphericity of the cluster and has the same numerical value everywhere the outermost shell of the structure is completed and everywhere else c has a zero value. Of course, one expects that different kinds of atoms will assume different values of parameters a,b, and c in (3.6).

The relative binding energy gap for a cluster with N atoms compared to clusters with N+1 and N-1 atoms is given by (3.7)

$$\delta(N) = 2E_B(N)-[E_B(N-1) + E_B(N+1)]. \qquad (3.7)$$

As is apparent throughout the present work and the cited references, the average positions of the atoms (or their ion cores) in the clusters have a shell structure either for fermionic or bosonic atoms. In this respect the structure of the clusters, to some extent, resembles nuclear structure.

Thus, several well-documented nuclear phenomena, e.g. collective effects, are reasonably expected as cluster phenomena as well. Hence, small fermionic clusters of size N far from magic numbers are expected to be deformed. Furthermore, deformed (prolate or oblate) fermionic clusters are expected to rotate [18], and spherical (close to magic numbers) clusters are expected to vibrate. Besides these collective excitations, clusters can show single particle excitation either due to their atom or electron structure (partial levels of ionization). All these interesting phenomena are out of the scope of the present work, which mainly intends to support coexistence of electronic shells and shells of atoms in microclusters according to the statistics of the constituent atoms (Fermi or Boson statistics, according to the half integer or integer spin of the atoms or ions involved in a specific cluster).

4. Application of the Model

According to the discussion in Section 2 application of the model on alkali clusters is apparent. Specifically, it can be said that the mass spectrum of Fig.1(a) is due to fermionic neutral sodium atoms, while those of Fig.1(b)-(d) are due to both the electron structure and the bosonic ion cores of sodium, rubidium, and lithium atoms, respectively.

A detailed application of the model is presented in ref. [7]. Below, only a brief reference to this application of the model is given which shows that the model is applicable in all cases of clusters and is supported by all experimental data available to date. Specifically, in all cases, where neutral atoms constitute the cluster [2,11,17-19] the magic numbers are due to atoms alone (present model). Depending on the case, the structure of the cluster is proper for atomic fermions [2,19] or for atomic bosons [11,16-18]. In contrast, in all cases where delocalized electrons and left-over ion cores constitute the relevant cluster [8-10,20-22] the magic numbers are due to both the electrons (jellium model) and the ion cores (present model).

As a special case we refer to large alkali clusters [23] whose mass spectrum for sodium is shown in Fig.3. These clusters, due to their large

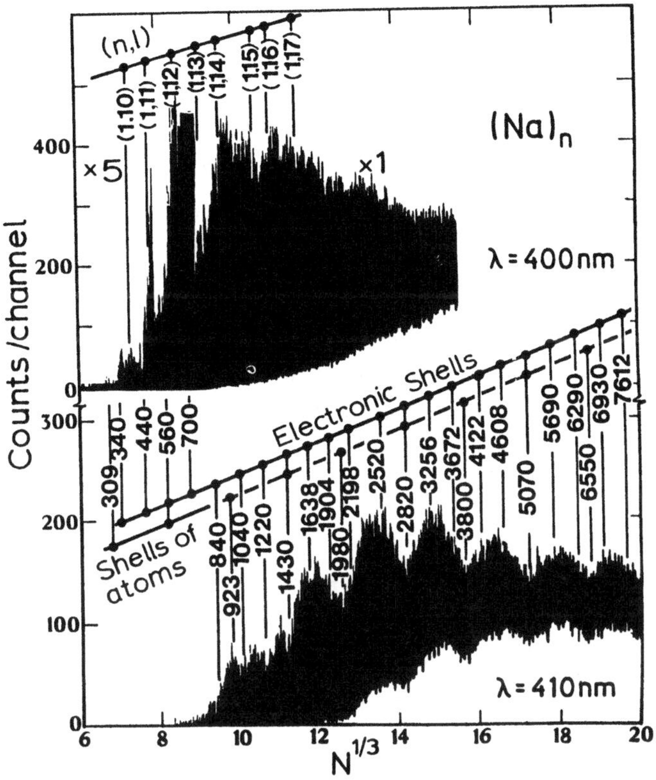

Fig. 3 Experimental mass spectrum of large sodium clusters.

size, are hot and thus they are under conditions favoring delocalization of the valence electrons. Hence, according to the present model, magic numbers due to delocalized electrons and magic numbers due to ion cores are expected. A complete development of this subject is given in ref.[24], here only the conclusion is stated. Specifically, all deep valleys in the spectrum, below and beyond 1500, are due to the ion cores, while much smaller valleys appearing in the structure of the spectrum are due to the delocalized electrons. Even completion of subshells (n,l), particularly the first ones, have an apparent contribution to the fine structure of the mass spectrum.

5. Extension of the Model to Nuclear Physics

By consulting ref.[15] one finds it interesting that the present model can be extended to Nuclear Physics when the nucleon finite size is taken into account. Results of the model on nuclear radii and binding energies are indeed impressive.

6. Concluding Remarks

In the present work and related references the novel idea that a neutral atom or its ion core can be considered as a composite particle is successfully introduced. Specifically, a neutral atom or its ion core is taken as an atomic fermion or as an atomic boson depending on whether the relevant number of electrons is odd or even, respectively. Thus, the magic numbers in clusters can be a result not only of the electron structure, but also of the atomic structure and structure of the ion cores.

Up to now only magic numbers due to delocalized valence electrons have been considered in the framework of the jellium models. This comes directly from the assumption usually made in the literature that the valence electrons are always delocalized (particularly in the case of simple metals, i.e., alkali) and also that the ion cores play almost no role in the formation of magic numbers. This assumption, however, has never been proved.

On the contrary, the present model considers that there are experimental conditions (e.g. born neutral, low temperature, small size clusters) for which the valence electrons are not delocalized. Thus, for these conditions the jellium models are not applicable and the magic numbers come strictly from the atoms themselves. The present model also assumes that in the case where the valence electrons are delocalized (e.g., born ionized, high temperature, large size clusters) the left-over ion cores form their own magic numbers besides those formed by the delocalized electrons.

Due to the fact that electrons and atomic fermions in theory possess the same magic numbers, i.e. those of fermions, some mass spectra in the literature explained as due to electron structure could be due to atomic structure (e.g., the spectrum of Fig.1(a)). Thus, further experimental verification of the present model is highly desirable.

References

1. Ekardt, W.: Phys. Rev. B **29**, 1558 (1984).
2. Knight, W.D., Clemenger, K., de Heer, W.A., Saunders, W.A., Chow, M.Y., Cohen, M.L.: Phys. Rev. Lett. **52**, 2141 (1984).
3. Knight, W.D., de Heer, W.A., Saunders, W.A.: Z. Phys. D-Atoms, Molecules and Clusters, 3, **109** (1986).
4. Ekardt, W., Penzar, Z.: Phys. Rev. B **38**, 4273 (1988).
5. Penzar, Z., Ekardt, W.: Z. Phys. D-Atoms, Molecules and Clusters, **17**, 69 (1990).

6. Anagnostatos, G.S.: Phys. Lett. A **154**, 169 (1991).
7. Anagnostatos, G.S.: Phys. Lett. A **157**, 65 (1991).
8. Saito, Y., Minami, K., Ishida, T., Noda, T.: Z. Phys. D - Atoms, Molecules and Clusters **11**, 87 (1989).
9. Bhaskar, N.D., Frueholz, R.P., Klimcak, C.M., Cook, R.A. **36**, 4418 (1987).
10. Saito, Y., Watanabe, M., Hagiwara, T., Nishigaki, S., Noda, T.: Jpn J. Appl. Phys. **27**, 424 (1988).
11. Echt, O., Sattler, K., Recknagel, E.: Phys. Rev. Lett. **47**, 1121 (1981).
12. Anagnostatos, G.S.: Phys.lett. A **124**, 85 (1987).
13. Anagnostatos, G.S.: Phys. Lett. A **133**, 419 (1988).
14. Mansikka-aho, J., Manninen, M., Hammarén, E.: Z. Phys. D - Atoms, Molecules and Clusters **21**, 271 (1991).
15. Anagnostatos, G.S.: Invited talk at the conference "Nuclear and Atomic Clusters 1991", Turku, 1991.
16. Mühlbach, J., Pfau, P., Sattler, K., Recknagel, E.: Z. Phys. B - Condensed Matter and Quanta **47**, 233 (1982).
17. Klots, T.D., Winter, B.J., Parks, E.K., Riley, S.J.: J. Chem. Phys. **92**, 2110 (1990).
18. Rayane, D., Melinon, P., Cabaud, B., Hoareau, A., Tribollet, B., Broyer, M.: Phys. Rev. D **39**, 6056 (1989).
19. Geusic, M.E., Morse, M.D., Smalley, R.E.: J. Chem. Phys. **82**, 590 (1985).
20. Katakuse, I., Ichihara, T., Fujita, Y., Matsuo, T., Sakurai, T., Matsuda, H.: Int. J. Mass Spectrom. Ion Processes **74**, 33 (1986).
21. Katakuse, I., Ichihara, T., Fujita, Y., Matsuo, T., Sakurai, T., Matsuda, H.: Int. J. Mass Spectrom. Ion Processes **69**, 109 (1986).
22. Begemann, W., Driehöfer, S., Meiwes-Broer, K.H., Lutz,H.O.: Z. Phys. D - Atoms, Molecules and Clusters **3**, 183 (1986).
23. Martin, T.P., Bergmann, T., Göhlich, H., Lange, T.: Chem. Phys. Lett. **172**, 209 (1990).
24. Anagnostatos, G.S.: To appear.

Magic Numbers of Coulomb and Lennard-Jones Crystals and Quasicrystals

Rainer W. Hasse

ESR Division, Gesellschaft für Schwerionenforschung GSI,
D-6100 Darmstadt, Germany

With the help of molecular dynamics computer simulations we study the equilibrium configurations at low temperature of large systems of strongly correlated particles either under the influence of their mutual long-range Coulomb forces and a radial harmonic external confining force or of the short-range Lennard-Jones potential. The former is a model for charged particles in ion traps and the latter for clusters. For the Coulomb plus harmonic force, the particles arrange in concentric spherical shells with hexagonal structures on the surfaces. The closed shell particle numbers agree well with those of multilayer icosahedra (mli). A Madelung (excess) energy of −0.8926 is extracted which is larger than the bcc value.
For the Lennard-Jones force we employ various initial configurations like multilayer icosahedra or hexagonal closed packed (hcp) spheres. Cohesive (volume) and surface energies per particle are extracted and compared to the energies of scaled mli quasicrystals and of spherical scaled crystals with N up to 36 000. It is shown that relaxed mli are the dominant structures for $N < 5\,000$ and hcp spheres for larger particle numbers. For $N < 22\,000$, hcp crystals have about the same closed shell numbers as mli quasicrystals but smaller ones for $N > 22\,000$. The same magic numbers obtain with other short range Mie potentials.

1. Introduction

The magic numbers of Mackay's multilayer icosahedra (mli) [1],

$$N = (2M+1)\,[\tfrac{5}{3}M(M+1)+1] \quad , \tag{1}$$

notably N =13, 55, 147, 309, 561, 923,... have been verified experimentally as closed shell particle numbers in metal clusters, see e.g. refs. [2, 3]. The subshell numbers are given by $10N^2 + 2$ =12, 42, 92, 162, 242,... Here it is shown that these magic numbers are more general in the sense that they not only are identical to those of stacked cuboctahedra but also show up in Coulomb systems and in spherical hcp matter under the action of the Lennard-Jones (LJ) force or of other Mie-type short-range forces. Due to the long range nature of the Coulomb force Coulomb quasicrystals tend to be as spherical as possible. This is not in contradiction to the edged nature of the mli because here the plane surfaces of the mli become curved.

2. Coulomb plus harmonic forces

The strongly correlated infinite one-component plasma (OCP) crystallises in an bcc lattice if the plasma parameter, i.e. the ratio of Coulomb to thermal energy, reaches the value [4] $\Gamma \approx 171$. Finite systems, on the other hand, exhibit hexagonal structures on the surfaces [5, 6], however imperfect due to the incompatibility between a perfect lattice and a curved surface and due to the incommensurability of two adjacent shells. A change of structure from the plane hexagonal into the bcc one occurs only if the dimensions of the system become as large as about 100 interparticle distances [7].

In order to study large but finite Coulomb systems we extend the calculations on small systems by Rafac et al. [8] and employ the molecular dynamics (MD) technique to solve the classical equations of motion under the external harmonic confining force

$$\mathbf{F}^i_{\text{conf}} = -K\mathbf{r}_i,$$

and the Coulomb force,

$$\mathbf{F}^i_{\text{Coul}} = -q^2 \sum_{j\neq i} \frac{\mathbf{r}_i - \mathbf{r}_j}{|\mathbf{r}_i - \mathbf{r}_j|^3}.$$

The initial momenta are chosen at random as to give an initial kinetic energy corresponding to $\Gamma \approx 1$ and the initial coordinates of N ions usually are chosen as those of the (N–1)-ion system with one ion added at random. The system then is followed in time thereby cooling by reducing the momenta until thermal equilibrium has been reached; for details see ref. [6]. In this chapter distances are measured in units of the Wigner-Seitz radius $a_{\text{WS}} = (q^2/K)^{1/3}$ and energies in units of q^2/a_{WS}. The total energy, i.e. the sum of confining and Coulomb energies per particle, then reads

$$\epsilon = \frac{1}{2N}\sum_{i=1}^{N} \mathbf{r}_i^2, + \frac{1}{N}\sum_i \sum_{j<i} |\mathbf{r}_i - \mathbf{r}_j|^{-1}.$$

The excess energy is then defined by subtracting the homogeneous value of $\bar{\epsilon} = \frac{9}{10}N^{2/3}$.

The resulting magic numbers are listed in Table I where the excess energy is minimal as compared to the neighbouring particle numbers and the excess energy is shown in Fig. 1. For systems with $N < 64$, there are strong shell effects in the excess energy with peaks and minima if a new shell opens, in particular the transitions at $N = 12, 60, 146$. One notes that the smaller magic numbers are even as compared to those of the mli because the center particle is always missing. In particular, the second shell starts at $N = 13$, the third one at $N = 61$, and the next ones at or around N = 147, 309, 565, 900, 1400, 2100, respectively.

Table 1:
Structures, rms radii and excess energies of closed shell N-particle systems

N	Structure	R_{rms}	ϵ_{excess}
12	12	1.6002	-0.87663
60	48+12	2.9335	-0.88498
146	93+41+12	4.0054	-0.88745
308	163+93+42+10	5.7404	-0.88919
561	255+161+93+42+10	6.3418	-0.88994
899	356+247+154+92+38+12	7.4361	-0.89059
1414	491+365+244+163+94+44+13	8.6600	-0.89124
2057	641+491+363+257+158+93+44+10	9.8209	-0.89132
2837	805+634+480+356+246+173+*143**	10.9384	-0.89172
3871	992+801+639+*1439**	12.1379	-0.8915†

* The remaining shells cannot be resolved
† At small temperature

An infinite Coulomb system can take advantage of all long range interactions in order to arrange in an bcc lattice and to minimize the Madelung energy by summing up all long range contributions. A finite system, on the other hand, can explore only the short and intermediate range parts of the Coulomb interaction. It is well known that systems with short range interactions arrange in fcc or hcp lattices with coordinations (the number of almost equal nearest neighbour distances) of 12 but with higher Madelung energies. A finite Coulomb system, hence, will arrange in such a way as to maximize the coordination i.e. to achieve the maximum possible number of equilateral triangles. In Fig. 2 is shown the front hemisphere of the outer shell of the 5000-particle system. Here the overall hexagonal structure is well pronounced, however with dislocations and pentagonal point defects. The particles in the next inner shell most often sit below the line connecting two particles rather than below the center of the triangle.

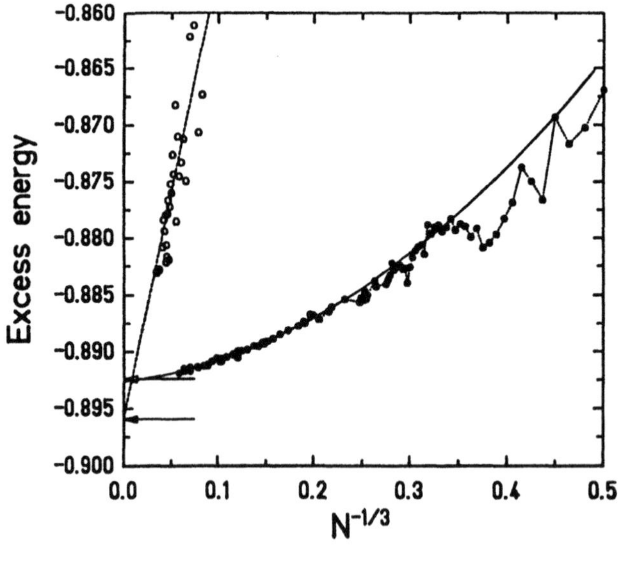

Figure 1:
Excess (Madelung) energy of Coulomb quasicrystals. The dots are MD data and the full line is the best fit. The open arrow points to the asymptotic value and the full arrow to the Madelung energy of the infinite bcc OCP. Open circles are results of spherical bcc matter together with a fit (dashed line).

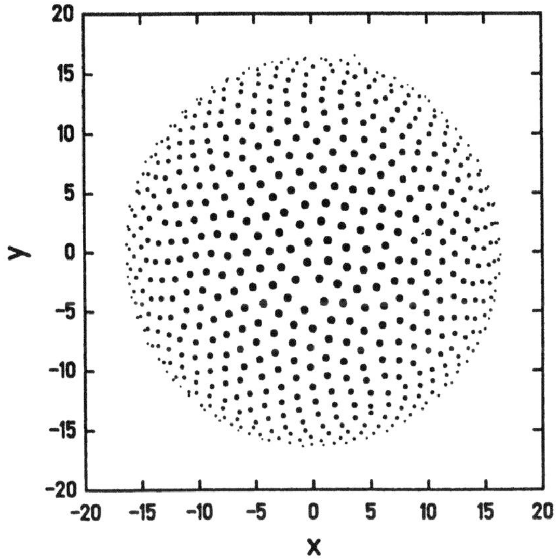

Figure 2:
The front hemisphere of the outer shell of the 5000 ion system. Note the haxagonal structure with approximately equilateral triangles and the point defect below the center.

A fit for $N > 300$ gives

$$\epsilon_{\mathrm{MD}} = -0.8926 \pm 0.0001 + 0.0088 N^{-1/3} + 0.096 N^{-2/3}$$

with an asymptotic Madelung energy of -0.8926 which is higher than the one of the infinite bcc OCP of -0.895929, thus indicating that even at very large particle numbers the hexagonal surface structure is still dominating over the bcc structure. This is due to the small surface energy of the hexagonal lattice on the curved surfaces. The surface energy extracted from the MD data is the average over the whole surface and also includes relaxation and reconstruction of the surface that minimize the total energy. To evaluate the order of magnitude of surface relaxation we calculated the excess energies of unrelaxed spherical fragments containing up to 25 000 particles, sliced out of infinite ideal bcc matter with the same density, see the open circles in Fig. 1. Extrapolation of these data to the bcc Madelung energy (dashed line) gives a surface energy coefficient of ≈ 0.4. It crosses the fit of the MD data at $N \approx 4 \times 10^6$. This number of particles at which the infinite bcc lattice takes over energetically is compatible with the estimate of Dubin [7].

3. Lennard-Jones force

The largest magic number identified in cluster experiments is around 21 300 [3]. They are well in agreement with those obtained from purely classical geometrical packing of N particles for instance into mli or cuboctahedra (mlc). However, the energetical stability of large crystals and quasicrystals under short-range forces has only been studied extensively by Raoult et al. [9]. Here icosahedral, octahedral and mono-twinned fcc, decahedral, tetrakaidecahedral, hexakaiicosahedral (hki), dodecahedral pentakaitetrakontahedral (dpk) and other truncated structures are considered.

We extend these calculations to very large particle numbers and, as an approximation to the effective two-body force in clusters, we employ the short range LJ potential,

$$V_{\mathrm{LJ}}(r) = \epsilon_0 \left[\left(\frac{\sigma}{r}\right)^{12} - 2\left(\frac{\sigma}{r}\right)^{6} \right].$$

In this chapter we will use $\epsilon_0 = \sigma = 1$. The energy per particle, hence, is given by

$$E_{\mathrm{LJ}} = \frac{1}{N} \sum_{i}^{N} \sum_{j<i} V_{\mathrm{LJ}}(|\mathbf{r}_i - \mathbf{r}_j|) \quad . \tag{2}$$

Due to the very short range nature of the LJ force a start with initial random coordinates never yields a stable configuration. Therefore we employ mli, mlc, spheres of hexagonal closed packed (hcp), face centered (fcc) or body centered cubic (bcc) matter, or even stacked rhombic dodecahedra (srd) proposed by Kepler, see ref. [10]. Systems with particle numbers up to $N = 6\,525$ are completely relaxed by MD and those with $2 < N < 36\,000$ are studied by uniform scaling and energy minimization. Systems with nonmagic numbers of particles were started with a magic core and the extra particles at random in the next shell; for details see ref. [11].

With inverse power-law potentials scaling effects can be calculated by scaling the dimensions in eq. (2) and minimising with respect to the scaling parameter, $s_0 = (e_{12}/e_6)^{1/6}$ to yield the minimum energy

$$E_{\mathrm{LJ}}^{\mathrm{min}} = -e_6^2/e_{12}.$$

For infinite bcc, fcc and hcp matter, the inverse-power sums e_n are known, see [12], to give the scaling parameters and cohesive energies of Table 2.

Table 2:
Cohesive and surface energies and scaling factors of different Lennard-Jones crystals and of infinite mli and mlc quasicrystals.

Structure	V_{LJ}	S_{LJ}	s_0
hcp	-8.611065[1]	15.5[2]	0.971228[1]
fcc	-8.610201[1]	15.4[2]	0.971234[1]
mlc	-8.59[3]	15.6[3]	0.974[3]
hki	-8.545[4]	14.23[4]	
mli	-8.54[3]	14.18[3]	0.953[3]
dpk	-8.538[4]	14.20[4]	
bcc	-8.237292[1]	15.1[2]	0.979204[1]

[1] Analytical values
[2] From minimizing scaled large spheres
[3] Estimated from MD relaxation
[4] Lower limits extrapolated from ref. [9]

Under the LJ force, hcp matter is lowest in energy, followed by fcc and bcc matter. We repeated those scaling calculations for spheres of such matter with up to 36 000 particles in order to also obtain the surface energy coefficient in the expansion

$$E_{\rm LJ} = V_{\rm LJ} + S_{\rm LJ} N^{-1/3} ...$$

which are also listed in Table 2. Relaxation of finite systems is achieved by solving the coupled classical equations of motion with standard MD as described above.

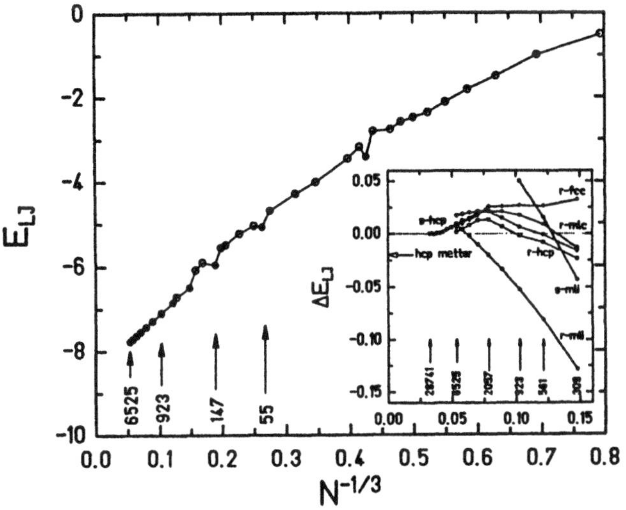

Figure 3:
Energies per particle of LJ crystals and quasicrystals. Full (open) circles are relaxed MD results with mli initial configurations at magic (nonmagic) particle numbers. In the insert are shown various energies (r: relaxed by MD, s: scaled and minimized) with the scaled-hcp average of $-8.591+15.035N^{-1/3}$ subtracted. The open arrow points to the hcp matter value.

Table 3:
Closed shell particle numbers N of shell M and their energies of relaxed mli, mlc, hcp and fcc structures and of scaled mli, hcp and fcc.

M	N	$E^{\rm MD}_{\rm mli}$	$E^{\rm MD}_{\rm mlc}$	$E^{\rm MD}_{\rm hcp}$	$E^{\rm MD}_{\rm fcc}$	$E^{\rm scaled}_{\rm mli}$	$E^{\rm scaled}_{\rm hcp}$	$E^{\rm scaled}_{\rm fcc}$
1	13	-3.4098					-3.4098	-3.1466
2	55	-5.0772	-4.8778	-4.9114	-4.7660	-5.0373	-4.9052	-2.4041
3	147	-5.9623	-5.8121	-5.8636	-5.7888	-5.8938	-5.8576	-4.0776
4	309	-6.4959	-6.3805	-6.3907	-6.3347	-6.4102	-6.3834	-5.3022
5	561	-6.8492	-6.7595	-6.7707	-6.7522	-6.7530	-6.7621	-6.1041
6	923	-7.0994	-7.0295	-7.0490	-7.0202	-6.9965	-7.0409	-6.6062
7	1 415	-7.2854	-7.2312	-7.2261	-7.2454	-7.1781	-7.2184	-6.9529
8	2 057	-7.4291	-7.3875	-7.3956	-7.3831	-7.3185	-7.3880	-7.1823
9	2 869	-7.5432	-7.5121	-7.5204	-7.5168	-7.4304	-7.5137	-7.3676
10	3 871	-7.6361	-7.6138	-7.6257	-7.6209	-7.5216	-7.6193	-7.5085
11	5 083	-7.7128	-7.6982	-7.7133	-7.7037	-7.5973	-7.7071	-7.6219
12	6 525	-7.7775	-7.7696	-7.7832	-7.7808	-7.6611	-7.7774	-7.7121
13	8 217					-7.7156	-7.8404	-7.7849
14	10 179					-7.7628	-7.8933	-7.8515
15	12 431					-7.8041	-7.9420	-7.9000
16	14 993					-7.8404	-7.9807	-7.9441
17	17 885					-7.8726	-8.0158	-7.9865
18	21 127					-7.9013	-8.0477	-8.0238
19	24 739					-7.9272	-8.0748	-8.0554
20	28 741					-7.9505	-8.1003	-8.0851
21	33 153						-8.1223	-8.1086

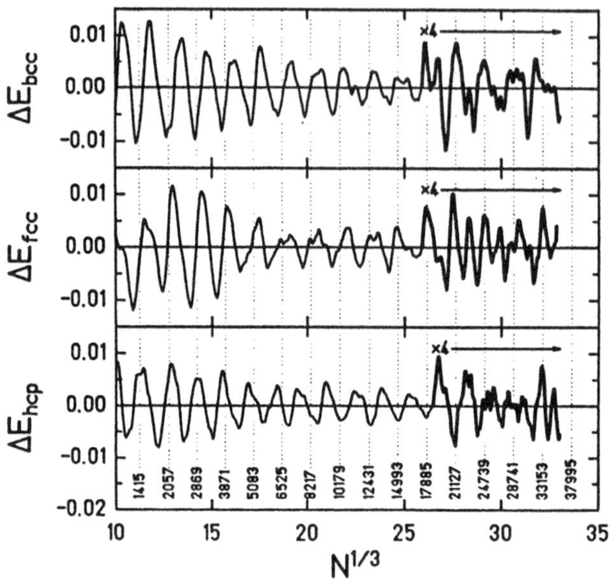

Figure 4:
Shell energies of scaled spherical LJ crystals, smoothed and the mean subtracted. The heavy parts are magnified.

The resulting energies are listed in Table 3 and shown in Fig. 3. For small systems it can be seen that the energy of mli with the magic numbers 13, 55, 147 and 309 attains minimal values as compared to surrounding nonmagic clusters. Relaxed icosahedrons are lowest in energy up to $N = 3\,871$ with either multilayer or stacked cuboctahedrons being much higher. In addition, since simple spherical hcp crystals always have lower energy than mlc and since even fcc crystals are lower in most of this interval of particle numbers, we exclude the posssibility of the existence of cuboctahedral clusters. Due to their bcc structure the same holds for rhombic dodecahedra.

Despite the fact that mli have a very low surface energy, the volume energy and, hence, the total energy becomes too large for large particle numbers. Around particle numbers of 5 083, hence, there is a transition to hcp crystals which, however, have larger surface energy but lower volume energy. Similar reasoning holds for the hki and dpk. The corresponding values for the cohesive and surface energies of Table 2 have been obtained by extrapolating the results of ref. [9]. They have to be taken as lower limits. However, from Table 2 it can also be seen that although hki have a slightly larger surface energy than dpk, their cohesive energy is slightly less. Hki, hence becomes the more stable configuration for $N > 40\,000$.

In order to follow the magic numbers beyond $N = 6\,525$ the shell energies of scaled (but not relaxed) spherical simple cubic (sc), bcc, fcc and hcp crystals have been calculated. By subtracting the mean and smoothing we obtain the shell energies of Fig. 4. All of them exhibit characteristic shell oscillations due to the fact that there exist spherical crystalline configurations with minimum (maximum) number of surface particles at the same radius, hence with small (large) surface energies. However, for large systems sc, bcc and fcc do not show the characteristic shell spacing of eq. (1) whereas hcp does for $N =$ 10 179, 12 431, 14 993, 17 885 and 21 127 at the correct positions (also, at 8 217 there is a local minimum). The next two minima appear at $N \simeq 23\,600$ and 27 500 rather than at 24 793 and 28 741. Here the magic numbers were not yet identified experimentally. For larger particle numbers the shell energy becomes rather small and one expects that magic numbers cease to exist.

In order to assure that the effect of magic numbers in hcp spheres is not an artifact of the LJ force we repeated the scaling calculations with the Mie potential $r^{-4n} - 2r^{-2n}$ ($n=3$ is the LJ potential) with $n=$ 1...4. For the short range potentials $n = 2, 3, 4$ we found the minima of the energy essentially at the same magic numbers as discussed above. In the long range case $n=1$, on the other hand, the shell oscillations are very irregular and one cannot associate magic numbers. However, we cannot exclude that macroscopic structures formed of hcp matter other than the sphere might have even less surface energy.

References

[1] A.L. Mackay, Acta Crystallogr. **15** (1962) 916

[2] H. Göhlich, T. Lange, T. Bergmann and T.P. Martin, Phys. Rev. Lett. **65** (1990) 748

[3] T.P. Martin, T. Bergmann, H.Göhlich and T. Lange, *in* Proc. 5th Int. meeting on small particles and inorganic clusters, Konstanz 10-14 Sept. 1990, Z. Phys. **D 19** (1991) 25

[4] W. Slattery, G. Doolen and H. DeWitt, Phys. Rev. **A 21** (1980) 2087

[5] R.W. Hasse and J.P. Schiffer, Ann. Phys. **203** (1990) 419

[6] R.W. Hasse and V.V. Avilov, Phys. Rev. **A 65** (1991) 4506

[7] D.H.E. Dubin, Phys. Rev. **A 40** (1989) 1140

[8] R. Rafac, J.P. Schiffer, J.S. Hangst, D.H.E. Dubin and D.J. Wales, Proc. Natl. Acad. Sci. USA **88** (1991) 483

[9] B. Raoult, J. Farges, M.F. De Feraudy and G. Torchet, Phil. Mag. **B 60** (1989) 881

[10] I. Stewart, New Scientist **13 July** (1991) 29

[11] R.W. Hasse, Phys. Lett. **A** in print

[12] C. Kittel, Introduction to solid state physics (Wiley, New York, 5th ed. 1976) p. 80.

CLUSTERS AS SOLITONS ON THE NUCLEAR SURFACE

Săndulescu, A. Ludu and W. Greiner
Institut für Theoretische Physik, Johann Wolfgang Goethe-Universität,
Frankfurt am Main, Germany

In the present paper, by introducing nonlinear terms in the hydrodynamical equations, we show that stable solitons exist on the surface of a sphere. Contrary to the Bohr-Mottelson model we assume that the outside nucleons do not polarize the double magic core ^{208}Pb.
The soliton itself contains the polarization effect. The total potential energy consists of four terms : surface, centrifugal, coulomb and shell energies. The last term was introduced phenomenologically proportional with the overlap between the final nucleus (considered to be a sphere) and the initial nucleus with quite different shapes than the shapes described by the usual multipole expansion of the nuclear surface in spherical harmonics. This term causes a new minimum in the total potential energy. Due to the fact that alpha and cluster decays are spontaneous decays we choose this minimum to be degenerate in energy with the ground state minimum. This description leads to a new coexistence model : the usual shell model and a cluster-like model described by a soliton moving on a sphere. Due to the large barrier between the two minima the amplitude of the cluster-like state is much smaller than the usual ground state. The ratio of the square of the two amplitudes gives the preformation probabilities of the corresponding clusters at the nuclear surface. The comparison with the present existing experimental data shows an excellent agreement.
Consequently let us consider a small perturbation propagating on the surface of a sphere of radius R. For an axial symmetry, with the symmetry axis in the direction of the perturbation, the problem reduces to a small perturbation propagating on a circle with the shape $r = R + \eta(\theta,t)$. We make three assumptions : first that the amplitude of the perturbation η_0 is small compared with the radius R so that we

can introduce a small parameter $\xi = (r-R)/R \ll 1$, second that the core is unperturbed up to the radius $r = R-h$ with $h \ll R$ and third that we have the case of an ideal, imcompressible and irrotational fluid which leads to a field of velocities $V = \nabla\Phi$ given by a scalar potential Φ satisfying the Laplace equation $\Delta\Phi=0$. As we shall show later the irrotational flow gives rise to some new type of collective oscillations of the fluid, coupling the tangential and the radial oscillations of the fluid at the surface.

In order to describe the dynamics of such a perturbation we have to find the equations for shapes and velocities.

Now it is natural to ask for solutions of the Laplace equation written as a power series into the small parameter $\xi = \frac{r-R}{R}$

$$\Phi(r,\theta,t) = \sum_{n\geq o} \xi^n f_n(\theta,t \tag{1}$$

It is easy to show that in second order of ξ the radial $v = \Phi_r$ and tangential $u = \Phi_\theta/r$ velocities depend only on two independent functions $f_{o,\theta}(\theta,t)$ and $f_1(\theta,t)$ denoted in the following by $g(\theta,t)$ and $j(\theta,t)$ respectively.

The first equation is given by the time derivative of the radial coordinate on the surface Σ :

$$v|_\Sigma = \frac{dr}{dt}|_\Sigma = \eta_t + \frac{1}{r} u \eta_\theta \tag{2}$$

By imposing the condition that $v = 0$ at $r = R-h$ we obtain in the first order in η, a relation between j and g_θ :

$$j = -\frac{h}{R} g_\theta \tag{3}$$

The second equation is given by the Euler equation :

$$p_m \left[\frac{\partial V}{\partial t} + (V.\nabla)V\right] = -\nabla P + \nabla\Phi_e \tag{4}$$

where ρ_m is the constant mass density, P the pressure and $\nabla\Phi_e$ the repulsive electrostatic interaction. The Poisson equation for the Coulomb potential is

$$\Delta\Phi_e = -\frac{\rho_{el}}{\epsilon_0} \tag{5}$$

where Φ_e represents the electrostatic potential generated by a charged distribution ρ_{el} and ϵ_0 the vacuum dielectric constant. The pressure of the liquid at the surface Σ is in the first order in η

$$P|_\Sigma = \frac{\sigma}{R}\left[1 - \frac{1}{R}(\eta+\eta_{\theta\theta})\right] \tag{6}$$

and the electrostatic potential Φ_e up to the second order in η

$$e|_\Sigma = \frac{\rho_{sph}R^2}{3\epsilon_0}\left[1 - \frac{\eta}{R} + \frac{\eta^2}{R^2}\right] - \frac{\rho_{sol}\,\eta^2}{2\epsilon_0} \tag{7}$$

From the continuity equation and the condition of irrotationality (rotV = 0) we can put the Euler equation (4) into a gradient form which leads to following equation

$$g_t + \frac{1}{R^2}gg_\theta - \frac{\sigma}{\rho_m R^2}(\eta_\theta + \eta_{\theta\theta\theta}) - \frac{\rho_{sol}\rho_{sph}R}{3\epsilon_0\rho_m}\eta_\theta + \frac{\rho_{sol}}{\rho_m\epsilon_0}(\rho_{sol} - \frac{\rho_{sph}}{3})\eta\eta_\theta = 0 \tag{8}$$

Eq.(2), by using rel. (3), gives in the second order in η :

$$R\eta_t + \frac{h}{R}g_\theta + \frac{1}{R}(\eta g)_\theta + \frac{h}{R^2}\eta g_\theta = 0 \tag{9}$$

The system formed by eqs. (8) and (9) is a system of partial nonlinear third order differential equations in the unknown functions $\eta(\theta,t)$ and $g(\theta,t)$. In order to solve this system we make the transformation

$$g = \chi\eta + \psi(\theta,t), \tag{10}$$

where χ is an arbitrary real parameter and $\psi(\theta,t)$ an arbitrary function and the functional transformation :

$$\psi(\theta,t) = \frac{k\chi\eta - \chi\eta^2/R - R\int\eta_t\theta + \psi^{(1)}}{k - \eta/R} \tag{11}$$

Choosing the arbitrary function $\psi^{(1)}$ so that :

$$\psi^{(1)}_t - \frac{\beta}{\chi}\psi^{(1)}_\theta + \frac{1}{R^2}\Psi^{(1)}\Psi^{(1)}_\theta + (\frac{\chi}{R^2} + \frac{\beta}{kR\chi})(\eta\psi^{(1)})_\theta = 0 \tag{12}$$

we obtain a Korteweg de Vries equation for η :

$$A\eta_t + B\eta\eta_\theta + C\eta_{\theta\theta\theta} = 0 \tag{13}$$

where

$$A = \chi + \frac{R^2\beta}{h\chi}\ ;\ B = \frac{\chi^2}{R^2} + \alpha + \frac{2\beta}{h}\ ;\ C = -\gamma$$

with the constants $\alpha = \frac{\rho_{sol}}{\rho_m \epsilon_0}(\frac{\rho_{sph}}{3} - \rho_{sol})$, $\beta = -\frac{1}{\rho_m R}(\frac{\sigma}{R} + \frac{R^2 \rho_{sph}\rho_{sol}}{3\epsilon_0})$, $\gamma = \frac{\sigma}{\rho_m R^2}$ and $k = \frac{h}{R}$. One of the solutions of K.dV equation in the following

$$\eta = \eta_0 \mathrm{sech}^2(\frac{\theta - Vt}{L}) \tag{14}$$

which represents the soliton characterized by the half-width $L = (12C/B\eta_0)^{1/2}$ and the angular velocity $V = B\eta_0/3A$. By assuming $\psi^{(1)} = f(\theta - Vt)$, i.e. $\frac{\partial}{\partial t} = -V\frac{\partial}{\partial\theta}$, the solution of eq. (12) is

$$\psi^{(1)} = 2R^2\left[V + \frac{\beta}{\chi} - (\frac{\chi}{R^2} + \frac{B}{kR\chi})\eta\right] \tag{15}$$

Introducing this expression in eqs. (3) and (10) we obtain the final forms for the unknown functions j and g and, consequently, the final form for the field of velocities.

The total potential energy must describe the transition from the ground state of the initial spherical nucleus of radius R_0 to a soliton moving on the surface of a layer situated above the ground state of the daughter nucleus of radius R, also considered as a sphere. Evidently the total potential energy must be separated in two parts with two different descriptions, one around the soliton shape and one around the ground state. Due to the fact that a soliton could have a very small amplitude and a large angular width close to π we may interpolate the potential energy of such a soliton in the range of the parameters where also the potential energy for the ground state is valid. In this way we can obtain a unique expression for the total potential energy. The corresponding dynamics is simply given by the mass of the emitted fragment if we assume that the final nucleus is a rigid core (ground state) unperturbed by the appearence of the soliton. The potential energy E_p, around the soliton shape, can be written as the sum of the surface energy E_S, the Coulomb energy E_C, the shell energy E_{shell} and the centrifugal E_{cf} energy.

$$E_S = 2\pi\sigma R^2 \int_0^{\pi} \left[2\frac{\eta}{R} + \frac{1}{R^2}(\eta^2 + \frac{\eta_\theta^2}{2})\right]\sin\theta d\theta \tag{16}$$

$$E_C = 8.6275\Big[0.07708(Z^2 \tilde{A}^{-1/3} + \lambda^2 Z_{cl}^2 A_{sol}^{-1/3}) + \tag{17}$$
$$+0.3854 Z^2 \lambda \tilde{A}^{-1/3} \int_0^{\pi} (2\frac{\eta+h}{R} + \frac{(\eta+h)^2}{R^2})\sin\theta d\theta\Big] - 0.665 Z_0^2 A_0^{-1/3}$$

where $\lambda = \rho_{sol}/\rho_{sph}$ and $\tilde{\chi} = A_0 - A_{sol}$.

$$E_{shell} = \frac{U_0 V_{over}}{V + [V_0-(V_{sol}+V_{layer})] - V_{over}} \tag{18}$$

where V_{over} denotes the volume of the overlap between the volumes of the initial V_0 and final V nuclei, V_{sol} is the soliton volume and V_{layer} is the layer volume on which the soliton is moving.

$$E_{cf} \simeq \frac{1}{3R} \int_0^{\pi} \frac{\eta}{(h+\eta)^4} \left[\tilde{\chi}\eta_\theta(h - \eta)^2 + (h^2\chi + hRV)\eta_\theta + \frac{1}{3}\chi h\eta\eta_\theta\right]^2 \sin\theta d\theta \tag{19}$$

For very compact solitons (small L) the centrifugal energy becomes

$$E_{cf} \simeq \frac{1}{3}\rho_m V_{cl} r_{CM}^2 V^2 \text{ with } r_{CM} = R(1 + \frac{\eta_0}{3R} + \frac{\eta_0^2}{15R^2}) \tag{20}$$

This already suggests that only one third of the outside nucleons of the rigid core (final nucleus) are participating in the soliton, the other two thirds forming the surface layer on which the soliton is moving. If we take this conclusion as one of the possible ways for the description of the cluster like amplitude in the potential minimum we can fix-up the parameter h by volume conservation

$$R_0^3 = \frac{1}{3} r_0^3 A_{cl} + (R + h)^3 \tag{21}$$

The expression of the potential energy around the ground state minimum can be written in terms of normal modes :

$$\eta(\theta,t) = \eta_0 \sum_{l \geq 2} \alpha_l(t) P_l(\cos\theta) \quad \text{with} \quad \alpha_l(t) = \int_{-\pi}^{\pi} P_l(\cos\theta)\eta d\theta \tag{22}$$

This leads to the following expression of the normalized potential energy in terms of normal modes :

$$E_p = \frac{1}{2} \sum_{l \geq 2} C_l \alpha_l^2 \text{ with } C_l = \frac{l-1}{4\pi}\left[E_S(0)(l+2) - 10E_C(0)\frac{1}{2l+1}\right] \tag{23}$$

The corresponding barriers along the minimum of the potential in the plane $\tilde{\chi} - \eta_0/R_0$ for different k-values corresponding to two different emissions ^{14}C and ^{24}Ne with different final nuclei are given in Fig. 1. The spectroscopic factors, as ratio of the experimental decay (λ_{exp}) over the theoretical one-body decay constant (λ_G) were first intro-

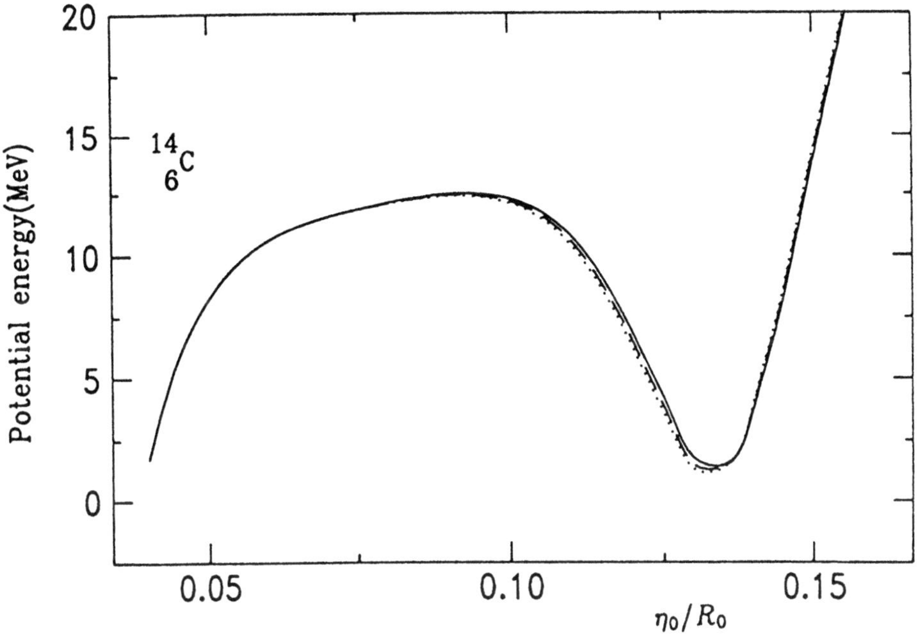

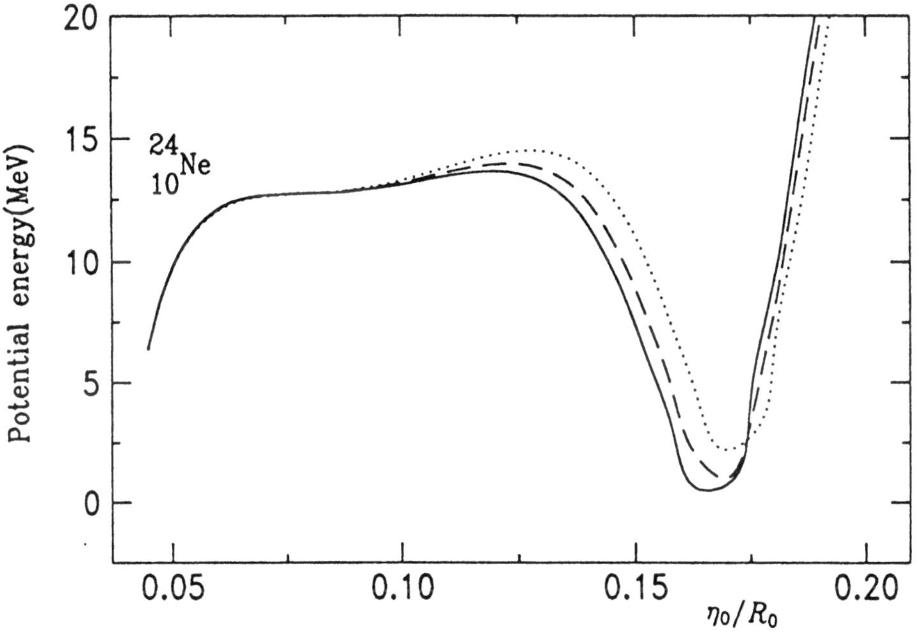

Fig. 1 The soliton potential barriers for ^{14}C-decay of ^{222}Ra(———), ^{224}Ra(– – –) and ^{226}Ra(. . .) and for ^{24}Ne-decay of ^{232}U(———), ^{234}U(– – –) and ^{236}Pu(. . .). We can see that with the decrease of the charge density of the final nucleus the soliton potential barriers are increasing leading to smaller spectroscopic factors S in complete agreement with the experiment (see Table).

duced in alpha decay in order to separate the nuclear structure effects from the decay constant as the product between the frequency of the collisions with the barrier wall and the barrier penetrability (λ_{Gamow}). Similar spectroscopic factors have been introduced for cluster decays.

We consider that without the introduction of the many body calculations in the shell model it is not possible to explain such a large enhancement of clusters on the nuclear surface. In the present paper we give an explanation of spectroscopic factors based on a collective model. By considering the solitons as nonlinear solutions of the hydrodynamic equations we obtained new shapes which show that some nucleons are grouping together to form the emitted cluster. In this way we have a new coexistence model consisting of the usual shell model and a cluster-like model (soliton).

In this soliton description the spectroscopic factors are given by the ratios of the two wave amplitudes in the corresponding wells, evaluated with the help of barrier penetrability between the two minima. The spectroscopic factors are given by the penetrabilities of the above potential barriers :

$$S = \exp\left[- \frac{2}{\hbar}\int_0^{\eta_{of}} (A_{red}(\eta_o)E_p(\eta_o))^{1/2} d\eta_o\right] \tag{24}$$

where η_{of} is the coordinate of the final state. For the reduced mass we take the expression :

$$A_{red} = \frac{A_{sol}(\eta_o)A}{A_o}; \quad A_o = A_{sol}(\eta_o) + A \tag{25}$$

In this way we calculated for the above decays the corresponding spectroscopic factors. The results are presented in the Table which consists in seven columns and nine lines each corresponding to a given cluster decay. In the first three columns we define the nuclei participating to the decay. In the fourth column we give the one-body decay constants (λ_G^{M3Y}) calculated with Michigan 3 Yukawa (M3Y) effective nuclear-nuclear forces [1].

TABLE Spectroscopic Factors for Exotic Decays

$^{A_i}Z_i$	$^{A_f}Z_f$	$Z_{cl}{}^{A_{cl}}$	$\lambda_G^{M3Y}(s^{-1})$	$S^{M3Y} = \frac{\lambda_{exp}}{\lambda_G^{M3Y}}$	S_{BW}	S_{sol}
^{222}Ra	^{208}Pb	^{14}C	2.8×10^{-1}	2.4×10^{-11}	2.9×10^{-10}	2.4×10^{-11}
^{224}Ra	^{210}Pb	^{14}C	4.1×10^{-6}	2.3×10^{-11}	2.9×10^{-10}	2.3×10^{-11}
^{226}Ra	^{212}Pb	^{14}C	3.2×10^{-11}	1.1×10^{-11}	2.9×10^{-10}	1.5×10^{-11}
^{228}Th	^{208}Pb	^{20}O	6.1×10^{-8}	1.2×10^{-14}	1.2×10^{-14}	1.3×10^{-14}
^{230}Th	^{206}Hg	^{24}Ne	1.6×10^{-8}	1.0×10^{-17}	1.3×10^{-17}	1.2×10^{-17}
^{232}U	^{208}Pb	^{24}Ne	1.6×10^{-4}	1.7×10^{-17}	1.3×10^{-17}	1.7×10^{-17}
^{234}U	^{210}Pb	^{24}Ne	7.4×10^{-10}	3.8×10^{-17}	1.3×10^{-17}	2.7×10^{-17}
^{234}U	^{206}Hg	^{28}Mg	4.4×10^{-7}	4.5×10^{-20}	1.6×10^{-20}	4.6×10^{-20}
^{236}Pu	^{208}Pb	^{28}Mg	5.0×10^{-3}	2.9×10^{-20}	1.6×10^{-20}	3.1×10^{-20}

In the fifth column we give the corresponding spectroscopic factors S^{M3Y} as ratios of the experimental decay constants λ_{exp} and the previous one body decay constants λ_G^{M3Y}. The sixth column contains the spectroscopic factors $S_{BW} = (6.3 \times 10^{-3})^{(A_{cl}-1)/3}$ obtained using a semiempirical heavy ion potential [2]. In the last column we give our spectroscopic factors given by eq. (24). First we mention that the agreement with the experimental values calculated with the M3Y forces is very good. The calculated spectroscopic factors differ from one isotope to another. This effect can be explained on one side by the contribution of the Coulomb energy (through ρ_{sol} and ρ_{sph}) and on the other side by the different values of the atomic mass of the final nuclei, which give different positions and half-widths for the absolute minimum of the shell potential energies. The spectroscopic factors S_{BW} do not depend on charge densities. Also they differ by an order of magnitude for ^{14}C-decays. For large A the minimum is obtained for smaller values of η_0 and viceversa. The variation of A affects also the value of the parameter h which changes the structure of all the other three potential energies, thus modifying the height and length of the potential barrier, as one can see in Fig. 1.

REFERENCES

1. Săndulescu A, Gupta R K, Carstoiu F, Horoi M and Greiner W 1991, to be published
2. Blendowske R and Williser A 1988 Phys. Rev. Lett. 61 1930

2. Fission

Fission versus metastable decay series of rare gas cluster ions

T.D. Märk
Institut für Ionenphysik, Leopold Franzens Universität,
Technikerstr. 25, A 6020 Innsbruck, Austria

1. Introduction

If the energy of an electron (or photon) beam colliding with gas phase clusters is greater than a critical value (appearance energy), some of these neutral clusters will be ionized. The abundance and the variety of the cluster ions P_n^+ produced from a specific neutral precursor P_m depend on geometric, electronic and energetic properties of the neutral and ionized clusters. Whereas inelastic interaction of electrons with single atoms results only in changes of the electronic configuration, interaction of electrons with van der Waals (vdW) clusters involves - besides electronic excitation - (i) changes in the nuclear motion (vibrational and rotational excitation), (ii) multiple collisions of the incoming electron at different cluster sites, and (iii) subsequent intermolecular reactions within the cluster. All of this leads to the production of excited ions, i.e. deposition of excess energy into various degrees of freedom. Energy flow between different degrees of freedom leads to spontaneous decay reactions of these excited ions in time scales ranging from a few vibrational oscillations for prompt dissociations up to the μs time regime for metastable dissociations.

It is clear that these prompt and/or delayed dissociations will lead to a strong modification of the original neutral cluster distribution. Curiously enough, despite this fact, in a number of earlier studies cluster ion mass spectra were related on a one to one basis to neutral cluster distributions. Today, it is widely accepted /1-4/ that magic numbers - discontinuities observed in otherwise smoothly varying mass spectral distributions of cluster ions (a beautiful example demonstrating the especially stable icosahedral shell closure at around n = 147 is given in Fig. 1) - are due to variations in the ionization efficiency and the properties (structure, binding energy, stability) of the ensuing ions.

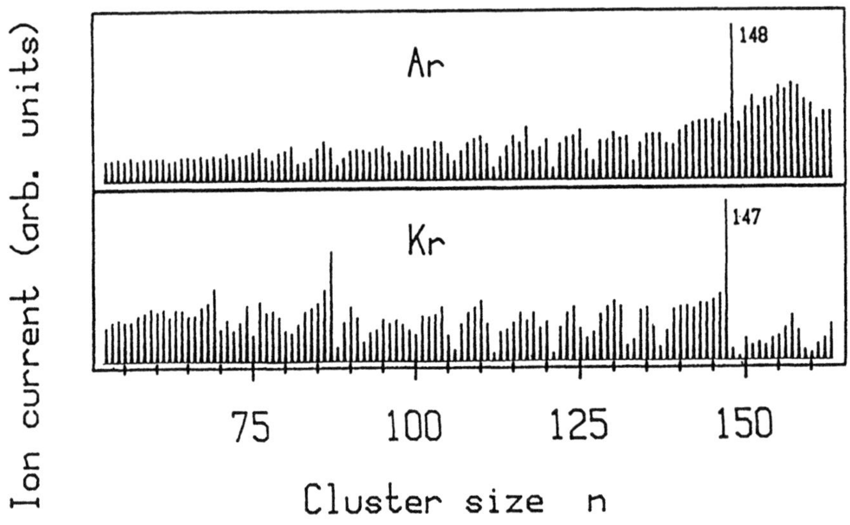

Fig. 1 Mass spectra (section) of argon and krypton cluster beams
after Ref. 5

There has been considerable progress in this field during recent years due to advances by several experimental and theoretical groups. Improvements in mass spectrometric techniques in conjunction with judicious use of theoretical concepts (statistical rate theory), now enable new insight to be gained regarding the nature of the dynamics, energetics and kinetics of cluster ion production and cluster ion reactions. Here I will summarize today's knowledge with particular attention to results from our laboratory on spontaneous decay reactions of excited atomic cluster ions, including monomer evaporation (vibrational predissociation), sequential metastable decay series, and excimer induced fissioning.

2. Statistical (vibrational) predissociation

The most common spontaneous cluster ion decay reaction in the metastable time regime is monomer evaporation,

$$P_n^{+*} \dashrightarrow P_{n-1}^{+} + P \qquad (1)$$

Electronic predissociation and barrier penetration has been named as likely mechanisms for small cluster ions. A typical example is the metastable decay of $Ar_2^+(II(1/2)_u)$ via $Ar_2^+(I(1/2)_g)$ into $Ar^+(^2P_{3/2})$ + Ar with a lifetime of 91 μs /6,7/. A particular variant of barrier penetration is tunneling through a centrifugal barrier (rotational predissociation), which has been proposed to account for the slow decay of small Ar cluster ions /6,8/.

Conversely, vibrational predissociation has been shown to be the dominant metastable dissociation mechanism for larger ($n > 10$) cluster ions. If a polyatomic cluster ion is complex enough, the random motion of an activated ion on its potential hypersurface will be complicated enough to increase its lifetime into the metastable time regime. This process has to be treated theoretically in the framework of statistical theories (RRKM, QET), where the unimolecular rate k (and other properties such as the release of translational kinetic energy T) are assumed to depend only on the internal energy E^* of the activated ion. Cluster ions produced by electron (or photon) impact ionization of a neutral cluster beam normally comprise, however, a broad range of energies E^* due to the broad range of energies deposited into the ions by the primary ionization process and the effect of subsequent unimolecular reactions. Therefore, parent ions are present with different energies and thus different decay rates. Moreover, in many cases, P_n^{+*} may decay by competing reactions and the produced daughter ions P_{n-1}^+ may not be stable and decay again by further decomposition reactions. This situation makes analysis of experimental data very difficult.

Nevertheless, considering the kinetics of monomer evaporation within the frame of QET and using the concept of an evaporating ensemble, Klots /9/ predicted the time dependence of an evaporating parent population, isolated and normalized to unity at a time t_o, to be given by a non-exponential decay function

$$[P_n^{+*}](t) = 1 - \frac{C_n}{\gamma^2} \ln \left[\frac{1}{t_o + (t-t_o)\exp(-\gamma^2/C_n)}\right] \qquad (2)$$

where C_n is the heat capacity of the cluster (in units of the Boltzmann constant k_B) and γ is the Gspann parameter, defined by

$$\gamma = \frac{E_{vap}}{k_B \cdot \sqrt{TT^*}} \qquad (3)$$

It contains the energy of evaporation E_{vap} and a geometric means of the before-and-after temperatures T and T^*, respectively. The Gspann parameter is very nearly independent of the size and composition of the cluster and only a weak function of time. The single unknown parameter in the Klots formula is the heat capacity. Choosing plausible values for γ and C_n there exists very good agreement (at least for larger clusters) between the predicted time (equ. (2)) dependence and the experimental findings (see Fig. 2) /10,11/.

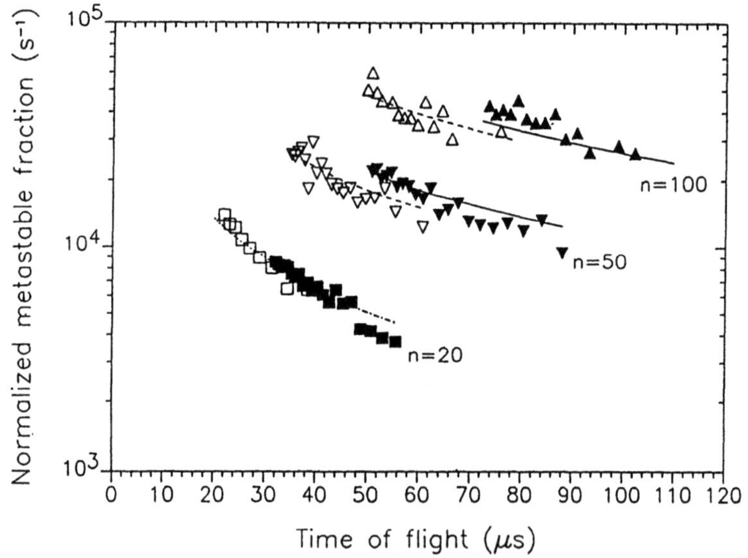

Fig. 2.Comparison of measured (open symbols refer to Ar clusters, filled symbols to Kr clusters) and predicted (equ. 2) normalized metastable fractions (metastable fraction divided by the length of the observational time window) after Ref. 10.

Moreover, equ. (2) may be also used to predict the dependence of the metastable fraction on cluster size n. There is good agreement /9/ in the general trend between existing experimental data and the predicted curve for atomic cluster ions such as Na, Cu, Xe and Ar. It is interesting to note that for certain cluster sizes the experimental values deviate from the predicted curve beyond quoted error bars (e.g. see Fig. 9 in Ref. 10). In most cases these anomalous small or large rates (metastable fractions) coincide in a mirror-like fashion with enhanced or depleted ion abundances in the ordinary mass spectrum (see Fig. 3). The reason for these mass spectral anomalies are additional structural stabilities ("magic numbers") not included in the continuum based model of Klots. According to calculations and experimental evidence (see Fig. 1) especially stable atomic clusters are obtained for n = 13,55,147,309 etc. atoms, their corresponding structure being icosahedral. Moreover, it is possible to relate anomalous small or large metastable rates of atomic cluster ions to corresponding larger or small binding energies using RRKM type treatments /12-15/.

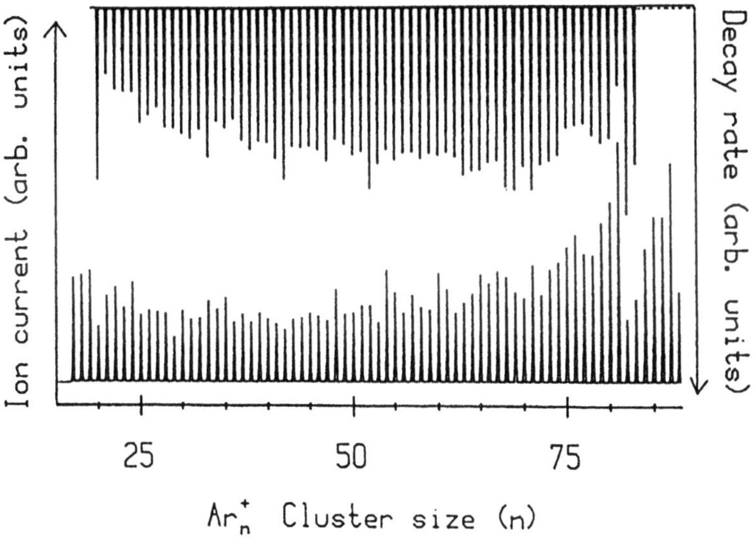

Fig. 3. Ordinary mass spectrum for argon cluster ions (20 ≤ n ≤ 83) and apparent metastable decay rates as a function of cluster size after Ref. 10.

3. Metastable decay series

Considering also thermochemical aspects Klots /9/ predicted that due to the influence of the surface energy, sequential evaporations of monomers should be a more likely cooling process for excited cluster ions than single step splitting off larger fragments. Using both field free regions of a double focussing sector field mass spectrometer as independent observational windows we were able to demonstrate recently that certain cluster ions P_n^{+*} (with P = Ar /16/ and N_2 /17/) decay by sequential decay series (and not single step fissioning), i.e.

$$P_n^{+*} \dashrightarrow P_{n-1}^{+*} \dashrightarrow P_{n-2}^{+*} \ldots P^+ \qquad (4)$$

evaporating a single monomer in each of these successive decay steps. Whereas in case of radioactive decay series the various decay rates are constants, in case of these cluster ions individual apparent decay rates are depending on time and on the parent ion (excitation) due to the fact that each ion (in the ensemble probed) may exist in a different excited state.

4. Excimer-induced fission

Recently, we have discovered a rather unusual metastable fragmentation channel occurring in argon cluster ions /18,19/. In contrast to the well known case of the single monomer evaporation reaction (1) due to vibrational predissociation, in the new decay reaction the average number of ejected Ar monomers rises from 2 for Ar_4^+ up to 10 for Ar_{30}^+ (see as an example the decay pattern of Ar_{10}^{+*} given in Fig. 4). After studying the dependence of the metastable fractions on (i) the electron energy, (ii) parent cluster size and the number of ejected monomers (see Fig. 4), and (iii) the time interval between ion formation and dissociation (yielding a fragmentation lifetime of 1.5 μs), we concluded that a metastable, electronically excited excimer Ar_2^* $(^3\Sigma_u^+)$ localized inside the cluster ion is responsible for the observed unusual decay pattern. The radiative decay of this excimer leads to repulsion of two Ar atoms in the ground state $Ar_2(^1\Sigma_g)$ and to a subsequent violet (non-statistical ? /20/) disintegration of the cluster.

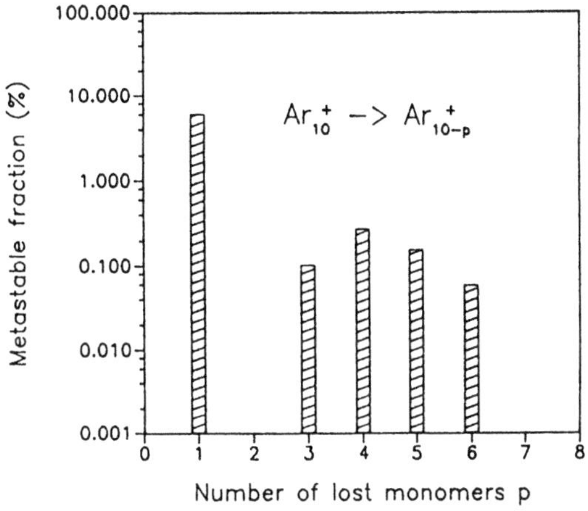

Fig.4. Metastable fraction versus p for the metastable decay reaction Ar_{10}^{+*} --> Ar_{10-p}^{+} + pAr after Ref. 19.

In order to further investigate this phenomenon we have recently extended our studies to metastable fragmentation of neon cluster ions /21/, in particular to the following reaction

$$Ne_4^{+*} \text{ --> } Ne_2^{+} + 2\ Ne \qquad (5)$$

Figure 5 shows the dependence of the fragment ion current resulting from the metastable decay reaction (5) on the

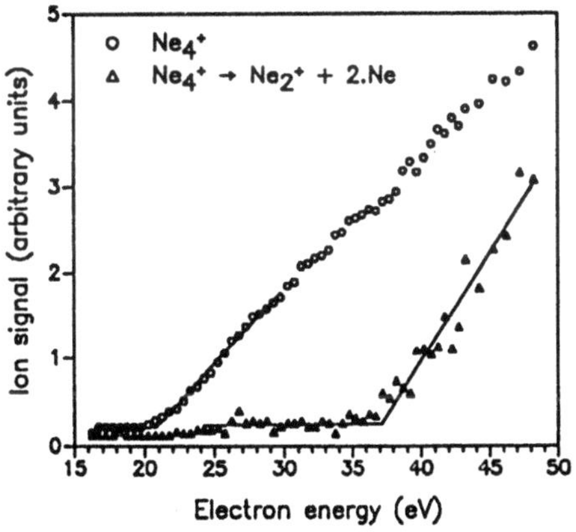

Fig. 5 Ion current as a function of the electron energy for the parent ions Ne_4^+ (o) and for the fragment ions Ne_2^+(Δ) produced by the metastable decay reaction Ne_4^+ --> Ne_2^+ + 2Ne. Full lines represent linear extrapolations of experimental data in the onset regions. After Ref. 21.

electron energy. For comparison, the electron energy dependence of the Ne_4^+ parent ion current is also shown. Linear extrapolation in the onset regions shows, that the appearance energy (AE) of the metastable decay (5) is 37.2 ± 0.8 eV, which is approximately 16.4 eV more than the AE of the parent ion Ne_4^+ (in this measurement AE(Ne_4^+) = 20.8 ± 0.3 eV).

The first electronically excited state of the Ne^+ ion is the $(2s2p^6)^2S_{1/2}$ state, lying 26.9 eV above the ground ionic state. The lowest excited states of the neutral Ne are the $(2p^53s)$ $^3P_2$, ^3P_o metastable states and the $^3P_1$, $^1P_1$ resonant states, lying 16.62 to 16.85 eV above the ground state. Taking into consideration the measured AE of the metastable fragmentations (5), we have to conclude therefore that as in the case of Ar electronically excited states of neutral Ne (i.e. their corresponding excimers) are involved in the fragmentation process (5), populated most likely by either the scattered or ejected electron after a successful ionization process (e + Ne --> Ne^+) inside the same cluster (see above). The properties (energy, lifetime) of these excimers as deduced in the present experiments are in excellent agreement with those reported in studies in liquid and solid rare gases /22/. Moreover, the present interpretation is in accordance with recent results on electron induced sputtering /23/, luminescence of neutral rare gas clusters /24/ and electron energy loss data on neutral Kr clusters /25/. In addition, Hertel and coworkers /26/ recently confirmed our results by reporting the direct observation of such an excitation and decay process during photoionization of Ar clusters using synchrotron radiation and threshold photoelectron photoion coincidence TOF analysis. The existence of this fission process provides an example of a system where statistical energy distribution does not occur upon initial excitation. This mode selective excitation of the excimer in the cluster ion via multiple collisions constitutes a beautiful example for the violation of vibrational energy equipartitioning in a large finite system due to the existence of an "isolated electronic state" /27/.

An interesting variant to this excimer production and decay process is the production of anion clusters via (i) multiple collision electron scavenging (auto-scavenging) or via (ii) core excited resonance attachment and subsequent simultaneous energy and charge transfer leading to dissociative anionization involving the first excited states of rare gas atoms in clusters containing molecules exhibiting zero energy (cluster) resonances, e.g. the production of $(XeSF_6)^-$ via the reaction sequence e + $Xe_n(SF_6)_m$ --> $Xe^*Xe_q(SF_6)_p^-$ --> $XeSF_6^-$ /28/ or the production of $(ArO_3)^-$ via e + $Ar_n.(O_2)_m$ --> Ar^{*-} . $Ar_q.(O_2)_m$ --> $(ArO_3)^-$ /29/.

Acknowledgements

Work partially supported by the österreichischer Fonds zur Förderung der wissenschaftlichen Forschung and the Bundesministerium für Wissenschaft und Forschung, Wien, Austria. It is a pleasure to thank all of my coworkers (see respective references throughout the text) for their valuable collaboration in elucidating the physics and chemistry of cluster ions.

References

1. T.D. Märk, Int. J. Mass Spectrom. Ion Proc., 79 (1987) 1; Z. Phys. D12 (1989) 263
2. A.J. Stace: In: Mass Spectrometry (M.E. Rose, Ed.) Royal Chemistry Specialist Report, London 9 (1987) 96
3. T.D. Märk, O. Echt: In: Clusters of Atoms and Molecules (H. Haberland, Ed.) Springer, Heidelberg (1991) in print
4. A.W. Castleman, Int. J. Mass Spectrom. Ion Proc., (1992) in print
5. T.D. Märk, Proc. 8th SASP, Pampeago (1992)
6. K. Stephan, A. Stamatovic, T.D. Märk, Phys. Rev. A 28 (1983) 3105; P. Scheier, A. Stamatovic, T.D. Märk, J. Chem. Phys., 89 (1989) 295
7. K. Norwood, J.H. Guo, C.Y. Ng. J. Chem. Phys., 90 (1989) 2995
8. E.E. Ferguson, C.R. Albertoni, R. Kuhn, Z.Y. Chen, R.G. Keesee, A.W. Castleman, J. Chem. Phys., 88 (1988) 6335
9. C.E. Klots, J. Phys. Chem., 92 (1988) 5864; Int. J. Mass Spectrom. Ion Proc., 100 (1990) 457
10. P. Scheier, T.D. Märk, Int. J. Mass Spectrom. Ion Proc., 102 (1991) 19
11. Y. Ji, M. Foltin, C.H. Liao, T.D. Märk, J. Chem. Phys., in print
12. P.C. Engelking, J. Chem. Phys., 87 (1987) 936
13. C. Brechignac, P. Cahuzac, J. Leygnier, J. Weiner, J. Chem. Phys., 90 (1989) 1492
14. P.G. Lethbridge, A.J. Stace, J. Chem. Phys., 89 (1988) 4062; 91 (1989) 7685
15. S. Wei, Z. Shi, A.W. Castleman, J. Chem. Phys., 94 (1991) 8604
16. P. Scheier, T.D. Märk, Phys. Rev. Lett., 59 (1987) 1813
17. P. Scheier, T.D. Märk, Chem. Phys. Lett., 148 (1988) 393

18. M. Foltin, G. Walder, A.W. Castleman, T.D. Märk, J. Chem. Phys., 94 (1991) 810
19. M. Foltin, G. Walder, S. Mohr, P. Scheier, A.W. Castleman, T.D. Märk, Z. Phys. D20 (1991) 157
20. This question is presently explored by classical molecular dynamics calculation in collaboration with Prof. U. Landman, Atlanta, USA
21. M. Foltin, T.D. Märk, Chem. Phys. Lett., 180 (1991) 317
22. N. Schwentner, E.E. Koch and J. Jortner, Electronic excitations in condensed rare gases, Springer, Berlin (1985)
23. R. Pedrys, D.J. Oostra, A. Haring, A.E. de Vries and J. Schou, Nucl. Instr. Meth. Phys. Res., B 33 (1988) 840
24. J. Wörmer and T. Möller: Z. Phys. D 20 (1991) 39; See also: E. Bodarenko, E.T. Verkhovtseva, Y.S. Doronin and A.M. Ratner, Chem. Phys. Lett., 182 (1991) 637
25. A. Burose, C. Becker and A. Ding, Z. Phys. D20 (1991) 35
26. H. Steger, J. de Vries, W. Kamke and I.V. Hertel, Z. Phys. D21 (1991) 85
27. C. Lifshitz, J. Phys. Chem., 87 (1983) 2304
28. T. Rauth, M. Foltin and T.D. Märk, J. Phys. Chem., submitted (1991)
29. M. Foltin, V. Grill and T.D. Märk, Chem. Phys. Lett., in print (1991)

Applications of the Liquid Drop Model to Metal Clusters

Winston A. Saunders[1] *and Nico Dam*[2]

1. Dept of Applied Physics, M. S. 128-95, California Institute of Technology, Pasadena, CA 91125 USA

2. Institut de Physique Expérimentale, Ecole Polytechnique Fédérale de Lausanne, PHB-Ecublens, CH-1015 Lausanne, Switzerland

The vibrations of a cluster surface, akin to those observed in liquid droplets and nuclei, are best described by a normal mode expansion in terms of spherical harmonics. The quadrupolar surface modes are especially important in determining the electronic and the vibrational properties of metal clusters. Here, we discuss the application of the Liquid Drop Model in understanding the fission of multiply charged metallic clusters and the thermal broadening of the plasmon line width.

Fission in metallic clusters has been studied in both the noble[1,2] and alkali[3] metals. Considering that these systems share a common description as free-electron metals in the bulk, it is of interest to find a picture which describes common aspects of their fission properties.

Adapting the liquid drop from nuclear physics to the description of metallic clusters is straight-forward.[4,5] In this approach, the energy of the droplet is assumed to be comprised of three terms,

$$E_T = E_b + E_s + E_c \tag{1}$$

where E_b is the bulk (volume) energy, Es is the surface energy, and E_c is the Coulomb energy. E_b is taken, within this approximation, to be the number of atoms in the cluster times the bulk cohesive energy of the metal. We adopt the standard parameterization of the cluster surface,

$$r_d = a\left(1 + \alpha_0 P_0(\cos\theta) + \alpha_2 P_2(\cos\theta)\right) \tag{2}$$

where $a = r_o n^{1/3}$, with r_o is the atomic radius and n is the number of atoms in the cluster. The α_l are the deformation parameters and the P_l are Legendre polynomials. The term α_2 in (7) is the quadrupolar surface deformation parameter. It can be shown that the deformation energy due to the increase in surface area upon deformation is[6]

$$E_s = \varepsilon_{so} n^{2/3}(1 + (2/5)\alpha_2^2 - (4/105)\alpha_2^3 - (38/175)\alpha_2^4) \tag{3}$$

to fourth order in α_2. Here, $\varepsilon_{so} = 4\pi r_o^2 \sigma$, where σ is the surface tension of the droplet. For the present purposes, curvature corrections[7] to the droplet surface tension are ignored and the bulk value is taken.

The Coulomb energy can be calculated by standard electrostatics.[8] The solution of Poisson's equation in spherical coordinates which converges for $r \rightarrow \infty$ is

$$\Phi(r, \theta) = \sum_{l=0}^{\infty} B_l r^{-(l+1)} P_l(\cos\theta) \tag{4}$$

where $\Phi(r, \theta)$ is the electrostatic potential. The B_l are constants to be chosen such that the boundary condition at $r = r_d$, (Eq. 2)

$$\Phi(r, \theta) = \Phi_0 = \sum_{l=0}^{\infty} B_l a^{-(l+1)}\left(1 + \alpha_0 + \alpha_2 P_2(\cos\theta)\right)^{-(l+1)} P_l(\cos\theta) \tag{5}$$

where Φ_0 = constant, is satisfied.

To fourth order in α_2, E_c for a conducting metallic droplet is given by

$$E_c = z^2 \varepsilon_{co} n^{-1/3}\left(1 - (1/5)\alpha_2^2 - (4/105)\alpha_2^3 + (53/245)\alpha_2^4\right) \tag{6}$$

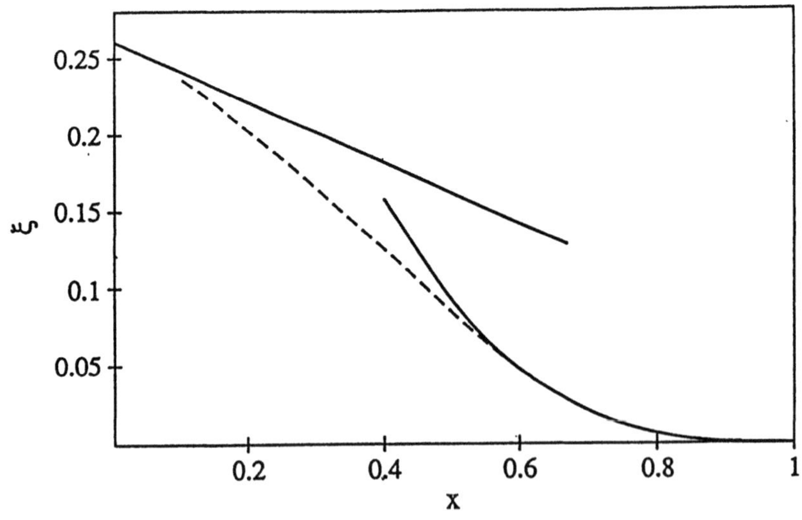

Figure 1. The fission barrier as a function of the fissionability parameter, x. In the limit $x \rightarrow 1$, the barrier follows the familiar Bohr-wheeler formula. In the intermediate range, the interpolated function is shown.

where $\varepsilon_{co} = e^2/(2r_o)$. Up to third order in α_2 this expression is the same as that for a uniformly charged (non-conducting) droplet.

This gives rise to the well known critical size for a charged droplet,

$$\left(\frac{z^2}{n}\right) \equiv \left(\frac{z^2}{n}\right)_c = 2\varepsilon_{so}/\varepsilon_{co} = 12(4\pi r_0^3/3)\sigma/e^2 \tag{7}$$

Based on (7) the fissionability parameter, the only parameter of the liquid drop model, is $x = z^2/n/(z^2/n)_c$. For $x = 1$ the droplet is unstable to spontaneous fission. For $x < 1$, the fission barrier can be derived from the expressions for the total energy of the cluster as a function of deformation in the limits of $x \ll 1$ and $(1 - x) \ll 1$ with relative ease. In intermediate cases, however, the series expansion technique used to derive (6) does not converge very strongly. As a result, an interpolation scheme between the two limits is used to derive the barrier. This is shown in Figure 1. Here, the fission barrier $E_f = \varepsilon_s n^{2/3}\, \xi$

The function shown in Fig 1 can be used to calculate the fission barriers for metallic clusters. The graph for Na clusters is shown below, in Figure 2. The crossing point, where the fission barrier and the evaporation energy are equal, is found to be at $n = 30$, in agreement with the available experimental results. In more detailed measurements on Au clusters, the model compares quantitatively to measurements of the fission rate relative to the neutral evaporation rate, Γ_f/Γ_e.

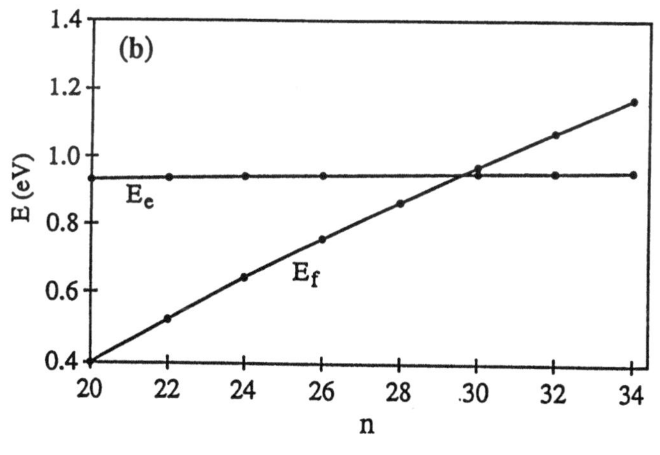

Figure 2. The fission barrier and evaporation energies for doubly charged Na clusters near the size of $n = 30$. The crossing point predicted by the model agrees with the experimentally observed crossing point to good accuracy.

The liquid drop model can also be applied to understand the broadening of the surface plasmon linewidth of metal clusters. The vibrational motion of the cluster surface increases the linewidth of absorption due to its sensitivity to the cluster shape. By taking an ensemble average of the instantaneous shapes of the clusters, weighted by a Bolzman factor, one can derive an elementary theory of the line shape for the clusters. The model predicts a significant broadening of the plasmon, but almost no shift in its position with increasing temperature The results of the calculation are shown in Figure 3. Experimental points are derived from measurements of the plasmon

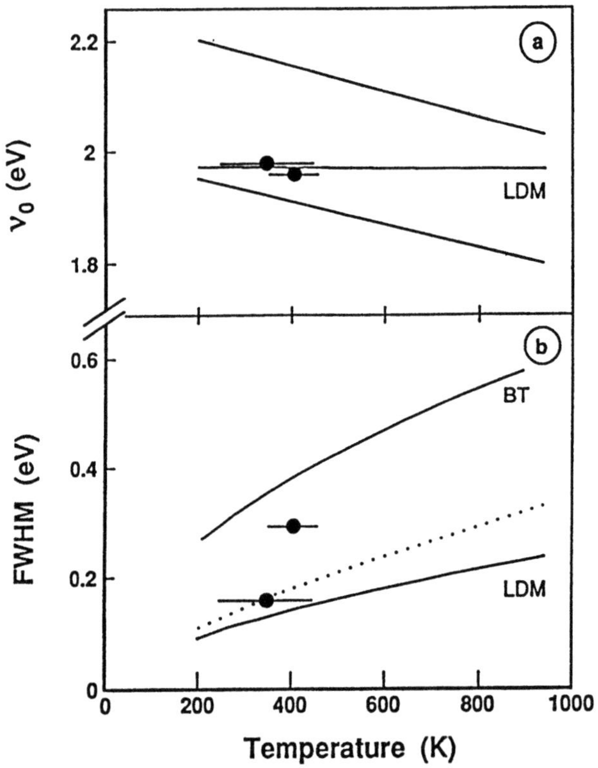

Figure 3. The plasmon position (a) and linewidth (b) as a function of temperature. Theoretical curves from other authors are shown for comparison. Experimental points are shown with errors in the estimate of the temperature.

linewidth[9,10] and the determination of the cluster temperature within the context of the evaporative ensemble.[11] Agreement between the model (LDM) and the experimental data are reasonable. In particular, this provides supportive evidence
that to a large degree the width of the plasmon is determined by the quadrupolar surface fluctuations. Other important effects include the electron-phonon interactions, which lead to the thermal decay of the plasmon and limit it's lifetime. The experimental data are also compared to the model of Bertsch and Tomanek (BT).[12] The temperature dependence predicted by Yannouleas *et al.*[13] is similar to that predicted by the LDM.

In conclusion, the liquid drop model provides a reasonable description of the vibrational properties of small metallic clusters. It is quite a striking result that a model rooted in a macroscopic picture and based on macroscopic parameters can describe a system of only a few atoms with reasonable accuracy. We expect further refinements of the model to yield greater insight into the properties of metallic clusters.

References

1. I. Katakuse, H. Ito, and T. Ichihara, Int. J. Mass Spectrom. Ion Proc. 97, 47 (1990)
2. W. A. Saunders, Phys. Rev. Lett 64, 3046 (1990). W. A. Saunders, Phys. Rev. Lett 66, 840 (1991)
3. C. Bréchignac *et al.* Phys. Rev. Lett 64, 2893, (1990)
4. S Sugano, A. Tamura, and Y. Ishii, Z. Phys. D12, 213 (1989).
5. E. Lipparini and A. Vitturi, Z Phys. D18, 222 (1990)
6. W. J. Swiatecki, Phys. Rev. 104, 993 (1956).
7. R. C. Tolman, J. Chem. Phys. 17, 333 (1949).
8. W. A. Saunders (to be published)
9. N. Dam and W. A. Saunders, Z. Phys. D19, 85 (1991). N. Dam and W. A. Saunders (to be published).
10. C. Brechignac, *et al.*, Chem. Phys. Lett. 164, 433 (1989).
11. C. E. Klots, Z. Phys. D20, 105 (1991).
12. G. F. Bertsch and D. Tomanek, Phys. Rev. B40, 2749 (1989).
13. C. Yannouleas, J. M. Pacheco, and R. A. Broglia, Phys. Rev. B41, 6088 (1990).

The Dynamical Model of the Atomic Cluster Fragmentation

V. A. Rubchenya
V. G. Khlopin Radium Institute, Sankt-Petersburg, USSR

1. Introduction

The investigation of the stability of atomic clusters and different channels of their dissociation is of great interest both from the point of view of cluster production and detection and from point of view of studying their structure. It is well known that multicharged atomic clusters X_N^{Z+} are observed as a rule only above the threshold size N_{cr} [1-4]. It was named "Coulomb's explosion" when coulomb repulsion between positive holes makes the cluster unstable as to fragmentation [5]. The energetic criterion is used for the stability criterion, that is the multicharged cluster is unstable when the decay on fragments is exothermic process [6,7]. However the metal clusters with $N < N_{cr}$ have also been observed. The fragmentation of clusters from the ground and excited states is determined not only by the energy release but also by energy potential barrier. The exact calculation of Born-Oppenheimer's energy surface for Be_3^{2+} [8] have shown the existence of the considerable barrier which hinders the decay from the ground state. In heated clusters the channels of fragmentation and evaporation of monomers will compete therefore the observed yields of clusters can depend upon the methods of production and registration. In paper [9] the calculations of potential barrier for the very asymmetric fragmentation and evaporation energy of monomer for Na_N^{2+} are performed and equality of the barrier height and evaporation energy is used as the stability criterion.

There are strong similarities between the properties of atomic clusters and atomic nuclei in spite of a distinction of kind in the forces binding these different systems. The structure of the electron states of alkali clusters has much in common with the structure of the quasiparticle states of nuclei. [10]. The vibration modes [11], similar to the liquid drop oscillations of the atomic nuclei can exist in atomic clusters. The atomic cluster fragmentation regularities are also similar to the nuclear fission basic characteristics which are determined by the properties of the charged liquid droplet [1,4]. It is possible to proceed these analogies and try to apply ideas and methods used to describe the atomic nuclei fission for the analysis of the atomic cluster stability and the fragmentation process.

In this report the dynamical model of the heated atomic cluster fragmentation is proposed to investigate the competition between the fission and evaporation of monomers and to estimate the lifetimes of heated atomic clusters.

2. Model

Let us investigate the decay of the excited atomic clusters. As the possible channels of the decay we'll take into consideration the symmetric and asymmetric fission and the evaporation of monomers. The fission is a large-scale collective motion process therefore we divide cluster degrees of freedom in collective and internal variables. The interaction of the collective variables with internal degrees of freedom can be taken into account by the inclusion of the friction forces into the equations of motion upon collective variables. The conception of dissipation when considering the decay of the excited system comfortably to the fission of atomic nuclei was first introduced by Kramers [12] as early as 1940 , but until recently it has not found experimental confirmation. In recent years the experimental results were obtained in the fusion-fission reactions showing the important role of dissipation and fluctuations during the decay process of hot compound nuclei [13].

One has to specify collective variables describing the process of two-body fragmentation, which provide description of a continuous sequence of shapes interpolating between one sphere or spheroid and two spheres or coaxial spheroids at distance. This variables will be taken as classical, generalized coordinates obeying to the stochastic equations of motion of Langevin type.

$$\dot{Q}_i = (B^{-1}(Q))_{ik} P_k ,$$

$$\dot{P}_i = F_i(Q) - \frac{1}{2}\frac{\partial}{\partial Q_i}(B^{-1}(Q))_{lk} P_l P_k - \eta_{il}(Q) (B^{-1}(Q))_{lk} P_k + R_i(t), \quad (1)$$

where Q, P are the collective variables and corresponding momenta, B_{ik} is the inertial tensor, $\vec{F}$ is conservative force, η_{ik} is the friction tensor, $\vec{R}$ is the random force. We will assume that non-collective degrees of freedom constitute a thermal bath characterized at each time by a temperature T. During the process of the collective motion the neutral or charged monomers can be emitted. The emission width of particles having spin s, the orbital momentum l and kinetic energy in the range of ε, $\varepsilon + \Delta\varepsilon$ is written as

$$\Gamma(E,N,Z,I,l) = \frac{2s+1}{2\pi\,\rho} \sum_{I_d=|I-l|}^{I_d=I+l} \int_{\varepsilon}^{\varepsilon+\Delta\varepsilon} T_l(\varepsilon)\, \rho_d(E-H_e-\varepsilon)\, d\varepsilon , \quad (2)$$

where ρ and ρ_d are the level densities of the decaying and daughter clusters, I is the spin of cluster, T_l is the transmission coefficients, H_e is the binding energy of monomer. It is assumed that the evaporation of particles take place without changing of collective coordinates and the parameters in equations (1) are redefined after particle emission.

The transmission coefficients T_l can be calculated using the approximation of parabolic barrier

$$T_l\ (\varepsilon) = \left[1 + \exp\left(- \frac{\varepsilon - V_l}{\hbar\omega_b} \right)\right]^{-1}, \quad (3)$$

where V_l is the barrier height, $\hbar\omega_b$ is the frequency of the reversed oscillator potential approximating the potential barrier.

$$\hbar\omega_b = \frac{\hbar}{2\pi} \left[\frac{N}{A\ (\ N-1)} \left(\frac{d^2 V_l}{dr^2} \right)_{r_b} \right]^{1/2}, \quad (4)$$

where A is the atomic weight of a monomer.

Unlike the case of the atomic nuclei at present we have no experimental data available on the level densities of atomic clusters. So we shall use the model description applied in nuclear physics [14].

$$\rho = \rho_{el}\ k_{vibr}, \qquad (5)$$

here ρ_{el} is the level density of electron subsystem, k_{vibr} is the vibration enhancement coefficient. In the case of metal clusters having delocalized electrons for the electron level density we shall use Fermi-gas formula

$$\rho_{el}(E_{ex}) = \frac{1}{48E_{ex}} \exp(2\sqrt{aE_{ex}}), \qquad (6)$$

where E_{ex} is the excitation energy, a is the level density parameter determined by the density of the single-particle states near the Fermi level

$$a = \frac{\pi^2}{6}\, g(\varepsilon_F). \qquad (7)$$

For alkali metal clusters the level density parameter is approximately equal to

$$a = 0.1759\, r_s^2\, N^{2/3}(N-Z)^{1/3}\ , eV^{-1}, \qquad (8)$$

where r_s is the radius parameter determined the radius of spherical cluster in the jellium model $R = r_s N^{1/3}$. The vibration enhancement coefficient is equal to the statistical sum for the phonon excitations.

$$k_{vibr} = \prod_{\lambda} (1-\exp(-\omega_\lambda/T))^{-g_\lambda}, \qquad (9)$$

where ω_λ and g_λ are the phonon frequencies and statistical weights. The expression (6) is valid for the gas of non-interacting fermions. If the pairing interaction takes place in the metal clusters [15] then it can be taken into account in the first approximation by introduction of the effective excitation energy $U_{eff} = U - E_{cond}$, where E_{cond} is the energy of transition of electron subsystem from the superconducting state into normal one.

The system of stochastic equations (1) can be solved by two methods. It can be transformed into differential equation of Fokker-Planck type for the probability distribution function or this system can be solved by Monte-Carlo method. In numerical calculations we shall use second method. In this case the value of random force is generated at every time step Δt and the particle emission probability with decay constant $\lambda = \Gamma / \hbar$ is also calculated. If a particle escapes then the parameters of equations are redefined. The history can come to end in two cases. First, when the excitation energy of a cluster becomes less than the binding energy of a monomer and/or less then the fission barrier. Secondly, when the values of collective variables Q_i are equal or greater to the values $(Q_i)_{sc}$ determined the position of fission barrier or scission configuration. The scission configuration is close to the configuration of two touching spherical or coaxial spheroidal fragments.

3. The results of calculations

Here we shall give results of the model calculations for sodium clusters in one-dimensional approximation for the collective motion. In one-dimensional case at constant transport coefficient the system (1) turns into equations

$$\dot{Q} = B^{-1} P , \qquad (10)$$
$$\dot{P} = - \frac{\partial U}{\partial Q} - \beta P + R (Q,t) ,$$

where $\beta = B^{-1}\eta$ is the reduced friction coefficient. It is usually assume that the random force averaged over small time interval Δt

$$R (\Delta t) = \int_t^{t+\Delta t} R (t) dt , \qquad (11)$$

has normal distribution with parameters:

$$\langle R(\Delta t) \rangle = 0 , \quad \langle R^2(\Delta t) \rangle = 2 B \beta T \Delta t . \qquad (12)$$

Here time interval Δt must be considerably less than the characteristic time of collective motion $\Delta t \ll \tau_{coll}$.

In calculations given below we shall suppose that the inertial parameter is equal to reduced mass of fragments N_f and $N-N_f$.

$$B = B_r \left(\frac{dr}{dQ} \right)^2_{Q=Q_{sc}} , \qquad (13)$$

$$B_r = 239.25 \; A \; N_f \; (N - N_f) \; N^{-1} \quad \hbar^2/eV{\cdot}Å^2 , \qquad (14)$$

where B_r is the reduced mass for relative motion of fragments. The coefficient of transition from B_r to B in (13) is determined by the applied parameterization of the cluster shape. The free energy is written in liquid-drop model approximation.

$$\begin{aligned} U\,(Q,T) = {} & E_{surf}(0)\;(1 + \alpha\;T^2)\;(\;B_{surf}(Q) - 1\;) + \\ & + E_{coul}(0)\;(1 + \gamma\;T^2)\;(\;B_{coul}(Q) - 1\;) , \end{aligned} \qquad (15)$$

where B_{surf} and B_{coul} are the dimensionless function of collective variables. Temperature dependence is included in the form which is used for hot nuclei. The surface energy of spherical cluster is expressed in the form

$$E_{surf}(0) = 7.843 \cdot 10^{-4} \; r_s^2 \; N^{2/3} \; \sigma \; , \; eV, \qquad (16)$$

where r_s is expressed in and surface tension coefficient σ is expressed in dyn/cm. The Coulomb energy is calculated in the jellium model for uniformly charged sphere.

$$E_{coul}(0) = 8.64 \;\; Z^2/r_s N^{1/3} \; , \; eV \; . \qquad (17)$$

The shapes of axial-symmetric cluster are described in the parameterization proposed by Pashkevich [16] where the Cassini ovaloids in a lemniscate coordinate system are used.

In the theory of fission the fissility parameter $Z^2/\;N$ is widely used and its critical value in liquid-drop model is equal to

$$(Z^2/ N)_{cr} = 1.8158 \cdot 10^{-4} \, r_s^3 \, \sigma \, . \qquad (18)$$

The cluster temperature at each point of potential surface is determined by expression

$$E_{ex}(Q) = (3N-6) \, k_B \, T(Q) \qquad (19)$$

Below we'll give some results of calculation of monomer evaporation times and fission times. The monomer evaporation time is defined by formula

$$\tau_{ev} = \sum_{i=1}^{n} \hbar / \Gamma_i \, , \qquad (20)$$

where n is the average multiplicity of evaporated monomers. It was assumed that the cluster spin is equal zero and only one partial wave with l=0 is taken into account. In these calculation three phonons with energies $\hbar\omega_2$=50meV, $\hbar\omega_3$=75meV and $\hbar\omega_4$=100meV was used. Another parameters was chosen as follows r_s=2.9 , V_0 = 0.1eV and $\hbar\omega_b$=0.01eV. The calculated evaporation times and average kinetic energies of neutral monomers emitted from neutral Na_N clusters at excitation energy E_{ex}=5eV is shown in Table 1 for two values of binding energy.

Table 1. Evaporation times of Na_N clusters at E_x=5eV.

	H_e = 1 eV		H_e = 1.6 eV	
N	τ_{ev}, 10^{-10}s	$\bar{\varepsilon}$, eV	τ_{ev}, 10^{-10}s	$\bar{\varepsilon}$, eV
20	0.15	0.166	2.69	0.196
40	0.27	0.151	2.51	0.105
60	0.38	0.136	2.18	0.096
80	0.66	0.120	1.62	0.096
100	0.88	0.107	1.78	0.092

The evaporation time very strongly depends on the level density of heated clusters therefore the investigation of monomer evaporation is needed to obtain the equation of state of atomic clusters.

The situation with fission mode of decay of heated clusters more complex. The fission rate depends on inertial parameter, friction coefficient and conservative force. At the small cluster charge the well defined saddle point on the potential surface exists only for small cluster size $N \approx N_{cr}$. From eq. (18) one can obtain the estimation of surface tension coefficient. For Z=2 and N_{cr}=27 [3] we obtain $\sigma \approx 40$ dyn/cm. At such small value of the surface tension coefficient the fission probability of heated clusters is very high. In Table 2 is presented the fission times for different cluster sizes and charges. The values of $\beta=10^{10}$ s^{-1}, H_e= 1.6 eV [5] and σ = 110 dyn/cm are used.

Table 2. The fission times of Na_N^{Z+} clusters at excitation energy E_{ex}=5eV and H_e=1.6eV.

	Fission Times [10^{-10} s]	
N	Z = 0	Z = 2
20	0.76	0.98
40	1.67	2.15
60	3.00	3.45
80	5.11	5.72

4. Conclusions

Thus the dynamic model of decay of heated atomic cluster is formulated. The evaporation width of monomers is calculated in the framework of statistical theory of atomic nuclei. The level density of heated metal atomic cluster is calculated at assumption that there are the electron and phonon modes of excitation. The collective motion of fission mode is considered as stochastic process which is described by Langevin type equation. The model includes some phenomenological parameters which can be obtained from experimental data The calculation of evaporation and fission times are performed for Na heated clusters. It was obtained that these time intervals for Na_N at N < 100 are order of 10^{-10}s.

References

1. D. Kreisle et al, Phys. Rev. Lett., **56**(1986)1551.
2. C. Brechignac et al, Phys. Rev. Lett., **63**(1989)1368.
3. C. Brechignac et al, Phys. Rev. Lett., **64**(1990)2893.

4. W. A. Saunders, Phts. Rev. Lett., **64**(1990)3046.
5. D. Tomanec et al, Phys. Rev. B, **28**(1983)665.
6. M. P. Iniguez et al, Phys. Rev. B, **28**(1986)2152.
7. B. K. Rao et al, Phys. Rev. Lett., **58**(1987)1188.
8. N. Khanna, F. Reuse, J. Buttet, Phys. Rev. Lett., **61**(1988)535.
9. F. Garcias et al, Phys. Rev. B, **43**(1991)9459.
10. C. Yannouleas et al, Phys. Rev. Lett., **63**(1989)255.
11. G. F. Bertsch, D. Tomanec, Phys. Rev. B, **40**(1989)2749.
12. H. A. Kramers, Physica, **7**(1940)284.
13. P. Grange, S. Hassani, H. A. Weidenmuller, Phys. Rev. C, **34**(1986)2063.
14. A. Bohr, B. R. Mottelson, Nuclear Structure, v. 1, 1974.
15. F. Iachello, E. Lipparini, A. Ventura, Preprint U. T. F. 233, 1991, Povo (Trento)-Italy.
16. V. V. Paschkevich, Nucl. Phys. A, **169**(1971)275.

Fission of Metallic Clusters

I. Katakuse[1] and H. Ito[2]

[1]Department of Earth and Space Science, Faculty of Science, Osaka University, Toyonaka, Osaka 560, Japan
[2]Institute of Physics, College of General Education, Osaka University, Osaka, 560, Japan

The stability of metallic clusters can be understood in terms of a shell model or jellium model. The shell model was proporsed to explain the stability of nuclei[1)]. Therefore, we can expect that there are many similar characteristics between clusters and nuclei. The dissociation of a doubly charged cluster to two singly charged clusters can be compared to the fission of nuclei. Such fission-like dissociations have been observed in sodium[2)], gold[3)] and silver[4)] clusters.

Alkali metal and noble metal clusters are typical metallic clusters. They are metallic clusters even in small size. However, clusters composed of some kinds of metals are nonmetallic when they are in small size. The transition from Van der Waals to metallic clusters was observed in mercury clusters[5)]. Small size clusters of mercury are Van der Waals-like clusters. As cluster size increases, they become metallic clusters. A similar transition is expect in Mn clusters. The size distribution of Mn clusters is quite similar to those of rare gas clusters[6)]. As Mn bulk is metal, the transition will occur at a certain size of clusters.

We expect that the bond types (metallic, covalent, Van der Waals, or others) in clusters are presumed from the patterns of fission-like dissociation. We have investigated dissociation patterns of doubly charged Pb and Ta clusters. From the patterns, the Pb clusters are supposed to be nonmetallic when they are in small size. In the case of Ta clusters, the transition from a nonmetallic to metallic bond is supposed to occur around cluster size of 10.

Experimental

Clusters were produced by the bombardment of Xe^+ ions on the metal sheet. The Xe primary ions were obtained from a compact discharge type ion gun[7]. The ion energy was about 7keV and the current 7μA. Spot size was about 1mm in diameter at the sample tip. The acceleration potential of secondary ions was 5kV($=V_0$) under normal conditions. Normal mass spectra were obtained using a double focucing mass spectrometer of a normal geometry by scanning the magnetic field. The strength of the electric analyzer was set to pass the ions having the kinetic energy of eV_0.

Now we consider dissociation of a doubly charged cluster, n^{++}, to two singly charged clusters, d^+ and $(n-d)^+$, which decomposes between the main slit and the entrance of the electric analyzer. We adopt here the notation n^+ and n^{++} for $(M)_n^+$ and $(M)_n^{++}$ for the simplicity.

$$n^{++} \rightarrow d^+ + (n-d)^+$$

The kinetic energy of the d^+ is $(d/n)*2eV_0$, where V_0 is acceleration voltage. Usually the daughter ions cannot pass the electric analyzer, because the kinetic energy is not equal to eV_0. If the kinetic energy of daughter ions is equal to eV_0 by changing the acceleration voltage $V=(n/2d)*eV_0$, they can pass the electric analyzer. By scanning the acceleration

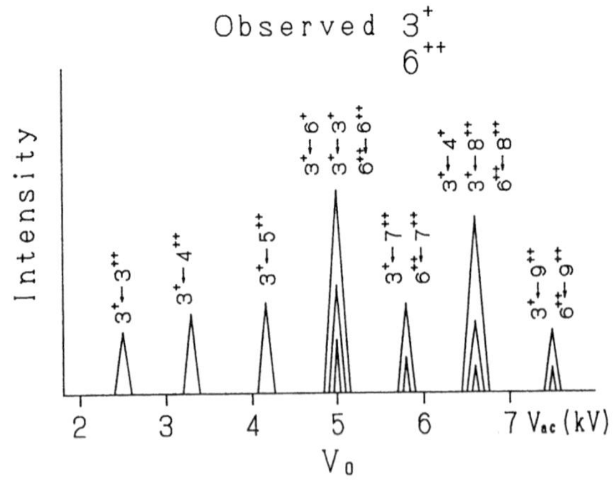

Fig.1 Dissociation spectrum shown schematically. When the V_{ac} is lower than 5kV, only fission fragments are detected.

voltage and by setting the magnetic field to pass a certain size of clusters, daughter ions decomposed from parent clusters with different sizes are detected successively. Such a dissociation spectra is shown schematically in Fig.1. In this case, the size of observed clusters is 3. If the acceleration voltage is lower than 5kV, fission fragments can be detected without the interference of strong ions decomposed from singly charged parent ions. Usualy, ions decomposed from singly charged clusters is overwhelmingly strong than the fission fragments. With the same manner, similar spectra of the m^+ are obtained by changing the magnetic field, where the m^+ can pass the magnet. In order to show the dissociation probability of the n^{++} (n=constant) to the d^+ more clearly, the dissociation spectra are rearranged according to the n^{++}.

Results and Discussion

Figure 2 is dissociation spectra of the 6^{++} and 11^{++} of lead clusters. The distinguished difference from those of silver clusters is the high probability of the asymmetric fission, that is, an atomic ion evaporation. The dissociation probability of silver clusters as a function of the daughter size shows rather flat structure. This may come from the small variation of the $\Delta E_1(d)$ as a function of daughter size, which

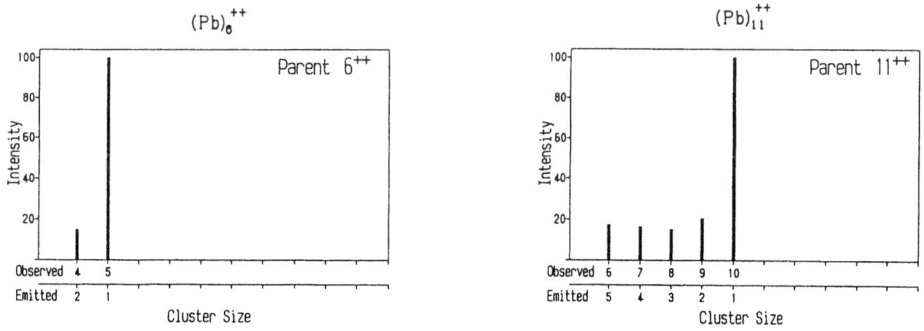

Fig.2 Dissociation probabilities of the 6^{++} and 11^{++} of lead clusters. The dominant channel is the evaporation of an atomic ion.

is the energy difference between the initial and final states of the fission of silver clusters. The calculated $\Delta E_1(d)$ of a liquid drop model by Makamura[8] is almost independent on the size of fragment clusters. The shell effect correction is rather large compared with the size effect. As the clusters produced by the ion sputtering method is highly excited, it is supposed that the barrier of the intermediate state of the fission is not so important.

The high probability of the evaporation of an atomic ion means that the bond is not metallic. It is presumably a covalent-like bond. The energy necessary to cut the covalent bonds is proportional to the number of bonds. The number of bonds to be cut increases as the cluster size to separate becomes larger and the separation energy becomes maximum at the symmetric fission. Therefore, it is difficult to occur symmetric fission. Figure 3 shows a normal mass spectrum of negatively charged lead clusters[9]. A similar spectrum of positive clusters was obtained. In both spectra, the irregularity of the ion intensity disappears around the 30-mers. The transition may occur in these size.

The dissociation spectra of the 6^{++}, 10^{++} and 15^{++} are shown in Fig. 4. As seen in the figure, the main dissociation channel of the 6^{++} is the evaporation of an atomic ion. The symmetric and asymmetric fissions are competing in the reaction of the 10^{++}. The symmetric fission become dominant at the 15^{++}. The transition of the bond may occur at around the 10^{++}.

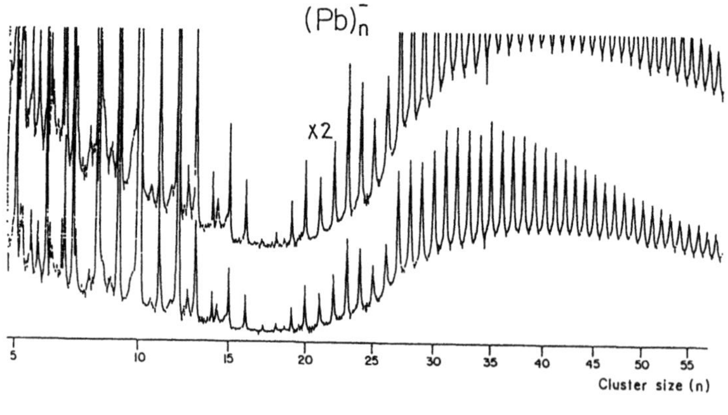

Fig. 3 Normal spectrum of negatively charged lead clusters.

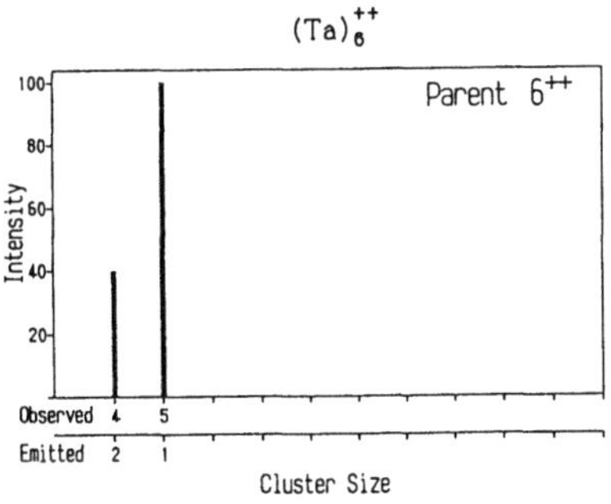

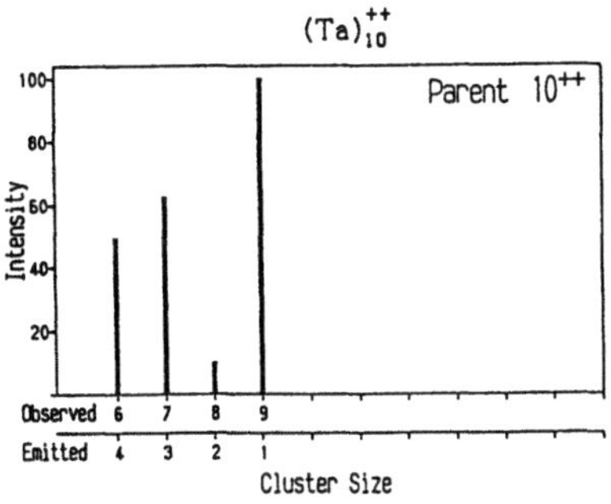

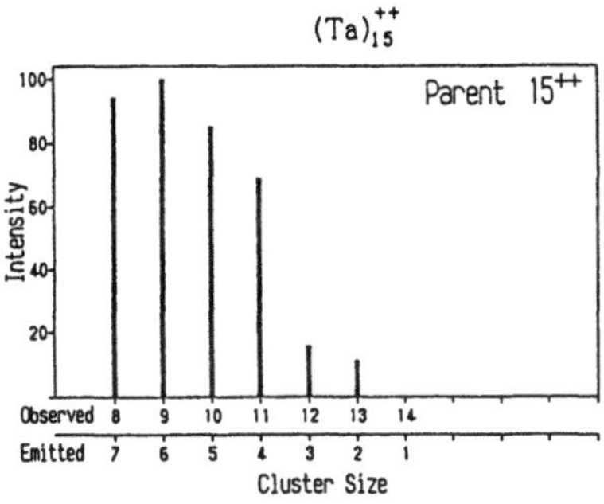

Fig. 4 Dissociation spectra of $(Ta)_6^{++}$, $(Ta)_{10}^{++}$ and $(Ta)_{15}^{++}$.

In conclusion, the dissociation spectra of doubly charged Pb and Ta clusters were obtained. The bond of Pb clusters is covalent-like. The transition from the covalent-like to metallic bond is observed around in 10-mers of Ta clusters.

This work was supported by Matsuo Foundation and Grant-in-Aid for Scientific Research, the Ministry of Education, Science and Culture (No. 03246103)

References

1) M.G. Meyer and J.H.D. Jensen, Elementary Theory of Nuclear Shell Structure, Wiley, New York, 1955.
2) C. Brechignac, Ph. Cahuzac, F. Carlier and M. de Frutos, Phys. Rev. Lett.,64, 2898 (1990).
3) W.A. Saunders, Phys. Rev. Lett., 64, 3046 (1990).
Phys. Rev. Lett., 64, 2898 (1990).
Ion Processes, 99, 207 (1990).
4) I. Katakuse, H. Ito and T. Ichihara, Int. J. Mass Spectrom.
Labastie, J.P. Wolf and L. Woste, Phys. Rev. Lett., 60, 275 (1988).
6) Y. Saito and I. Katakuse, Z. Phys. D, 19, 189 (1991).
7) I. Katakuse, T. Ichihara, H. Nakabushi, T. Matsuo and H. Matsuda, Mass Spectrosc. (Japan), 31, 111 (1983).
8) M. Nakamura, Z. Phys. D, 19, 149 (1991).
9) I. Katakuse, T. Ichihara, H. Ito, T. Matsuo, T. Sakurai and H. Matsuda, Int. J. Mass Spectrom. Ion Processes, 91, 93 (1989).

Dissociation of Doubly-Charged Alkali-Metal Clusters

J. A. Alonso, J. M. López
Departamento de Física Teórica. Universidad de Valladolid.
E-47071 Valladolid. Spain.

F. Garcias
Departamento de Física. Universidad de las Islas Baleares,
E-07071 Palma de Mallorca, Spain.

M. Barranco
Departamento de Estructura y Constituyentes de la Materia.
Universidad de Barcelona, E-08028 Barcelona, Spain.

Abstract

We have studied the competition between fission and evaporation as fragmentation channels of excited doubly charged Sodium clusters. We have used an Extended Thomas-Fermi method and the jellium model. A preliminary account of the influence of shell effects is also given.

1 Introduction

Since the experimental discovery [1] that the electrostatic repulsion in isolated doubly (or multiply) charged clusters X_N^{2+} may lead them to fragment into smaller agregates

$$X_N^{2+} \rightarrow X_n^+ + X_{N-n}^+ \tag{1}$$

the question of what is the critical size N_c below which multiply charged clusters can not be observed has attracted a considerable interest. The fission process of simple metal clusters has been substantially clarified by a series of experiments studying the dissociation of excited doubly charged alkaline [2, 3] and noble metal clusters [4, 5]. The experiments have shown the competition between fission and monomer evaporation. Atom evaporation is dominant for large clusters, but asymmetric fission becomes competitive as N decreases, X_N^{2+} being undetectable in the mass spectra below the critical size N_c. The competition is qualitatively ilustrated in fig. 1. If N is large the cluster preferentially evaporates an atom because its binding energy ΔH_e is smaller than the fission barrier F_m.

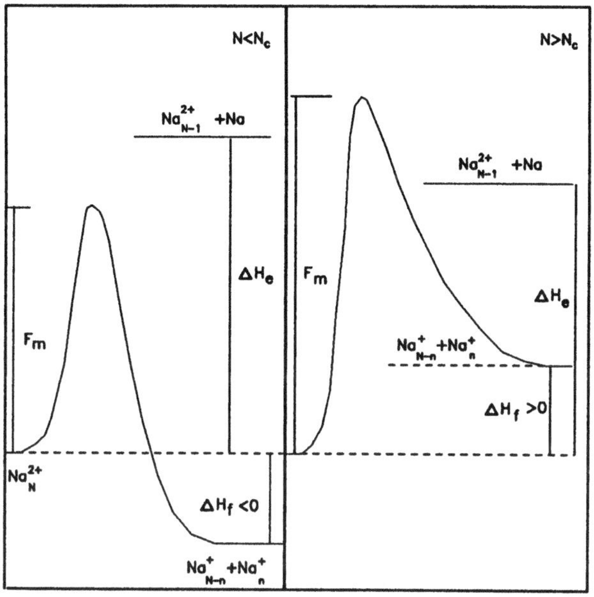

Figure 1: Schematic representation of the competition between fission and evaporation. ΔH_e and ΔH_f are the heats of evaporation and fission respectively. F_m is the maximum of the fission barrier and B_m the maximum of the opposite (capture) barrier.

On the other hand, F_m is lower than ΔH_e below N_c and the excited cluster preferently undergoes fission. In fact, N_c is defined [2] as the size for which F_m and ΔH_e become equal. The latest experiments [3, 5] also indicate that, at least for K and Ag, the most probable fission channels are influenced by shell closing effects (K_3^+, Ag_3^+ and Ag_9^+ emissions are found to be dominant channels). In summary, the two key ingredients for a theory aiming at explaining the critical size N_c must be : (i) consideration of fission barriers, and (ii) introduction of electronic shell effects.

2 Density Functional Theory for Cluster Fission

With reference to fig. 1 we can express the fission barrier height F_m as the sum

$$F_m = B_m + \Delta H_f \tag{2}$$

where B_m is the maximun of the barrier $B(d)$ for the opposite (capture) process in which X_n^+ and X_{N-n}^+ react to give X_N^{2+}, and ΔH_f is the heat of fission

$$\Delta H_f = E(X_{N-n}^+) + E(X_n^+) - E(X_N^{2+}). \tag{3}$$

In a previous paper [6] we have used the density functional formalism and the jellium model to calculate ΔH_f and B_m. The computation of ΔH_f is easy if we treat the parent and product clusters (at infinite separation) by the spherical jellium model. The same occurs in the evaluation of the evaporation energy of the monomer

$$\Delta H_e = E(X_{N-1}^{2+}) + E(X) - E(X_N^{2+}). \tag{4}$$

An Extended Thomas-Fermi (ETF) energy functional was used in these calculations. The exchange and correlation energies were treated within the local density approximation, and the kinetic energy of the electron gas was written as (using Hartree atomic units):

$$T[\rho] = \int d^3r[\frac{3}{10}(3\pi^2)^{2/3}\rho^{5/3} + \frac{\lambda}{8}\frac{(\nabla\rho)^2}{\rho}] \tag{5}$$

where the first term is the Thomas-Fermi term and the second is the first gradient correction. The value $\lambda = 1$, originally introduced by Von Weizsäcker, was used in ref [6].

In contrast, the computation of the barrier $B(d)$ requires the evaluation of the electron density, and the corresponding energy , for the deformed cluster undergoing fission. For this we have used a deformed, fully self-consistent, ETF model. The initial configuration of the fissioning system is a deformed cluster composed of $N-2$ electrons moving in the mean field created by two tangent jellium spheres corresponding to cluster sizes $N-n$ and n respectively. The other cluster configurations along the dissociation path have been obtained by increasing the separation d between the two jellium spheres representing the emerging fragments.

Using this method we have studied [6] the fission of Na_N^{2+} through the most asymmetric channel: $n = 1$ in eq. (1). A comparison of the calculated fission barriers with the energy of monomer evaporation (eq. (4)) gives a critical size $N_c = 40$, which is not far from the experimental value $N_c^{exp}(Na) = 27$ [2].

3 Further Results

A key ingredient in the calculation of fission barriers is the value of λ in eq. (5). From an empirical point of view, a value $\lambda = 0.5$ has been found to be appropriate for describing other properties of simple metal clusters [7]. Our first task is then to investigate the sensitivity of the fission barrier and N_c to the value of λ. The results are given in fig. 2. F_m is lower for $\lambda = 0.5$. The heat of evaporation also decreases for $\lambda = 0.5$, and the two curves intersect each other at the critical value $N_c = 36$, which reduces a little the discrepancy with experiment. The remaining discrepancy can be attributed to the fact that the most favourable fission reaction is often not the Na^+ channel.

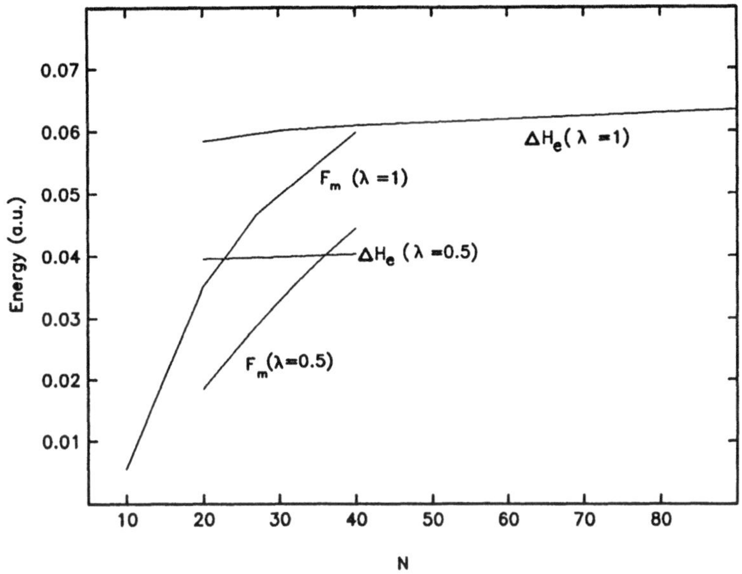

Figure 2: Fission barrier height for the most asymmetric channel (F_m) and heat of monomer evaporation (ΔH_e) versus parent cluster size N.

Concerning the fission of K_N^{2+} and $(K_{N-1}Na)^{2+}$ clusters, the experimental situation is rather clear [3]: K_3^+ is the preferred fission channel. The same occurs for Ag_N^{2+}, with the addition of the Ag_9^+ channel. Notice that those two fission channels involve a closed-shell fragment, with 2 and 8 electrons respectively. This is a shell-effect that can not be accounted for by the ETF theory, which predicts the most asymmetric (X^+) channel to the most probable one. A more correct description of cluster fission should be based on the wave-mechanical Kohn-Sham (KS) version of density functional theory. We take here an intermediate step in this direction, which consists in using the KS method to evaluate the heats ΔH_f and ΔH_e (we stress that the calculation of these two magnitudes only involves spherical clusters) but we still use the ETF method to evaluate B(d), as before (again the ETF calculation is performed with $\lambda = 0.5$).

We have calculated the fission barriers for the two asymmetric channels Na^+ and Na_3^+. Which of the two channels has the lowest fission barrier depends on the parent size. Our results are plotted in fig. 3. The barrier for the trimer ion channel strongly oscillates with the parent size. The oscillation is due to shell effects included in ΔH_f and the large magnitude of the oscillation is due to the spherical jellium description of the parent and products (at infinite separation). A similar overestimation of the oscillation with size is well documented in the literature for other properties of spherical-jellium clusters.

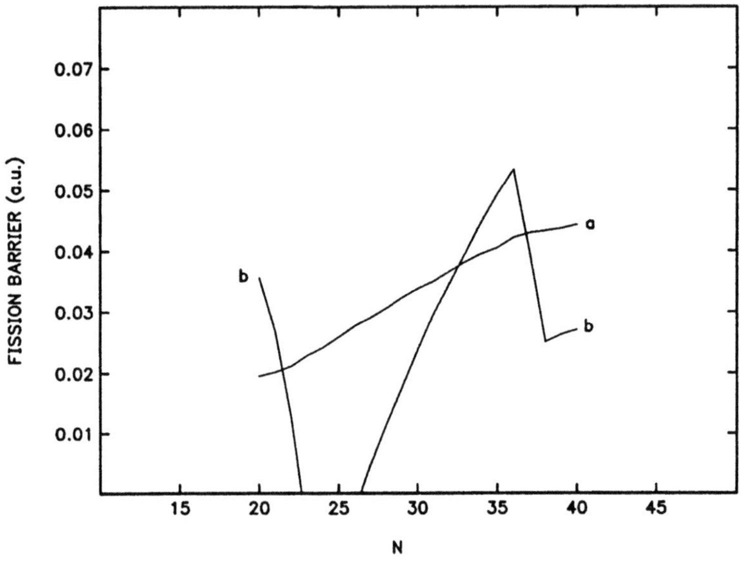

Figure 3: Fission barrier height as function of parent size for the mixed ETF-KS theory. (a) Na^+ channel. (b) Na_3^+ channel.

To compare these predictions with experimental data, we are only aware of the results of Bréchignac and coworkers [2]. In contrast to the case of K_N^{2+} or Ag_N^{2+}, the experimental situation concerning the fission behaviour of Na_N^{2+} clusters (data exist only in the range $N = 24 - 28$) is more complicated. The channels Na^+, Na_3^+, Na_5^+ and Na_7^+ have been detected. In short, if we restrict ourselves to the competition between the Na_3^+ and Na^+ channels the experiments show that the trimer ion channel is more favorable for $N = 24$ and $N = 27$, whereas the Na^+ channel is more favorable for $N = 28$. Thus, there is a non trivial competition between the two channels. A transition from the trimer to the monomer channel occurs in the calculations at $N = 32$; in the experiments this transition apparently occurs at $N = 28$. Evidently, consideration of other channels (Na_5^+, Na_7^+) is necessary for a better theoretical description.

Finally we turn again to the competition between fission and evaporation. To be consistent within our theory we have considered the evaporation channel with the lowest evaporation energy. The KS spherical jellium model gives Na_2 as the preferred evaporation channel for all Na_N^{2+} (this is due to the overestimation of the internal bonding in the dimer which can be corrected by introducting the discreteness of the ions (as opposed to the continuous character of the positive charge distribution of the jellium model) [8]. The critical number that we obtain for this mixed ETF-KS theory is $N_c = 36$, the same value obtained in the pure ETF theory (with $\lambda = 0.5$).

In summary, we have presented a theory for the fission of doubly charged clusters of simple metals which includes the effect of the fission barrier and a preliminary account of the influence of electronic shell effects. The theory gives a reasonable agreement with experiment for the critical size for the observability of doubly charged clusters. Improvements of the present theory should go in two directions. Concerning the electronic aspect, one should calculate the fission barriers with a full Kohn-Sham theory. Concerning the geometrical description of the system, the spherical jellium model should be trascended. These geometrical improvements can go from an ellipsoidal description of parent and separated fragments to a general treatment of the deformation in the fissioning cluster, and finally to a full geometrical description of the granularity of the system based on ionic pseudopotentials.

Acknowledgments

This work has been supported by DGICYT (Grants PB-89-0332 and PB-89-0352-C02-01). We are grateful to A. Rubio for helpful comments.

References

[1] K. Sattler, J. Mühlbach, O. Echt, P. Pfau and E. Recknagel, Phys. Rev. Lett. **47**, 160 (1981).

[2] C. Bréchignac, Ph. Cahuzac, F. Carlier and M. de Frutos, Phys. Rev. Lett. **64**, 2893 (1990).

[3] C. Bréchignac, Ph. Cahuzac, F. Carlier, J. Leygnier and A. Sarfati, Preprint (1991).

[4] W. A. Saunders, Phys. Rev. Lett. **64**, 3046 (1990).

[5] I. Katakuse, H. Ito and T. Ichihara, Z. Phys. D**19**, 101 (1991).

[6] F. Garcias, J. A. Alonso, J. M. López and M. Barranco, Phys. Rev. B**43**, 9459 (1991).

[7] Ll. Serra, F. Garcias, M. Barranco, J. Navarro, L. C. Balbás, A. Rubio and A. Mañanes, J. Phys.: Condens. Matter. **1**, 1039 (1989).

[8] M. P. Iñiguez, J. A. Alonso, A. Rubio, M. J. López and L. C. Balbás, Phys. Rev. B**41**, 5595 (1990).

3. Reactions

Projectile Break-up in Heavy Ion Collisions: Molecular Dynamics for Clusters of Alpha Particles

K.Möhring

Hahn Meitner Institut Berlin, D-W1000 Berlin 39, Germany

This contribution will discuss a dynamical model [1,2] for projectile break-up in heavy ion reactions in the Fermi energy domain. In the spirit of the seminar, I will mainly discuss the concepts of the model rather then its actually quite satisfactory confrontation with the experimental observations. However, at variance to the title of the seminar, I will discuss "cluster" concepts as used in the context of nuclear, not yet of atomic cluster physics.

We studied the collision of a bound cluster of alpha particles with a heavy target and investigated the various possible exit channels like total fusion, incomplete fusion, projectile break up and combinations of the latter two.

Motivation: The task was a description of heavy ion collisions at bombarding energies where collisions on the one hand no longer are binary in character, on the other hand are still too gentle to forget mean field effects altogether. More specifically, as there is a large body of experimental data on angular and energy distributions for specific ejectiles, in particular projectile like fragments as well as on correlations and coincidences between various ejectiles, a reasonable parametrization of such data was asked for. So we had to design a many body dynamical model which allows the prediction of highly differential cross sections for specific ejectiles.

In the literature, one finds several attempts to study the time evolution of the nuclear many body system *quantummechanically* [3,4,5]. Combining mean field dynamics with a proper collision term, they allow for valuable insight into the collision dynamics. But, as they evaluate the nucleon *single particle* density, the formation of well defined nuclei after the collision and the evaluation of corresponding cross sections cannot be covered within these schemes.

We, therefore, have to retreat to *classical* many body dynamics where all the many body correlations are incorporated automatically. Quantum corrections might then be introduced in some way or the other. The conceptually most advanced ansatz is probably the "Quantum Molecular Dynamics" [6,7]. Here quantum effects are incorporated by dressing every nucleon with a Gaussian in phase space allowing for some control of quantum uncertainty as well as of Pauli blocking. However, even accepting that this allows for a satisfactory description of the nuclear many body problem, the evaluation of highly differential cross sections is well beyond todays computational possibilities. Specific exit channels demand a very detailed scanning of the initial phase space, for a system of typically 200 constituents a prohibitive task.

This was the reason to try a drastic reduction in the numbers of degrees of freedom, or constituents, to study explicitly: For the projectile we consider only α - particle constituents and for the target, we disregard any explicit internal structure altogether. Obviously, this restricts the applicability of the model to sufficiently asymmetric collisions of light projetiles like ^{20}Ne or ^{32}S with a heavy target like ^{197}Au, and the study of not too slow ejctiles in the forward hemisphere. Still it seems the first dynamical model allowing the discussion of fusion, differential and coincidence cross sections on a consistent footing.

Friction: Excitation of any degree of freedom not followed explicitly, in particular energy and angular momentum transfer to the target, is simulated by friction forces between the α - particles and the target. The concept of friction was very succesfully introduced to describe binary heavy ion collisions at low bombarding energies, in particular to parametrize the phenomenon of deep inelastic scattering [8,9]. It is interesting to note that this concept works also in the present context of higher bombarding energies and a much more detailed model.

Initialization: The first step in performing a classical molecular dynamics calculation is the preparation of a proper initial phase space distribution: To model a projectile cluster of N nuclear physics α - particles in its ground state, we fill the N - particle phase space under the constraints of given binding energy E, total momentum $\boldsymbol{P}$and center of mass position $\boldsymbol{R}$. This is equivalent (assuming $\boldsymbol{R} = \boldsymbol{P} = 0$ for the sake of convenience) to an integration

$$\int d\boldsymbol{r}_1..d\boldsymbol{p}_N\, \delta(\sum_1^N \boldsymbol{r}_i)\, \delta(\sum_1^N \boldsymbol{p}_i)\, \delta(E - \sum_1^N \frac{p_i^2}{2m} - V(\boldsymbol{r}_1,..,\boldsymbol{r}_N)) f(\boldsymbol{r}_1,..,\boldsymbol{p}_N)$$

over proper functions f.

After eliminating $\boldsymbol{p}_N$ by means of the momentum conserving δ function, the argument of the energy conserving δ function is a quadratic form of the remaining momenta and can be transformed onto normalized principle axis. The remaining integral over momenta,

$$\int d\boldsymbol{p}'_1...d\boldsymbol{p}'_{N-1} 2m\, \delta(2m(E-V) - \sum_1^{N-1} \boldsymbol{p}_i'^2) \,... \,,$$

runs over the surface of a $3(N-1)$ dimensional hypersphere, and can be Monte Carlo integrated by sampling (3 N - 4)-tupels of Gaussian distributed random numbers [10].

We stay with a weighted integral over r-space,

$$\int d\boldsymbol{r}_1...d\boldsymbol{r}_N\, \delta(\sum_1^N \boldsymbol{r}_i)\, [2m(E_{bind} - V(\boldsymbol{r}_1,...,\boldsymbol{r}_N))]^{\frac{3N-5}{2}} \,... \,,$$

which very efficiently can be evaluated using Metropolis importance sampling [11].

This *microcanonical* procedure leads to gives a projectile cluster which, provided enough statistics, is perfectly stable in time, when evolving under its own internal dynamics. We find that over a time interval several times larger than a typical collision time, the chance to loose a constituent into the continuum, is negligible. Note that we do *not* construct the cluster in its *classical* ground state, but with prescribed realistic binding energy, allowing for a finite internal kinetic energy in similarity to the quantum mechanical ground state. Note also that the form of the α - α potential, cf. fig. 1, automatically prohibits spacial overlap of two α -s in a bound cluster.

Table 1 shows the rms radius and the separation energies of such a 5 - α cluster with the binding energy of ^{20}Ne in rather surprising agreement with the experimental values.

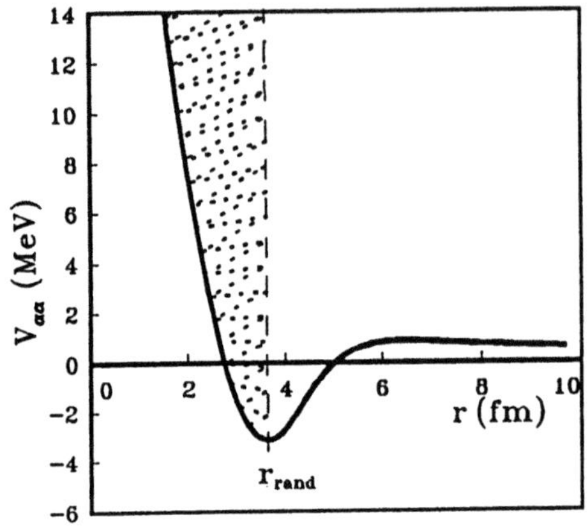

Figure 1: The α -α potential. The hatched area marks the range of the random force.

	Model	Exp.
R_{rms} :	3.04	2.99 fm
E^1_{sep} :	5.51	4.73 MeV
E^2_{sep} :	6.20	7.16 MeV
E^3_{sep} :	4.77	7.27 MeV
E^4_{sep} :	2.69	-0.09 MeV

Table 1: RMS radius and separation energies E_{sep} for the least bound, second least bound etc. α - particle of ^{20}Ne. (The differences in $E^{3,4}_{sep}$ reflect the fact that the classical model allows for a bound two-α -cluster.)

Statistics: Every point of the initial projectile phase space, together with an impact parameter, serves as an initial condition for solving the equations of motion for N α - particles colliding with the target. To give an idea of the necessary statistics: For 400 MeV ^{20}Ne on ^{197}Au the prediction of angular distributions of specified ejectiles demands some 10^4 trajectory calculations. At least 10^5 events are necessary to produce reasonably stable results for coincidences of two ejectiles. This demonstrates that the severe restriction in the number of explicit degrees of freedom is indeed necessary.

The random force: First we tried such a classical molecular dynamics calculation just with given conservative and friction forces. For the α - α potential, as displayed in figure 1, we rely on ATDHF [12], for the α - target interaction we use a single folding potential and a friction force as introduced in [8]. The result of such calculations was that the projectile was decomposed into its constituents with a chance definitely much higher than compatible with the experimental observation, as demonstrated by the left part of fig. 2.

Analyzing this discrepancy we came to the following conclusion: The classical α - α scattering for the potential $V_{\alpha\alpha}$ is strongly forward peaked, in fact much more than the experimental one, cf. fig.3. It follows that hardly any momentum transfered to an α in a cluster can be randomized between the constituents, any kicked α rather has to leave the cluster.

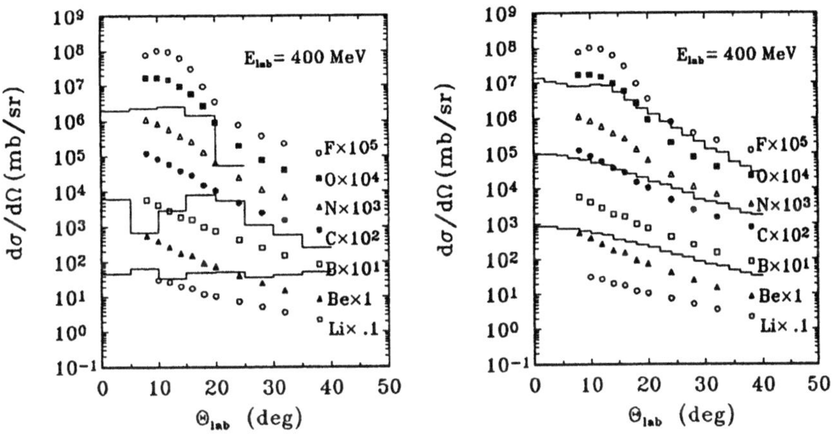

Figure 2: Angular distributions of various ejectiles from 400 Mev ^{20}Ne on ^{197}Au . Symbols represent the experimental data [13], histograms the calculations without (**left**) and with (**right**) the random α - α force for, from top to bottom, ^{16}O , ^{12}C and ^{8}Be .

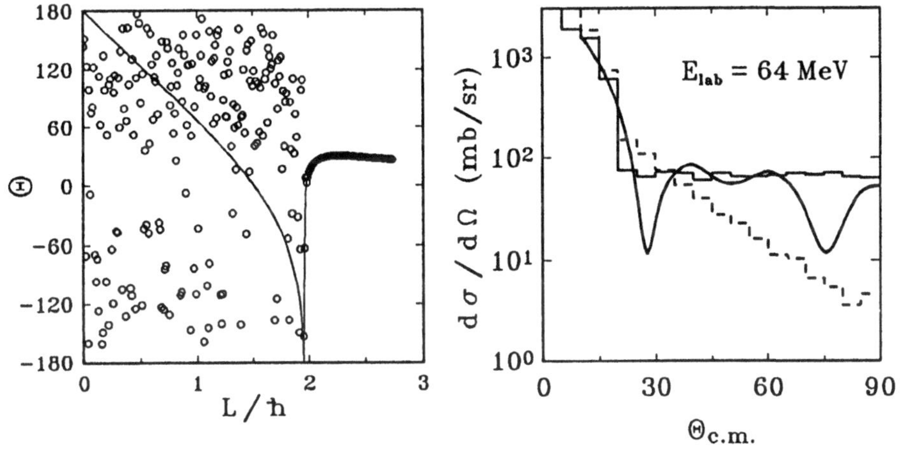

Figure 3: Left: Classical deflection "function" for α - α scattering at $E_{lab} = 3.54 MeV$. The circles represent the result for 250 trajectories with the random force included. **Right**: Differential cross section for $E_{lab} = 64\,MeV$. The smooth curve represents the experimental cross section [14], the histograms the model results with (solid) and without (broken) random force.

More specifically, if we compare to a quantummechanical description of α - α scattering, the classical model fails in reproducing the wide spread angular distributions for low partial waves. To cure this defect, we introduce a random force causing isotropic elastic scattering. It acts with a given frequency whenever the distance of two α - particles shrinks below a critical value r_{rand}, cf. fig. 1. Fig. 3 demonstrates how this additional interaction widens the classical correlation of scattering angle Θ and angular momentum L for small L and improves the α - α scattering cross section at larger angles. Correspondingly we now obtain much more realistic cross sections for sizable clusters, as can be seen from figure 2, right part.

So it seems that the dynamical model is sensitive to a distinctly quantummechanical behavior of α - particles. Work for a better understanding of quantum corrections to classical many body dynamics is under way, but still far from being conclusive. Also, it is not yet ruled out that the added random force rather simulates the fact that in realistic nuclei α structures are more or less washed out. In any case, it should be well understood that the introduction of the random force and in particular the specific chosen form, is just heuristic and waits for proper theoretical foundation.

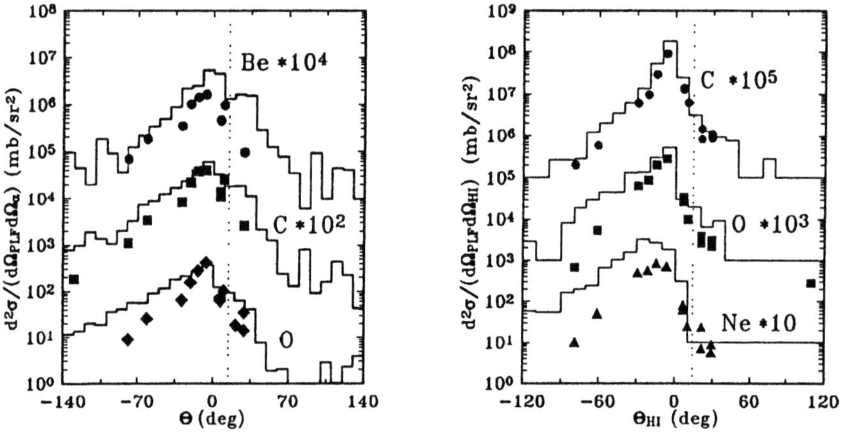

Figure 4: Angular distributions of α particles (**left**) and charged particles with $Z \geq 3$ (**right**) in coincidence with specific fragments at 14.3^o from 840 MeV ^{32}S on ^{197}Au . Symbols represent the experimental data [15], histograms the model results.

Taking the model at face value it can consistently and almost quantitatively reproduce a large variety of experimental findings and offers a simple dynamical scenario to understand these data. Representative examples for coincidences are shown in fig. 4.

The author is most indepted to his coworkers D.H.E. Gross, T. Srokowski and T. Schmidt. He gratefully acknowledges many fruitful discussions with his experimental colleagues H. Homeyer, H. Fuchs, W. Terlau and C. Schwarz.

[1] K.Möhring, T.Srokowski, D.H.E.Gross, and H.Homeyer, Phys.Lett. **B 203** (1988) 210

[2] K.Möhring, T.Srokowski, and D.H.E.Gross, Nucl. Phys. **A 533** (1991) 333

[3] J.Aichelin and G.Bertsch, Phys.Rev. **C 31** (1985) 1730

[4] H.Kruse, B.V.Jacak, J.Molitoris, G.D.Westfall, and H.Stöcker, Phys.Rev. **C 31** (1985) 1770

[5] C.Gregoire, B.Remaud, F.Sebille, L.Vinet, and Y.Raffray, Nucl.Phys. **A 465** (1987) 317

[6] J.Aichelin, G.Peilert, A.Bohnet, A.Rosenhauer, H.Stöcker, and W.Greiner, Phys.Rev. **C 37** (1988) 2451

[7] J.Jaenicke, J.Aichelin, N.Ohtsuka, R.Linden, and A.Faessler, preprint HD **TVP 90 1** (1990)

[8] D.H.E. Gross and H. Kalinowski, Phys.Rep. **45 C** (1978) 175

[9] J.Randrup, Ann.Phys. **112** (1978) 356

[10] D.Knuth, *The art of computer programming (2nd edition)* (Addison Wesley, Reading, Mass 1981)

[11] J.M.Hammersley and D.C.Handscomb, *Monte Carlo Methods* (Methuen's Statistical Monographs, Methuen + Co, London 1964)

[12] K. Goeke, F. Grümmer, and P. G. Reinhard, Ann.Phys. **150** (1983) 504

[13] Ch.Egelhaaf, M.Bürgel, H.Fuchs, A.Gamp, H.Homeyer, D.Kovar, and W.Rauch, Nucl.Phys. **A 405** (1983) 397

[14] P.Darriulat, G.Igo, and H.G.Pugh, Phys.Rev. **137 B** (1965) 315

[15] C. Schwarz, *PhD thesis*, Freie Universität Berlin 1991

Nuclear Heavy-Ion and Atomic Cluster-Cluster Collisions

R. Schmidt[1,2], G. Seifert[2] and H.O. Lutz[1]

[1] *Fakultät für Physik, Universität Bielefeld, W-4800 Bielefeld 1, F.R.G.*
[2] *Institut für Theoretische Physik, TU Dresden, O-8027 Dresden, F.R.G.*

1. Introduction

Phenomena in atomic nuclei often have corresponding analogues in atomic clusters. Well known examples are shell closing effects and collective excitations of nucleons in nuclei and electrons in clusters, or the fission process of nuclei and clusters. Obviously, collisions between atomic clusters represent the atomic counterpart to nuclear heavy-ion collisions.

Whereas the study of heavy-ion collisions (HIC) represents a domain of nuclear research since more than 20 years, the investigation of atomic cluster-cluster collisions (CCC) is becoming just now a very attractive field of cluster studies. Recently, a first experiment on CCC has been successfully performed.[1]

Preceding theoretical studies of CCC[2,3,4,5] have shown that remarkable analogies between nuclear HIC and CCC of metallic clusters may exist. In both cases, the reaction channels can be classified according to the classical impact parameter and impact energy. At low energy, complete fusion and quasielastic reactions dominate. At large energies and intermediate impact parameters two large, highly excited and deformed fragments are formed in the exit channel, largely preserving the identity of projectile and target; this behaviour is well known as deep inelastic collisions (DIC) in nuclear physics. To illustrate the similar behaviour of HIC and CCC, we compare in Fig. 1 the collisional dynamics of typical DIC-events in nuclear and cluster collisions. In the case of nuclei a sequence of the nucleonic density distributions (vertical direction), as obtained from time-dependent Hartree-Fock calculation[6], is shown for $^{136}Xe + ^{209}Bi$ collisions (collision axis is horizontal) at an impact energy of 1130 MeV. It is compared with the collisional dynamics between two Na_9-clusters at an energy of 10 eV, obtained from molecular dynamics (MD) simulations of CCC combined with

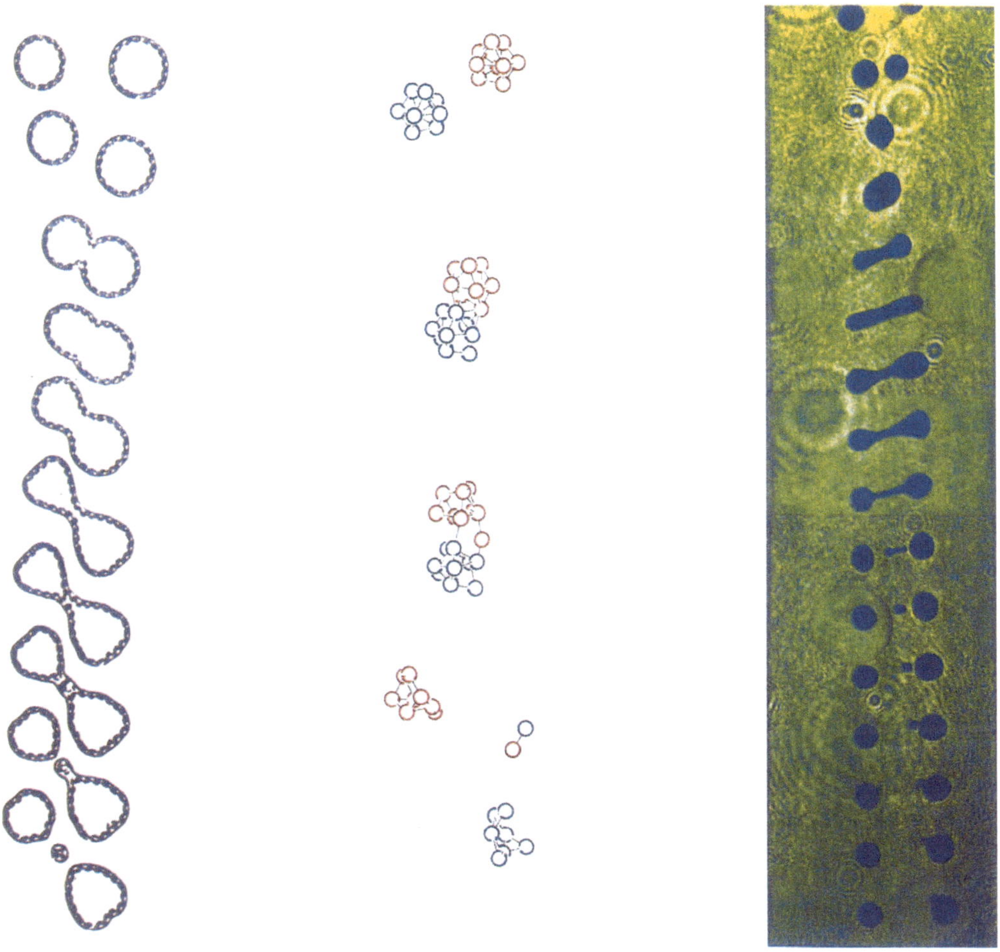

Fig. 1: Snapshots of the collisional dynamics between atomic nuclei, atomic clusters and liquid droplets

density functional theory (DFT) in local density approximation (LDA).[2] In addition, we have presented in Fig. 1 the direct optical visualization of collisions between two propanol-2 droplets with typical radii of $\sim \mu$m and relative collision velocity of $\sim$ m/s, obtained recently with a stroboscopic laser technique.[7]. In all three cases, the colliding aggregates stick together after the approach phase. Afterwards the systems rotate as a whole, form a neck and finally decay into two large and deformed products; from the neck regions a small particle is emitted.

The close relations in the collisional dynamics between nuclei and droplets are of course not completely unexpected. Since more than 50 years it has been well known that, e.g., the gross features of the fission process of nuclei can be well understood in the framework of the liquid drop model (LDM), where the delicate balance between surface (deformation) energy and Coulomb repulsion determines the stability of nuclei against (binary or ternary) fragmentation. In collisions, an additional centrifugal energy term has to be regarded. In fact, it has been shown recently[4] that the stability of nuclei and liquid droplets against *centrifugal* fragmentation can be treated with exactly the same stability condition, based on specific (simplifying) assumptions of the dissipation of collisional energy in the entrance and exit channel and formulated in the so-called rotating liquid drop model (RLDM).[8] The resulting critical angular momenta l_{cr}, below which the fused aggregates remain stable, can be related to the (high-energy) fusion cross section σ_{CF} (or impact parameter for fusion b_{CF}) $\sigma_{CF} \equiv \pi b_{CF}^2 = \pi\lambda^2 l_{cr}^2$, leading in the dimensionless form to a universal scaling law for fusion

$$\left(\frac{\sigma_{CF}}{\pi R_{12}^2}\right)^{1/2} \equiv \frac{b_{CF}}{R_{12}} = \frac{\tilde{v}}{v} \qquad (1)$$

where $\tilde{v} = l_{cr}/\mu R_{12}$, and $R_{12} = R_1 + R_2$ denotes the contact radius of the colliding aggregates with radii R_1, R_2; μ is their reduced mass and v is the collision velocity. In Fig. 2 we compare the predicted universal scaling law for fusion (solid line) with available experimental data of the impact parameter for fusion (or fusion cross sections) for various nuclear HIC as well as collisions between water and propanol-2 droplets having different radii $R_1 = R_2$. The calculated critical angular momenta l_{cr} for nuclei and droplets, determing $\tilde{v}$ in Fig. 2, differ by more than 20 orders of magnitude! The excellent scaling of both collisional systems demonstrates in an impressively *quantitative* form that also in fusion nuclei behave like colliding droplets. In both cases, fusion occurs as the consequence of *dissipation* of relative kinetic energy into intrinsic excitations of the constituents (i.e. into heat); thereby, the systems completely loose their "memory" on the entrance channel, and the maximum fusion cross section is determined by the "macroscopic" balance between surface tension and centrifugal (in the case of nuclei also Coulomb) repulsion, largely independent of the underlying microscopic forces between the constituents. In this spirit, it has been possible to develope an extended version of the model, thus calculating the fusion cross sections in CCC as function of cluster size, charge and impact energy.[5]

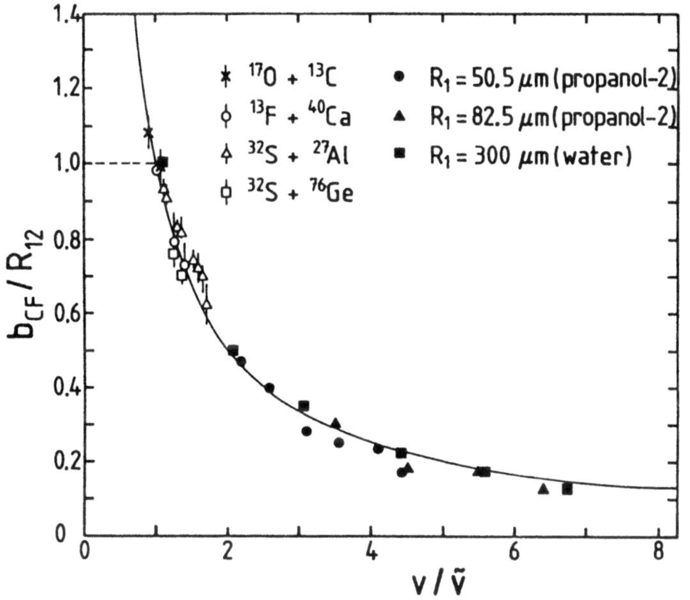

Fig. 2: Universal scaling law of the dimensionless impact parameter for fusion b_{CF}/R_{12} vs. impact velocity $v/\tilde{v}$ (solid line) as compared to experimental data for different nuclear HIC (points with error bars) and macroscopic droplet collisions (full points), from ref. 4.

From this macroscopic point of view, one may also expect close relations in other reaction mechanisms between HIC and CCC. In the work reported here we will demonstrate this for the deep inelastic reaction channel. It will become apparent that, in close analogy to nuclear DIC, deep inelastic CCC are characterized by *dissipation and fluctuation* of collective degrees of freedom coupled to a large, but finite number of intrinsic (atomic) degrees of freedom. CCC represent, therefore, a fascinating tool to study dissipation and fluctuation phenomena in *finite* atomic many-body systems.

2. The collision system and the MD-DFT method

We treat CCC in a fully microscopic description using the same MD-method combined with DFT in LDA as in ref. 2, 3 (for details see also the contribution of G. Seifert et al., this book), i.e., we explicitly follow the classical trajectories of each individual atom in the colliding clusters. From this we can define and derive the relevant macroscopic (collective) quantities: center-mass distance between the colliding clusters, kinetic energy and orbital angular momentum of their relative motion, deformation energy and total intrinsic angular momenta of the clusters (see below). To avoid any difficulties in the definition of the collective quantities, arising from the transfer of atoms between projectile and target, we choose a symmetric and double "magic" collision system, namely $Na_8 + Na_8$. In this case, the size-(mass-) asymmetry is frozen during the interaction due to strong shell effects and thus the reaction products of DIC are the same as in the entrance channel, i.e., $Na_8 + Na_8 \rightarrow Na_8 + Na_8$. A systematic study of the evolution of the size-asymmetry (as an important collective degree of freedom) in CCC will be given elsewhere.[9]. The center-mass (c.m.) collision energy is fixed to be $E_{c.m.} = 0.9$ eV and a typical intermediate impact parameter for DIC of $b = 10$ a.u. (1 a.u. $= 0.529 \cdot 10^{-10}$ m) is taken. It corresponds to an inital orbital angular momentum of the relative motion of $L_i = 1.2$ k$\hbar$ (1k$\hbar = 10^3\hbar$). The evolution of the corresponding collective (macroscopic) quantities is then followed as a function of time.

3. Relaxation phenomena in CCC

Before we will discuss the reaction mechanism of deep inelastic CCC in detail, let us briefly summarize the basic feature of nuclear DIC:

The most striking property of nuclear DIC consists in the strong dissipation of relative kinetic energy and orbital angular momentum. In a macroscopic picture, there are three mechanisms that contribute to the total kinetic energy loss ΔE of the relative motion. These processes can be well understood (and quantitatively described) assuming a velocity-proportional friction force which acts between the colliding nuclei, taking into account the deformation of the fragments in the exit channel (for a typical example of a classical friction model of HIC, see, e.g., ref. 10). The frictional force mediates the loss of radial ΔE_r and tangential ΔE_t kinetic energy of the relative motion. The radial part of the kinetic energy loss ΔE_r is stored into internal (nucleonic) heat energy. The tangential part ΔE_t results from the loss of relative orbital angular momentum into (collective) rotational motion of the scattered nuclei. The deformation of the reaction products, mainly induced by the neck formation before scission (see left column in Fig. 1), leads to an additional loss of energy ΔE_d. After the collision, this energy will be converted into intrinsic heat due to shape relaxation of the excited fragments. Thus, the total amount of dissipated energy consists of three terms:

$$\Delta E = \Delta E_r + \Delta E_t + \Delta E_d \qquad (2)$$

Besides the energy loss, the dissipation of orbital angular momentum into intrinsic angular momenta of the fragments is one of the most interesting relaxation phenomenon in DIC. A velocity-proportional frictional force predicts a maximum amount of angular momentum transfer; it can be estimated from the condition that the relative tangential velocity between the colliding partners vanishes. In this situation, the nuclei stick together, and from the conservation of the total angular momentum one simply obtains

$$\Delta L_{sticking} = \frac{J_1 + J_2}{J_{rel} + J_1 + J_2} L_i \qquad (3)$$

corresponding to the classical "sticking" limit of a friction force for the maximum transferred angular momentum. In eq. (3), L_i is the initial orbital angular momentum of the relative motion, J_{rel}, J_1, J_2 are the moments of inertia of the relative motion and the individual reaction partners, respectively. For two equal spheres and rigid-body moments of inertia, eq. (3) reduces to $\Delta L_{sticking} = 2/7 L_i$.

The angular momentum dissipation is connected with polarization and orientation phenomena. The final angular momenta of both fragments, induced by friction, should have the same directions as the initial orbital angular momentum L_i, i.e. perpendicular to the reaction plane. The strong polarization, however, is strongly decreased by the excitation of in-plane components of L, due to statistical fluctuations. For the same reason, the total amount of transferred angular momentum can considerably exceed the classical sticking limit (for a discussion of angular momentum dissipation and related orientation phenomena in HIC see, e.g., ref. 11, 12).

In the following, we will show, to what extent this macroscopic picture of the reaction mechanism of nuclear DIC applies also to atomic CCC.

3.1 Dissipation of energy and angular momentum

In Fig. 3, we present the calculated c.m. distance $|\vec{R}|$ between the colliding clusters, the total kinetic energy of their relative motion $\mu\dot{\vec{R}}^2/2$ and the orbital angular momentum L_z with respect to the c.m. betweeen both clusters as a function of time. The arrows indicate the scission point where the clusters leave their mutual interaction region. A typical interaction time is $\tau_{int} \approx 7 \cdot 10^4$ a.u. ≈ 1.7 ps (1 a.u. $= 2.4 \cdot 10^{-17}$ s). As can be seen from the middle part, the cluster are initially slightly accelerated, as expected from the attractive conservative part of their interaction potential (see contribution of O. Knospe et al., this volume). However, even during the approach phase, they loose in a very short time about 2/3 of their initial bombarding energy E_i. Within the same time, their relative orbital angular momentum L_z is strongly decreased. Its final value, L_f, lies in the vicinity of the sticking limit (dashed line) of the classical friction force estimated for two equal spheres. After the approach phase, both energy and angular momentum of the relative motion remain nearly constant during the entire duration of the interaction. This behaviour is in accord with the expectations of a strong friction mechanism leading to dissipation of energy and angular momentum.

In Fig. 4, the partitioning of the total energy loss ΔE into different contributions is shown. In the upper part, we plotted for comparison again the behaviour of the total kinetic energy $\mu\dot{\vec{R}}^2/2$ (solid line) and, in addition its radial part $\mu\dot{R}^2/2$ (dashed curve). The difference between the solid and the dashed curves gives, therefore, the tangential relative kinetic energy, i.e. roughly $L_z^2(t)/2J_{rel}(t)$. In the middle part of Fig. 4, the total intrinsic kinetic energy $\Delta E_r + \Delta E_t$ of the system is presented. It is defined as the difference between the total kinetic energy of all atoms and that of the relative motion of the clusters. We do not try to decide explicitly between ΔE_r and ΔE_t. To do that, a refined technique to separate rotational motion from vibrations taking into account the nonrigidity of the clusters is

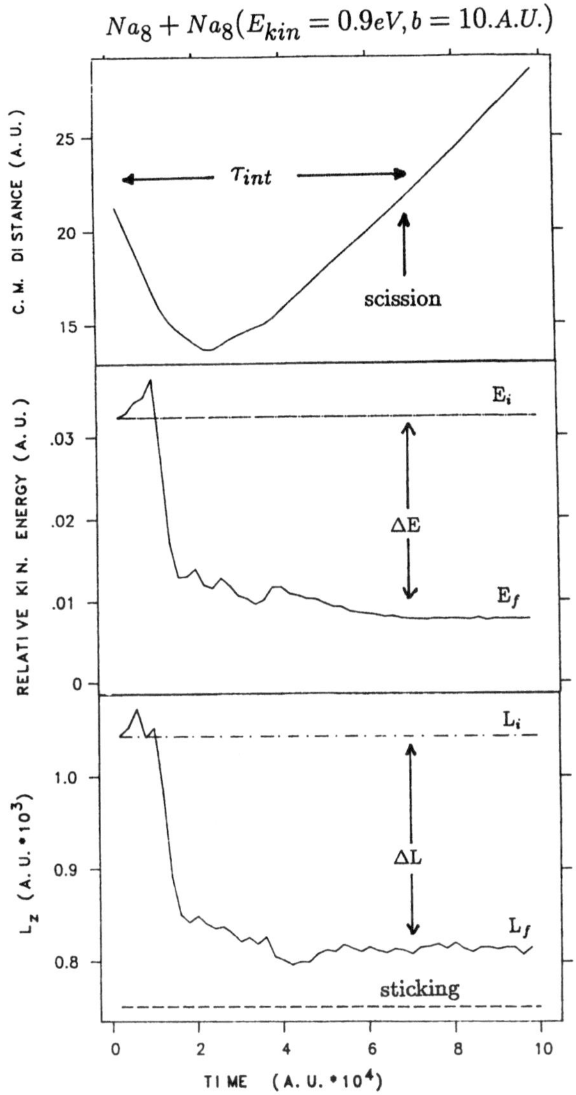

Fig. 3: Dissipation of energy ΔE and angular momentum ΔL in a deep inelastic collision of two Na_8 clusters: Distance between the mass centers of colliding clusters (top), total amount of relative kinetic energy (middle) and relative orbital angular momentum (bottom) as function of time (solid lines). Indicated are the interaction time intervall τ_{int}, initial (dashed lines) and final values of energy E_f and angular momentum L_f as well as the classical sticking limit for L_f of two colliding equal spheres interacting via a frictional force.

required[13] which, at present, is beyond the scope of our studies. The total intrinsic kinetic energy $\Delta E_r + \Delta E_t$ rapidly increases during the approach phase and then remains constant during the interaction. The large fluctuations result from the finite size of the system. In contrast, the potential energy increases continuously up to the scission point. The potential energy is defined as the difference betweeen the actual Born-Oppenheimer energy and that of the colliding (separated) clusters. During the approach phase this quantity is of course negative, reflecting the conservative part of the attractive interaction potential between the clusters.[9] However, if projectile and target collide in their ground states and the size-asymmetry is not changed during the interaction, this difference directly measures the energy which is needed to perturb the atomic ground state structure; thus, *after the scattering process*, it is identical with the (positive) deformation energy. From Fig. 4, one realizes that about 2/3 of the total ΔE is stored into $\Delta E_r + \Delta E_t$ and about 1/3 in ΔE_d.

3.2 Shape relaxation

As already mentioned, in nuclear HIC (and in nuclear fission) the deformed products undergo shape relaxation, thereby increasing their intrinsic heat energy. The shape relaxation process is important for the energy balance which determines secondary decay (or deexcitation) processes (e.g., particle evaporation) of the reaction products. It is, therefore, of interest to search for this process in CCC, too.

To do that, we have followed the dynamics of the scattered (completely non-interacting and free flying) two Na_8-clusters up to long time scales. In Fig. 5, their internal kinetic (heat) energy and Born-Oppenheimer (deformation) energy are presented up to times of about $4 \cdot 10^5$ a.u. $\approx$ 10 ps. After scission, besides the typical fluctuations of the finite systems, both quantities remain constant (their sum, of course, remains conserved at all times). However, at $t \gtrsim 1.3 \cdot 10^5$ a.u., the intrinsic excitation energy increases, at the same time the deformation energy decreases to zero, indicating a damped oscillation. This clearly demonstrates the shape relaxation process in CCC. For illustration, we have presented in Fig. 5, also snapshots of typical geometrical shapes of one of the Na_8 clusters: just after scission (1), at a moment where the potential energy approaches zero at $t \approx 2 \cdot 10^5$ a.u. (2) and in one of the positive maxima of the potential energy at $t \approx 3 \cdot 10^5$ a.u. (3). Evidently, the geometric structure of (2) is very similar to the ground state structure of Na_8.[14,15]

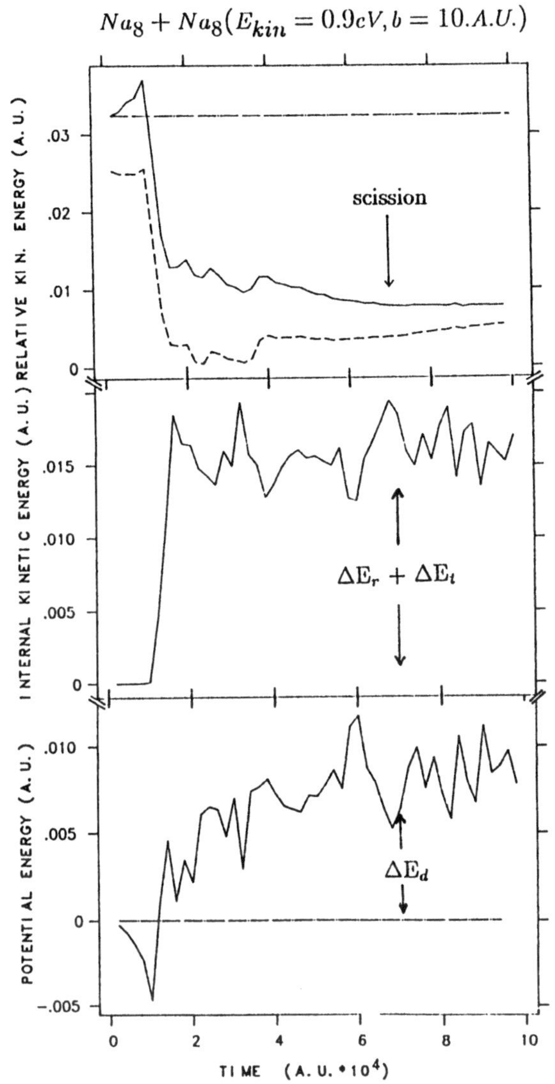

Fig. 4: Partitioning of the total kinetic energy loss $\Delta E = (\Delta E_r + \Delta E_t) + \Delta E_d$ into deformation energy ΔE_d (lower part) and radial plus tangential internal kinetic energy $\Delta E_r + \Delta E_t$ (middle part) of colliding clusters as function of time. In the upper part, the time dependence of the total (solid line) and radial (dashed line) kinetic energy of the relative motion are presented.

3.3 Polarization and orientation

A purely tangential friction force automatically leads to polarization of the reaction products perpendicular to the reaction plane because their intrinsic (transferred) angular momenta have the same direction as the relative orbital angular momentum. Any in-plane components of the intrinsic angular momentum transfer cannot be induced by friction. Of course, they must

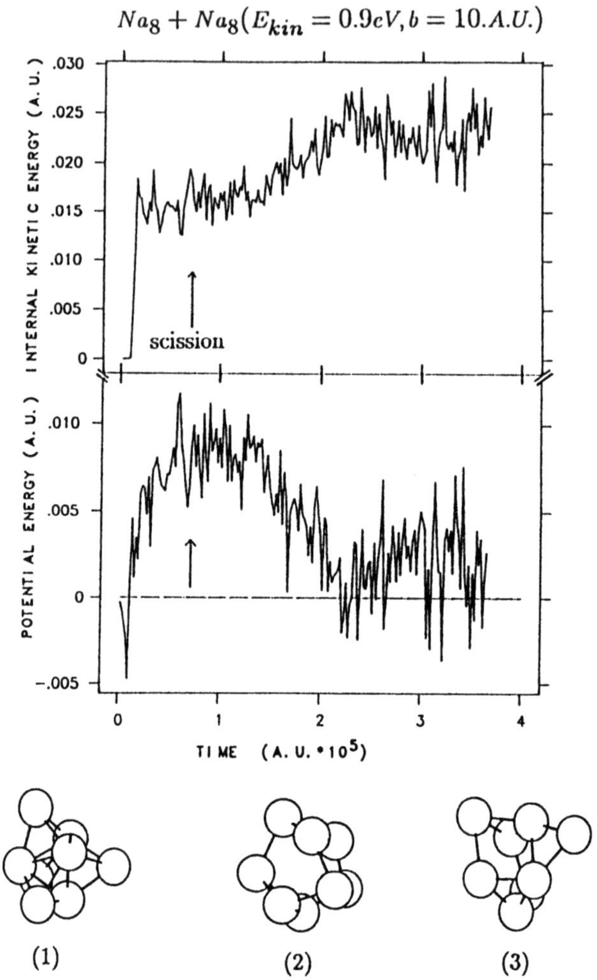

Fig. 5: Shape relaxation in CCC: Internal kinetic energy and deformation (potential) energy as function of time (different time scale as compared to Fig. 3). Note the increase of internal heat energy *after* the collision process (i.e., after scission). In the lower part typical geometrical shapes of one Na_8-cluster after the collision process are shown:
(1) $t = 1 \cdot 10^5$ a.u. (just after the collision)
(2) $t = 2 \cdot 10^5$ a.u. (minimum of the deformation energy)
(3) $t = 3 \cdot 10^5$ a.u. (second maximum of the deformation energy)

have equal magnitudes and opposite signs, due to conservation laws. In HIC, their excitation leads to a considerably increase of the final spins and, simultaneously, decreases their polarization perpendicular to the reaction plane.[11,12].

In Fig. 6, the calculated components of the intrinsic angular momenta (defined with respect to the c.m. of the individual Na_8-clusters) are shown as function of time. As expected from a frictional mechanism, their z-components $L_z^{(1)}$, $L_z^{(2)}$ (perpendicular to the reaction plane) have equal amount and the same sign as the relative orbital angular momentum; their sum $\Delta L_z = L_z^{(1)} + L_z^{(2)}$ lies in the vicinity of the classical "sticking" limit. However, large in-plane components (x, y directions) of the intrinsic angular momenta are created during the initial stage of the collisional process, too. Their absolute values can exceed twice that of the out-of-plane ones (see $L_y^{(1)}$, $L_y^{(2)}$). Thus, the total transferred angular momentum in this particular collisional event amounts to about 600 a.u. (1 a.u. $= 1\ \hbar$) and exceeds considerably the one resulting from dissipation of relative orbital angular momentum $\Delta L \equiv \Delta L_z \approx 200$ a.u. (cf. also Fig. 3). By the same reason, the polarization of reaction products is drastically reduced.

4. Summary

We have studied the reaction mechanism of deep inelastic CCC in microscopic detail. It has been shown that the finite, but large number of atomic degrees of freedom plays the role of a heat bath, responsible for the *irreversible* loss of energy and angular momentum of the relative motion. The kinetic energy of the relative motion is dissipated into intrinsic heat energy (chaotic atomic motion), collective rotational energy (rotation of the scattered clusters as a whole) and deformation energy (distortion of the ground state atomic structure of the scattered clusters). After the collision process, shape relaxation leads to an additional increase of the internal heat energy. The amount of dissipated angular momentum of the relative motion lies in the vicinity of the "sticking" limit of a classical friction force. The polarization of the scattered clusters perpendicular to the reaction plane is strongly reduced by the excitation of large in-plane components of the internal angular momentum.

Altogether, the deep-inelastic reaction mechanism of CCC resembles very much that known in nuclear HIC. In both cases, dissipation and fluctuation phenomena in finite systems show up. The results of the present study of CCC, based on a MD-DFT method, suggest strongly to treat these processes with the help of transport theories, which played always a key role in the understanding of nuclear HIC.

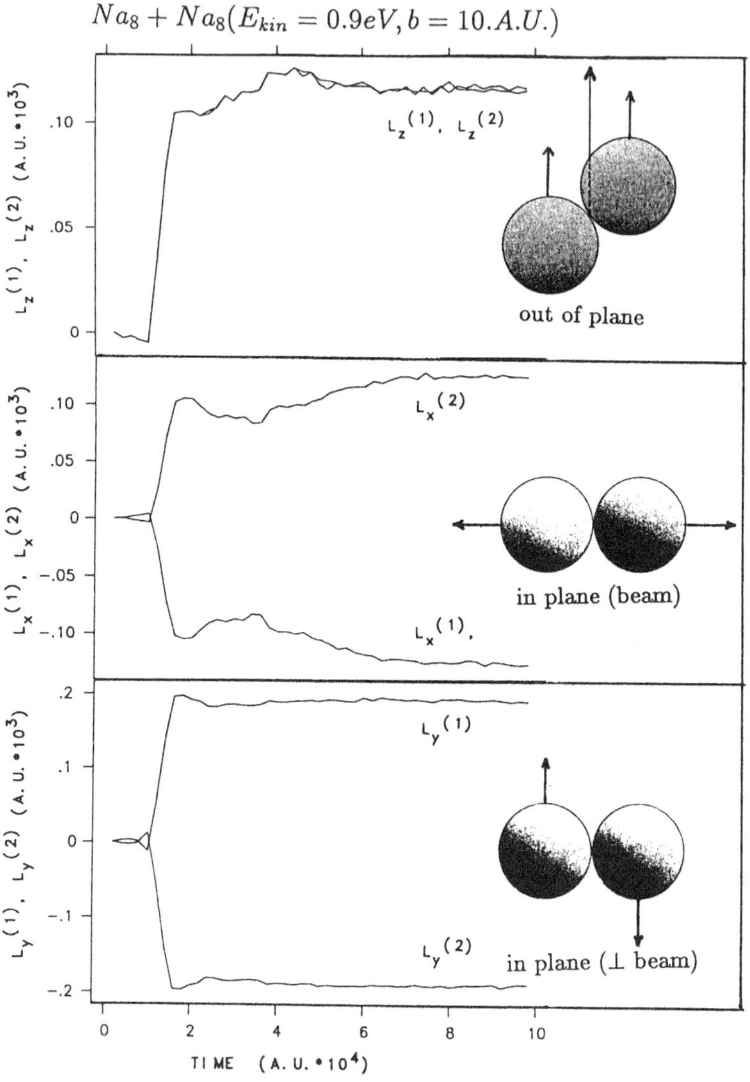

Fig. 6: Angular momentum dissipation and orientation phenomena in CCC. The components of the total internal angular momenta of both colliding Na_8-clusters (1), (2) perpendicular to the reaction plane $L_z(1)$, $L_z(2)$ and in-plane (x and y directions) as function of time. The small arrows indicate the directions of the internal angular momenta for the corresponding excited rotational modes, the large arrow in the upper part that of the relative orbital angular momentum.

Acknowledgement

This work has been supported by the Deutsche Forschungsgemeinschaft.

References

[1] E.E.B. Campbell et al., this volume
[2] R. Schmidt, G. Seifert and H.O. Lutz, Phys. Lett. A 158 (1991) 231
[3] G. Seifert, R. Schmidt and H.O. Lutz, Phys. Lett. A 158 (1991) 237
[4] R. Schmidt and H.O. Lutz, Phys. Rev. A (1992), in print
[5] R. Schmidt and H.O. Lutz, to be published
[6] A.K. Dhar, B.S. Nilsson, K.T.R. Davies and S.E. Koonin, Nucl. Phys. A 364 (1981) 105
[7] G. Brenn and A. Frohn, Exp. in Fluids 7 (1989) 441; Spektrum der Wissenschaft 12 (1990) 116 and priv. communication
[8] S. Cohen, F. Plasil and W.J. Swiatecki, Ann. Phys. 82 (1974) 557
[9] O. Knospe, R. Dreizler, R. Schmidt and H.O. Lutz, this volume, and to be published
[10] R. Schmidt, V.D. Toneev and G. Wolschin, Nucl. Phys. A 311 (1978) 247
[11] R. Schmidt and R. Reif, J. Phys. G 5 (1981) L181 and J. Phys. G 7 (1981) 775
[12] G. Wolschin, Nucl. Phys. A 316 (1979) 146
[13] J. Jellinek and D.H. Li, Phys. Rev. Lett. 62 (1989) 241, and Chem. Phys. Lett. 169 (1990) 380;
D.H. Li and J. Jellinek, Z. Phys. D 12 (1989) 177
[14] V. Bonačić-Koutecký, P. Fantucci and J. Koutecký, Phys. Rev. B 37 (1988) 4369
[15] U. Röthlisberger and W. Andreoni, J. Chem. Phys. 94 (1991) 8129

Molecular Dynamics Simulations of Cluster-Cluster Collisions

G. Seifert[1], R. Schmidt[1,2] and H.O. Lutz[2]

[1]*Institut für Theoretische Physik, TU Dresden, O-8027 Dresden, Germany*
[2]*Fakultät für Physik, Universität Bielefeld, W-4800 Bielefeld 1, Germany*

Introduction

Clusters may be viewed as a bridge between atoms and molecules and the bulk. There are also interesting analogies between clusters and atomic nuclei; their exploitation has been stimulated by the similarity of the nuclear shell structure and the electronic shell structure of alkaline metal clusters [1].

It is a challenge to search for such common features also in dynamical processes, like fragmentation, fission or collision events. Clearly, uncovering such close relations could cross-fertilize cluster as well as nuclear physics and act as an important stimulus for both fields.

Collision phenomena played always a key role in nuclear physics [2]. Thus, one may expect that studies of cluster-cluster collisions (CCC) will also be able to uncover new structural as well as dynamical properties of atomic clusters. In previous papers we have outlined various aspects of CCC on the basis of molecular dynamics (MD) simulations [4,5] and a phenomenological model [6] - see also the contribution of *R. Schmidt et al.* and *O. Knospe et al.* in this book. In the present paper we will mainly stress the interplay between geometric and electronic structure in collisions between neutral sodium clusters.

Alkaline metal clusters are characterized by delocalized electrons with a shell-like level structure. This leads to the appearance of "*magic* " numbers — in close analogy to atomic nuclei [1]. To uncover such relationship also in dynamical features, we performed calculations of Na_9+Na_9 and Na_8+Na_8 collisions. The Na_8 cluster is a "*magic*" one with filled 1S and 1P shells, whereas Na_9 is a "*non-magic*" cluster with filled 1S and 1P shells and one electron in the 1D shell. The possible product of a fusion between two Na_9 clusters is again a "*magic*" cluster (Na_{18} - 1S 1P 1D shells filled). In contrast, $Na_8 + Na_8$ is a collision between two "*magic*" clusters possibly forming the "*non-magic*" cluster Na_{16}.

The collisions were simulated by MD combined with density-functional theory (DFT) in the local density approximation (LDA), using an LCAO representation of the wave functions; such a combination allows a realistic account of the electronic structure. The main features of the method will be outlined in the following, and after that the results of the simulations will be discussed.

LCAO-Method

To simulate collision processes between alkaline metal clusters by MD-DFT one obviously needs a very time-efficient method. We applied a simplified scheme where the Kohn-Sham orbitals ($\psi(\vec{r})$) are written as linear combinations of atomic orbitals $\phi_\mu(\vec{r}-\vec{R}_j)$, centred at the nuclei j (LCAO-ansatz):

$$\psi(\vec{r}) = \sum_\mu C_\mu \phi_\mu(\vec{r}-\vec{R}_j)$$

In this way the Kohn-Sham equations are transformed into a set of algebraic equations (secular equations):

$$\sum_\mu C_\mu (h_{\mu\nu} - \varepsilon S_{\mu\nu}) = 0$$

with $h_{\mu\nu}$ and $S_{\mu\nu}$ the elements of the hamilton matrix and the overlapmatrix, respectively. These equations can be solved by diagonalizing the secular matrix. (In a previous paper [8] a variant of this method has been presented, which applies the orbital dynamics — as proposed by Car and Parrinello [3,7] — instead of diagonalization the secular matrix.) The $h_{\mu\nu}$ are the hamiltonian matrix elements of the basis functions with the Kohn-Sham hamiltonian $\hat{h}$:

$$\hat{h} = \hat{t} + V_{eff}(\vec{r})$$

where $\hat{t}$ is the operator of kinetic energy, and V_{eff} is the effective one particle potential, consisting of the electron-nuclear part (V_{ext}), the mean field electron-electron interaction contribution (Hartree potential - V_H) and the exchange-correlation part (V_{XC}) in LDA:

$$V_{eff} = V_{ext} + V_H + V_{XC}.$$

The overlap matrix elements ($S_{\mu\nu}$)are due to the nonorthogonality of the basis functions at different sites in the system. As an approximation we write V_{eff} as a sum of potentials of neutral atoms (V_j^0):

$$V_{eff} = \sum_j V_j^0.$$

Consistent with this approximation one has to neglect several contributions to the hamiltonian matrix elements $h_{\mu\nu}$ (see [9] and [10]):

$$h_{\mu\nu} = \begin{cases} \langle \phi_\mu | \hat{t} + V_j + V_k | \phi_\nu \rangle, & \text{if } \mu \cap \nu \epsilon \{j, k\}; \\ 0, & \text{otherwise.} \end{cases}$$

This treatment may be viewed as the LCAO variant of a cellular Wigner-Seitz method as applied for molecules by Inglesfield [13].

The forces on the ions (F_X^k) can be divided in an purely electronic (F_e^k) part and a contribution from the repulsion of the ion cores (F_r^k):

$$F_X^k = -\partial E / \partial X_k = F_{X,e}^k + F_{X,r}^k.$$

It is possible to write the electronic part of the force as a sum of orbital contributions:

$$F_{X,e}^k = \sum_i n_i F_{X,i}^k,$$

where n_i is the occupation number of orbital i. The LCAO representation leads to:

$$F_{X,i}^k = \sum_\mu \sum_\nu c_\mu^i c_\nu^i [-\partial h_{\mu\nu} / \partial X_k + \varepsilon_i \partial S_{\mu\nu} / \partial X_k].$$

The core repulsion of the ions is treated by a semiempirical repulsion potential:

$$F_{X,r}^k = -\partial E_r / \partial X_k$$

with

$$E_r = a \sum_{k,j} U(R_{kj})$$

$$U(R) = (R - R_1)^2$$

for $R \leq R_1$ and zero for $R > R_1$. The parameters ((a, R_1)) are obtained by fitting the calculated equilibrium distance and vibration frequency in a diatomic molecule to their experimental values. Such description of the forces is comparable to a similar partitioning of the cohesive energy,as used in calculations of structural properties of clusters by Tomanek and Schlüter [15]. The scheme proposed here may be viewed as a "hybrid" between *ab initio* molecular dynamics [7] - based on DFT - and the use of purely empirical potentials. It has the advantage over the latter of overcoming the transferability problem, where parametrized potentials based on experimental data can be difficult to transfer from one system to another. Furthermore it requires about two orders of magnitude less computation time than the *ab initio* molecular dynamics. For studies of CCC it is, however, restricted yet to neutral clusters. An extension of the LCAO-method to include the charge of colliding clusters is in progress [24].

The Verlet algorithm [12] was applied to the integration of Newton's equations for the atoms:

$$M_k \ddot{\vec{R}}_k = -\partial E / \partial \vec{R}_k.$$

A time step of 100 atu (1 atu=2.4 · 10^{-17} sec) has shown to guarantee the conservation of energy over the entire time (several 10^{-12} sec) of the simulations. The Hedin-Lundquist form [14] for V_{XC} has been used. The atomic s and p valence wave functions were considered in the LCAO ansatz. Each wave function is represented by a set of 12 Slater type functions - as to details see [9]. Considering the experimental values for the interatomic distance ($R_e = 5.82 a_B$) and the vibration frequency ($\omega_e = 159 cm^{-1}$) for Na_2 ([19]) one obtains $6.56 a_B$ and $0.49 eV/a_B^2$ for the parameters R_1 and a in the potential U.

The colliding clusters were prepared in a T=0 state. Molecular dynamics, combined with the technique of simulated annealing [16], was applied to find the ground state structure of the Na_8 and Na_9 clusters. The resulting ground state geometries for the initial clusters (Na_8, Na_9) are in good agreement with results given by *Bonačić-Koutecký et al.* [17] and *Röthlisberger and Andreoni* [18]. The method described here has also been applied for describing structural properties of phosphorus [11] and carbon [21] clusters, using MD and the technique of simulated annealing.

The two colliding clusters are then positioned at a fixed distance between their centres of mass along the collision axis; collisions were simulated with various values of centre-of-mass (c.m.) energy and impact parameters b.

Results and discussion

As it was discussed in [4], the most characteristic feature of the simulations consist in the preferential occurrence of a very restricted number of outgoing reaction channels with respect to the final product sizes. Each of these channels has a counterpart in nuclear heavy ion collisions with respect to the final mass number (i.e. cluster size) and the global as well as collective behaviour. Which one of the channels dominates depends on the impact parameter b (in $a_B; 1 a_B = 0.529 Å$) and the incident kinetic energy E. These channels are in the case of $Na_9 + Na_9$ collisions:

I - complete fusion ($E \leq 1 eV, b \leq 15. a_B$)

$$Na_9 + Na_9 \longrightarrow Na_{18}$$

II - incomplete fusion ($E \geq 1 eV$)

$$Na_9 + Na_9 \longrightarrow Na_{16} + Na_2$$

III - deep inelastic collision ($E \geq 1eV, b \sim 10...15a_B$)

$$Na_9 + Na_9 \longrightarrow Na_8 + Na_8 + Na_2$$

IV - quasielastic collision ($E \sim 0.5...5eV, b \geq 15.a_B$)

$$Na_9 + Na_9 \longrightarrow Na_8 + Na_{10}.$$

In the case of high impact energies ($E = 5eV$) and near-central collisions, fragmentation channels yielding more than three final products (including monomers) are observed.

In contrast to nuclear heavy-ion collisions, however, shell effects show up dramatically in all three types of cluster collisions leading to pronounced manifestations of "magic" numbers (Na_2, Na_8) in the exit channel of quasi-elastic and deep inelastic collisions. This manifestation of the "magic" numbers is illustrated in fig. 1 for typical events. In the case of Na_9+Na_9 (right hand side in fig. 1) the collision leads to a Na_8 and a Na_{10} cluster. However, no exchange of atoms appears in the collision of $Na_8 + Na_8$ (left hand side in fig. 1).

The difference between both collisions and the preferential occurrence of "magic" clusters (e.g., Na_8) in the exit channel of CCC can be quantified in terms of the so-called fragmentation potential (for definition and approximations see [22].

$$U(N_1) = E(N_1) + E(N - N_1) - 2E(N/2)$$

where $E(N)$ is the total energy of a cluster with N atoms. In ref. [22] the energy is calculated in the jellium approximation. Here we use our LCAO-scheme for $E(N)$ and in addition normalize the potential to the energy of the entrance channel $2E(N/2)$ in symmetric collisions. The potentials are shown in fig. 2 for $Na_8 + Na_8$ (upper part) and $Na_9 + Na_9$ (lower part). In spite of an even-odd alternation the shell closing effects can be seen clearly in both curves. The "entrance point" $N_1 = 9$ lies at a maximum in case of $Na_9 + Na_9$. This leads to an exchange of atoms, since $Na_8 + Na_{10}$ ($N_1 = 8$, $N - N_1 = 10$ is energetically favoured. In contrast, energy is needed for an exchange of atoms in the case of Na_8+Na_8 — the entrance point $N_1 = 8$ lies in a deep minimum of $U(N_1)$. For comparison, the potentials obtained by applying the liquid drop model [22], is also drawn in fig. 2 (smooth solid lines). From these curves on would expect a broad product distribution in collision processes. In contrast, the results of our simulations clearly indicate the dominant role of electronic structure effects in collisions of alkaline metal clusters. This holds for quasielastic collisions (see discussion above) as well as for deep inelastic collisions and for fusion processes (see [4]) and has been found, in additon, in dynamical studies of cluster fission, too [23].

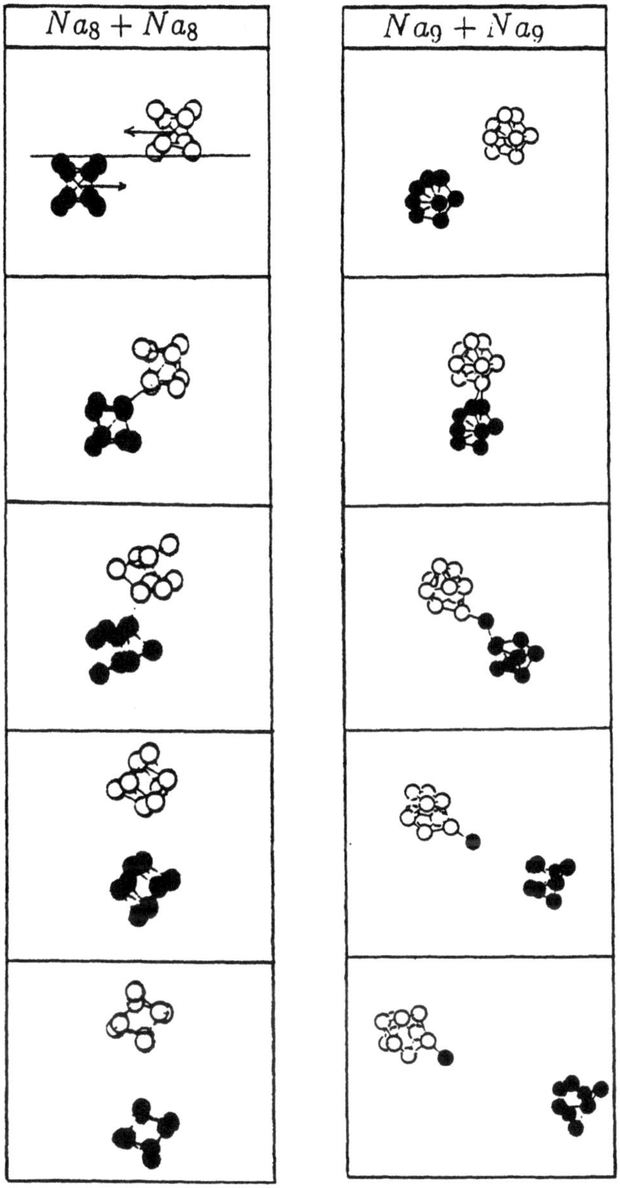

Fig. 1: Time evolution of typical events in MD simulations for $Na_9 + Na_9$ collisions (right hand side) and $Na_8 + Na_8$ collisions (left hand side).

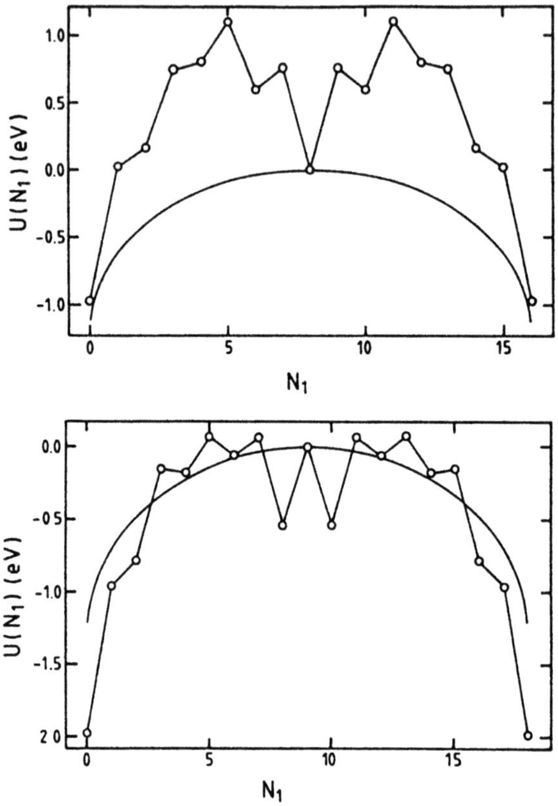

Fig. 2: Fragmentation potential $U(N_1)$ for $Na_8 + Na_8$ (upper part) and $Na_9 + Na_9$ (lower part) collisions

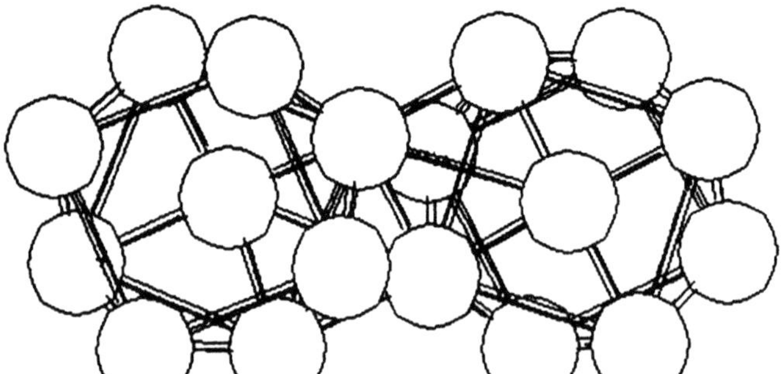

Fig. 3: Structure of the $Na_9 - Na_9$ molecule as obtained by simulated annealing.

In the case of fusion between $Na_9 + Na_9$ a long-lived "cluster-cluster molecule" can be formed — its structure after a simulated annealing procedure [16] is shown in fig. 3. Such a molecule is a stable entity on the potential hypersurface of Na_{18} [5]. Independent from our LCAO calculations the stability of such structure can be confirmed by applying an "ab initio" LDA scheme using plane wave expansions for the Kohn-Sham orbitals [20]. Such molecular cluster configuration is found initially also in the case of collisions between two "*magic*" Na_8 clusters. However, in contrast to $Na_9 - Na_9$ the $Na_8 - Na_8$, molecule-like configuration is unstable and decays after a short time ($\approx 1 \cdot 10^5$ atu) into a compact excited Na_{16} compound. Its appearance resembles "*nuclear molecules*" which show up as transition state in heavy-ion collisions and fission [25].

Acknowledgement

The authors thank Dr. R.O. Jones for his kind help. This work was supported by the Deutsche Forschungsgemeinschaft.

References

[1] See, e.g., Ekardt, W.: *Phys. Rev.* **B29** , 1558 (1984), Knight, W.D., Chou, M.Y. and Cohen, M.L.: *Phys. Rev. Lett* **52**, 2141 (1984)

[2] Bock, R. (ed.): *Heavy ion collisions*, Vols. 1-3 (North-Holland, Amsterdam, 1980)

[3] Car R. and Parrinello M.: *Phys. Rev. Lett.* **55**, 2471 (1985)

[4] Schmidt, R., Seifert G. and Lutz, H.O.: *Physics Letters* **A158**, 231 (1991)

[5] Seifert, G., Schmidt, R. and Lutz, H.O.: *Physics Letters* **A158**, 237 (1991)

[6] Schmidt, R. and Lutz, H.O.: *Phys. Rev.* **A** (1992) in print

[7] See, e.g., Car, R., Parrinello, M.: in *Simple Molecular Systems at Very High Density*, NATO Advanced Study Institute Series B **186**, 455 (1988). Jones, R.O.: *Angew. Chem.* **103**, 647 (1991)

[8] Seifert G., Bryja L. and Ziesche P.: in*Proc. 19th Int. Conf. on Electronic Structure of Solids*, Dresden 1989, p. 220 (Ed. P. Ziesche)

[9] Seifert G., Eschrig H. and Bieger W.: *Z. Phys. Chem. (Leipzig)* **267**, 529 (1986)

[10] Seifert G. and Eschrig H.: *phys. stat. sol.* **b127**, 573 (1985)

[11] Seifert G. and Jones R. O.: *Z. Phys.* **D20**, 77 (1991)

[12] Verlet L.: *Phys. Rev.* **159**, 98 (1967)

[13] Inglesfield J. E.: *Mol. Phys.* **37**, 873,889 (1979)

[14] Hedin L. and Lundquist B. I.: *J. Phys.* **C4**, 2064 (1971)

[15] Tomanek D. and Schlüter M. A.: *Phys. Rev.* **B36** , 1208 (1987)

[16] Kirkpatrick, S., Gelatt, C.D. and Vecchi, M.P.: *Science* **220**, 671 (1983)

[17] Bonačic-Koutecký, V., Fantucci, P. and Koutecký, J.: *Phys. Rev.* **B37**, 4369 (1988)

[18] Röthlisberger, U. and Andreoni, W.: *J. Chem. Phys.* **94**, 8129 (1991)

[19] Radzig, A.A. and Smirnov, B.M.: *Ref. Data on Atoms, Molecules and Ions*, (Springer-Verlag, Berlin, 1985)

[20] Seifert, G. and Jones, R.O.: (unpublished results)

[21] Blaudeck, P., Frauenheim, Th., Seifert, G. and Fromm, E.: *J. Phys. Condensed Matter* (submitted for publication)

[22] Knospe, O., Dreizler, R., Schmidt, R. and Lutz, H.O.: contribution in this volume and to be published

[23] Barnettt, R.N., Landman, U. and Rajagopal, G.: *Phys. Rev. Lett.* **67**, 3058 (1991) and contribution of U. Landman in this volume

[24] Schmidt, R., et al., to be published

[25] Fink, H.J., Greiner, W. and Scheid, W. in *Heavy Ion Collisions*, ed. by R. Bock, North-Holland, Amsterdam, Vol. **2**, 1980, p. 399, and contribution of W. Greiner this volume

COLLISION-INDUCED REACTIONS OF SIZE-SELECTED CLUSTER IONS OF Ar

T. KONDOW, S. NONOSE, J. HIROKAWA and M. ICHIHASHI

Department of Chemistry, Faculty of Science,
The University of Tokyo, Bunkyo-ku, Tokyo 113,
Japan

Abstract: The collision-induced reactions of Ar_n^+ (n = 2 - 15) with Kr and Ne were studied by use of a tandem mass-spectrometer equipped with octapole ion guides. The results showed that evaporation, charge transfer and fusion were dominant pathways in the reactions. The absolute cross sections and the branching ratios were obtained as functions of the size of the parent cluster ion and the collision energy. The size- and energy-dependences were explained in the scheme of the charge-induced dipole and the induced dipole-induced dipole scatterings. It was also shown that the conversion efficiency of the collision energy into the internal energy of Ar_n^+ was proportional to the cluster size. In the Ne collision, the upper limit of the conversion efficiency was estimated to be ~ 60 % at the collision energy of 0.2 eV.

1. Introduction

Collision-induced reactions of cluster ions have attracted much attention [1-5], because the reactions depend critically and characteristically on their cluster size. In particular, a specific nature of the cluster collisions could be demonstrated in collision processes involving ions of van der Waals clusters which are often made of a tightly bound ion core and a weakly bound solvation shell. This characteristic structure provides us a unique opportunity to explore reaction processes which are scarcely encountered in ordinary collision processes. In this regard, the collisional reactions of argon cluster ions, Ar_n^+, with Kr and Ne atoms were investigated as typical examples. It was found that the collision energy is transmitted to the internal degrees of freedom of Ar_n^+ system in a statistical manner and the large total cross sections observed are attributable to the independent contributions of every constituent atom of Ar_n^+ to the scattering process.

2. Experimental

An experimental setup is shown in Fig. 1. The cluster ions produced by electron impact on neutral clusters of Ar were mass-selected by a quadrupole mass spectrometer (Extrel, 162-8), and were allowed to collide

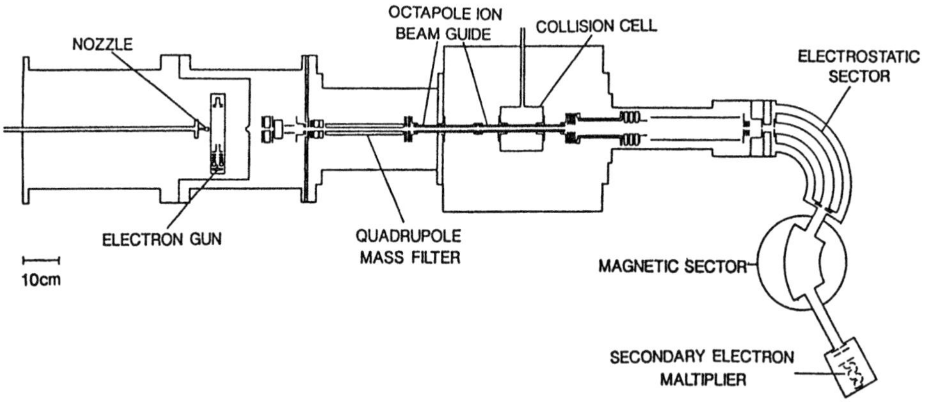

Fig. 1: A quadrupole-sector magnet tandem mass spectrometer with two-stage octapole ion guides.

with a target molecule in a collision region enclosed by an octapole ion guide. The product ions were mass-analyzed by a sector-magnet mass spectrometer (JEOL, JMS-D300). The ion source, the quadrupole mass spectrometer, and the collision region were floated up to 1 kV against the ground. The total cross section was measured by varying the pressure of the target gas in the range of 10^{-5} - 10^{-6} Torr. The background pressure was attained to be less than 5 x 10^{-7} Torr. The branching ratio of a given product ion was estimated from the intensities of the product ions. Because of a large uncertainty in the pressure reading, the observed cross sections were normalized against the reported cross sections for the process, Ar_2^+ + Rg -> Ar^+ + Ar + Rg, where Rg represents Kr or Ne [6].

3. Results and Discussion

3-1. Reaction Pathways

In the collision of Ar_n^+ with Kr, the product ions, $Ar_{n'}^+$ (n'<n) and $Ar_{n'}Kr^+$ (n' = 0, 1, 2 and 3) were observed, while in the Ne collision no charge-exchange species $Ar_{n'}Ne^+$ were detected. The measurements of the product ions show that the reaction of Ar_n^+ with Rg (= Kr or Ne) proceeds as follows:

$$Ar_n^+ + Rg \rightarrow Ar_{n'}^+ + (n-n')Ar + Rg \quad \text{(evaporation)} \quad (1)$$

$$Ar_n^+ + Rg \rightarrow Rg^+ + nAr \quad \text{(charge transfer)} \quad (2)$$

$$Ar_n^+ + Rg \rightarrow Ar_{n'}Rg^+ + (n-n')Ar \quad \text{(fusion).} \quad (3)$$

In the Ne collision, pathways (2) and (3) do not proceed in the collision energy range studied. The branching ratio of pathways (2)+(3) in the Kr collision is shown in Fig. 2, as a function of the cluster size, n. The branching ratio decreases smoothly, and levels off at n ~ 6 with the increase of n; these data points follow the dashed and dotted lines. This

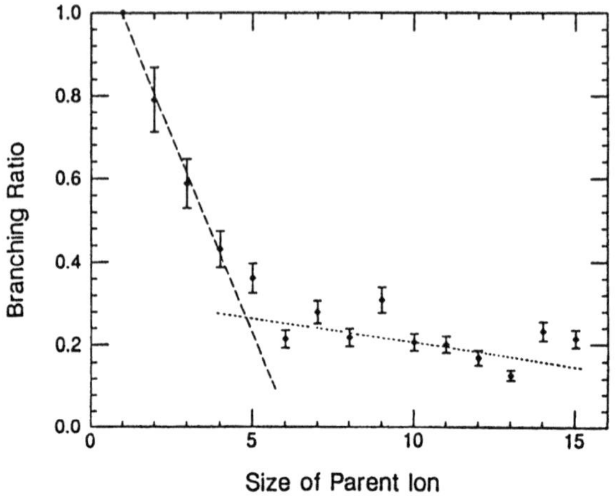

Fig. 2: The branching ratio of pathways (2)+(3) is shown as a function of the cluster size, n, in the reaction of Ar_n^+ with Kr. The collision energy is 1.0 eV.

finding indicates that pathway (2) dominates for n < 5 whereas pathway (3) does for n > 5; in pathway (3), the target, Kr, is conceivably fused with Ar_n^+ before Ar atoms are released. This phenomenon may arise from the specificity of the Ar_n^+ structure.

3–2. Collision–Energy Dependences of Cross Sections

Figure 3 shows the collision–energy dependences of the total cross section, the cross sections for pathway (1) and pathways (2)+(3), in the Ar_{11}^+ + Kr reaction system. As shown in Fig. 3, the total cross section decreases gradually while that of pathway (1) does not change appreciably, as the collision energy increases. On the other hand, the cross section for pathways (2)+(3) decreases rapidly with the collision energy. It is evident from this collision energy dependence that pathways (2)+(3) are exothermic and have no activation energy. The total cross section can be explained by assuming that every atom in Ar_n^+ behaves as a scatterer of the target atom in the scheme of charge–induced dipole [7] and induced dipole–induced dipole scattering. The cross section thus estimated is shown as the dashed line of Fig. 3. A similar tendency was observed for the Ne collision.

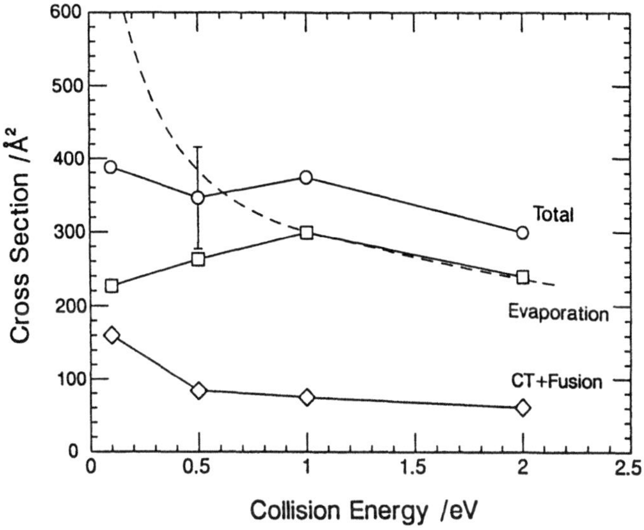

Fig. 3: Dependences of the cross sections on the collision energy for the Ar_{11}^{+} + Kr system. The solid and dashed lines represent the experimental and the calculated cross sections, respectively.

3-3 Cluster-Size Dependences of Cross Sections

The total cross section, the cross section for pathway (1) and pathways (2)+(3) are plotted against the cluster size, n, as shown in Fig. 4. The total cross section increases in proportion to n, and reaches 450 $Å^2$ at n = 15. The cross section for pathway (1) increases monotonously with the increase of n, while the cross section for pathways (2)+(3) does not change much with n. The size-dependence of the total cross section is also predicted by the model proposed in section 3-2. The cross sections for the Ne collision are also shown in Fig. 5. As is the case of the Kr collision, the total cross section is roughly proportional to n and is consistent with the prediction given by the model proposed.

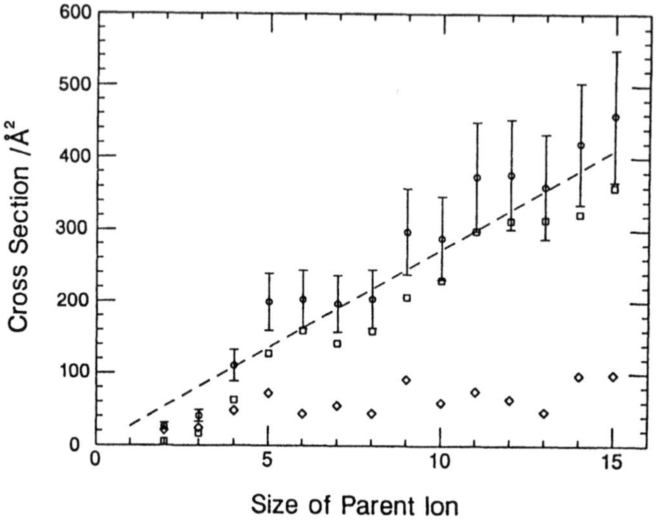

Fig. 4: The total cross section, the cross sections of pathway (1) and pathways (2)+(3) are shown as a function of the cluster size, n, in the reaction of Ar_n^+ with Kr; open circles, open squares, and open diamonds show the total cross section, the cross sections of pathway (1) and pathways (2)+(3), respectively. The dashed line shows the total cross section calculated on the basis of the model given in the text. The collision energy is 1.0 eV.

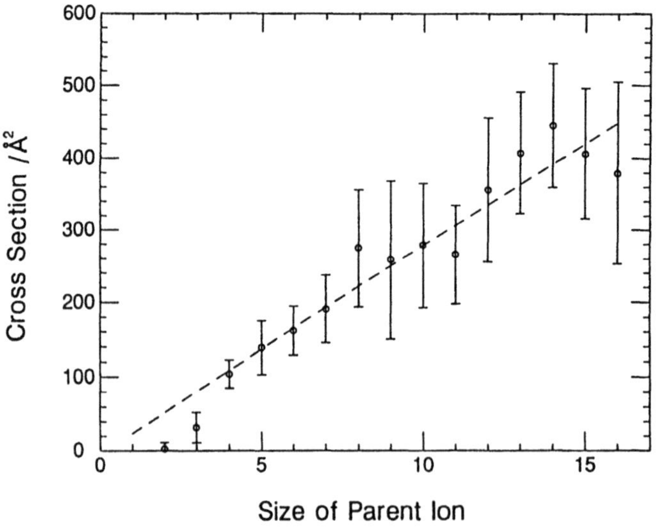

Fig. 5: The total cross section is plotted as a function of the cluster size, n, in the Ar_n^+ + Ne reaction system with the collision energy of 0.2 eV. The dashed line shows the total cross section calculated.

It was also found that the average number of Ar atoms evaporated from Ar_n^+ by 1.0 eV collision of Kr was 0.3n. If the average number is proportional to the increment of the internal energy of Ar_n^+, the conversion efficiency of the collision energy into the internal energy of Ar_n^+ turns out to be proportional to the cluster size or, in other words, the internal degrees of freedom. It is inferred from the above argument that the collision energy is partitioned statistically into the internal degrees of freedom. In the 0.2 eV collision of Ne, the average number was given by 0.25n for $n < 10$, and reached a constant number, 2.5, for $n > 10$. As the average energy of Ar evaporation from Ar_n^+ is approximately 0.05 eV per atom [8], the minimum energy given to the internal degrees of freedom by the Ne collision is calculated to be 0.125 eV, this is, about 60 % of the collision energy is converted to the internal energy.

A systematic study of the rare gas collision with Ar_n^+ will be undertaken to obtain an insight into the collision processes of the finite systems.

References

[1] L. Hanley and S. L. Anderson, J. Chem. Phys. 89, 2848 (1988).

[2] S. A. Ruatta, L. Hanley and S. L. Anderson, J. Chem. Phys. 91, 226 (1989).

[3] S. A. Ruatta, P. A. Hintz and S. L. Anderson, J. Chem. Phys. 94, 2833 (1991).

[4] S. K. Loh, D. A. Hales. L. Lian and P. B. Armentrout, J. Chem. Phys. 90, 5466 (1989).

[5] C. A. Woodward and A. J. Stace, J. Chem. Phys. 94, 4234 (1991).

[6] W. B. Maier II, J. Chem. Phys. 62, 4615 (1975).

[7] R. D. Levine and R. B. Bernstein, *Molecular Reaction Dynamics* (Oxford University Press, Oxford, England, 1974), 1st ed.

[8] N. E. Levinger, D. Ray, M. L. Alexander and W. C. Lineberger, J. Chem. Phys. 89, 5654 (1988).

Reactions and Fission of Clusters

Uzi Landman, R. N. Barnett, C. L. Cleveland, H.-P. Kaukonen and G. Rajagopal

School of Physics, Georgia Institute of Technology, Atlanta, Georgia 30332

Abstract

Classical and quantum molecular-dynamics simulations are used to investigate a new class of cluster-catalyzed reactions induced via collisions between reactants embedded in clusters, and the energetics and dynamics of fission of doubly charged sodium clusters, exhibiting shell-closing effects, double-hump barriers, and a precursor fission mechanism.

I. Introduction

Studies of reactions induced by collisions between clusters, between clusters and surfaces, or between atomic reactants colliding with clusters [1-3], and of the energetics and dynamics of cluster-fission [4-9], open new avenues in investigations of reaction and fragmentation dynamics [4-8], pathways of collisional energy distribution in materials aggregates, and a new class of reactions which are catalyzed by the cluster environment [1].

Recent studies of metallic clusters [9] (particularly of simple metals) unveiled systematic energetic, stability, spectral [10], and fragmentation [5,6] trends. Moreover, several properties of atomic clusters (e.g., electronic shell structure [9], and most recently supershells [11], portrayed by the occurrence of magic numbers in the abundance spectra and ionization potentials; the influence of shape fluctuations analyzed within the jellium model [9]; giant spectral resonances interpreted as evidence for collective plasma oscillations [5,12]; and fragmentation, fission, patterns of ionized clusters [5,6]) bear close analogies to corresponding phenomena exhibited by atomic nuclei, suggesting an intriguing universality of the physical behavior of finite size aggregates, though governed by interactions of differing spatial and energy scales [13].

In this short paper we summarize results of our recent theoretical investigations of reactions in clusters [1] and of cluster-fission [4,13], using classical and quantum molecular dynamics simulations, demonstrating the wealth of dynamical information obtained via such studies and certain analogies between phenomena occurring in atomic cluster and nuclear systems.

II. Reactions in Clusters

While much progress has been achieved in the experimental generation, size selection, characterization, and probing of the properties of clusters, and in the theoretical treatment of finite systems in both the classical and quantum regimes, relatively few investigations of the chemical properties of cluster systems, and in particular chemical reactions between colliding clusters or clusters colliding with atomic, molecular, or ionic reactants, have been made [1-3].

Recently we have proposed [1] a new mode, of nonelectronic nature, by which clusters may catalyze reactions (CCR). In this mode reactants are embedded in colliding clusters and the role of the cluster environment is that of a local heat bath (i.e., extended third body). In addition to this main subject, our studies relate to certain issues pertaining to the dynamics and kinetics of gas-phase recombination processes (particularly ion-ion recombination [14]).

The reaction systems that we have chosen to investigate are mixed clusters composed of small charged sodium-chloride fragments embedded in argon, colliding with a Cl^- anion. To investigate the dependence on size, several reaction systems were investigated: (i) $[Na_4Cl_3]^+ Ar_q$ for q = 12 and 32 and (ii) $[Na_{14}Cl_{12}]^{+2}Ar_{30}$. Furthermore, systematic dependencies on characteristic parameters of the reaction were studied [1] (i.e., initial temperature of the cluster, relative translational kinetic energy between the collision partners, and impact parameters).

The interactions between the ionic constituents are described in our studies by Coulomb and Born-Meyer repulsion potentials, parametrized according to Born and Huang [15]. For the interactions between the argon atoms and the Na^+ and Cl^- ions, the potentials developed by Ahlrichs et al. [16] were used, and the interactions between the rare gas atoms were described by 6-12 Lennard-Jones potentials with $\varepsilon = 120K$ and $\sigma = 3.4$ Å.

Given these interaction potentials, we have performed extensive MD simulations, where the classical particles' equations of motion (in the center of mass coordinate frame) were solved. For each case that we have studied, 100 reaction trajectories were simulated, where in each one a cluster configuration was randomly chosen from an ensemble of equilibrium configurations, to which a random rotation was applied. Finally, the impact parameters between the clusters and the Cl^- were selected randomly to appropriately cover uniformly the cross sectional area of the target cluster [1]. The initial distance between the reactants was 32 Å in each case.

In this paper we limit ourself to discussion of reactions of the system $[Na_4Cl_3]^+Ar_{12}$ + Cl^-. Prior to summarizing our results we note that the product cluster Na_4Cl_4 is characterized by a low-temperature isomer of approximate cubic structure and a high-temperature one exhibiting a ring-like (two-dimensional) structure [1,17].

First we note that collisions between a Cl^- and a cold (30K) $[Na_4Cl_3]^+$ cluster, even for an initial relative translational energy $E_R \simeq 0$, result in fragmentation and/or generation of high-potential-energy isomers (open chain structures). Upon collision of a Cl^- with a $[Na_nCl_m]^{+(n-m)}$ Ar_q cluster, a fraction of the relative translational kinetic energy between the collision partners, as well as the binding energy released upon reaction, is imparted to the compound (combined) cluster (i.e., the intermediate cluster composed of the two colliding entities). Since the binding energies between the Ar atoms and between them and the alkali-halide ions are significantly smaller than the binding energy between the Na^+ and Cl^- ions, the released excess energy could result in breaking of argon atom bonds to the cluster. Consequently, we would expect that the heat content of the combined, collision-intermediate cluster will be reduced via ejection of Ar atoms, resulting in effective cooling of the products. Under ideal circumstances (which depend on system parameters, such as activation barriers, binding strengths, vibrational relaxation times, initial temperature, and relative translational energy), one could expect that such a process may anneal the product to its lowest (ground) state (in our particular example, to the low-temperature cubic isomer).

The results of our studies show that:

(a) For a given initial vibrational temperature of the reactant cluster, the total average probability for associative reaction products exhibits a decreasing trend with increasing relative translational energy (E_R) between the reactants. Other collision events which occur for high E_R are stripping reactions and glancing collisions. In both the latter cases the products are vibrationally colder than for the corresponding associative reactions, illustrating a spectator mechanism.

(b) The averaged results for the branching ratios plotted vs E_R for the reaction $[Na_4Cl_3]^+Ar_{12}$ + Cl^-, for three initial vibrational temperatures of the target clusters [$T^o{}_{vib}$ = 31.6, 189, and 316K] are summarized in Fig. 1. These results show that for associative reactions with the reactant clusters at the same initial temperature, the branching ratio, P_C/P_R, between the probability to produce the ground state isomer (cube) to that of producing the higher-energy isomer (ring), decreases with increasing E_R. Examination of the individual probabilities (P_C and P_R)

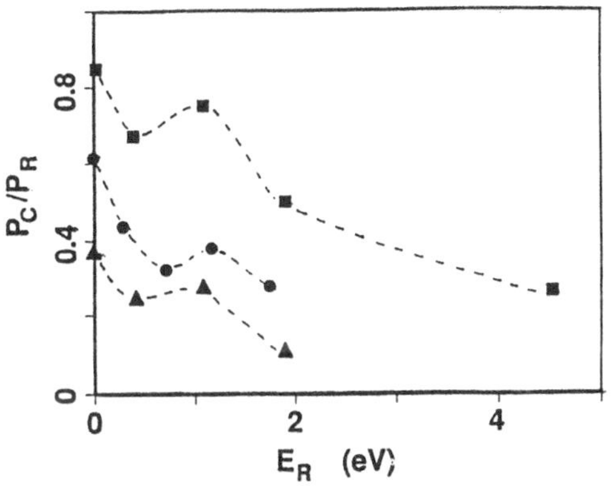

Figure 1: Summary of the branching ratios (P_C/P_R) plotted vs. relative translational energy between the reactants (E_R), for the reaction $[Na_4Cl_3]^+Ar_{12} + Cl^-$, for three initial vibrational temperatures of the reactant cluster T^o_{vib} = 31.6, 189, and 316K denoted by filled squares, circles, and triangles, respectively. The dashed lines are drawn to guide the eye. Energy in eV.

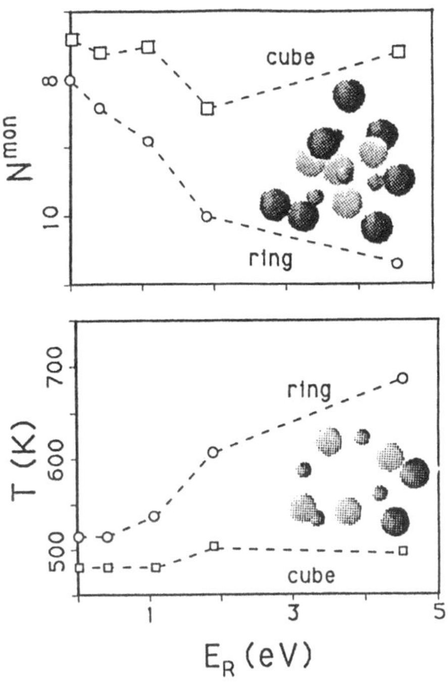

Figure 2: Number of Ar monomers released in the collision (N^{mon}, in a) and vibrational temperatures of the products (in b) for the reaction $[Na_4Cl_3]^+Ar_{12} + Cl^-$, with the initial vibrational temperature of the cluster T^o_{vib} = 31.6K, plotted versus the relative translational energy, E_R. Results corresponding to cube and ring-isomeric products are denoted by squares and circles, respectively. Typical cube and ring isomeric configurations of products are shown in (a) and (b), respectively (large grey, small dark, and large darker spheres correspond to Cl^-, Na^+ and Ar, respectively).

for producing products with the alkali-halide in the cube and ring isomeric forms indicates that the main trend correlating with the above conclusion is the systematic decrease in the cube-isomer products (P_C) with increasing E_R. The vibrational temperature of the lower-energy cube-isomeric reaction products is lower than the ring-isomers, and is quite insensitive to the initial relative kinetic energy between the reactants (see Fig. 2b).

(c) For a given size of the alkali-halide fragment, reactions in which the argon environment in the reactant cluster is larger yield a higher proportion of ground state isomer products.

(d) All associative reactions involve the release of Ar atoms, almost all in the form of monomers. The number of ejected Ar atoms depends on the size of the reactant clusters, the reaction conditions, and the reaction pathways (see for example Fig. 2a).

Studies of the dynamics of the associative reactions reveal that the major part of the energy release occurs in the very initial stage of the collision, resulting in the ejection of one or two translationally hot Ar monomers. The amount of energy carried by these initially ejected Ar atoms determines the reaction path; i.e., a larger energy loss at the initial stage of the reaction is more likely to result in a product with a ground state isomeric structure.

In this context we remark that in all our calculations the simulated trajectories were of 25 ps in duration. Test runs, where the simulations were continued to the nanosecond range, show that typically about two additional Ar atoms were evaporated, resulting in further small cooling of the product cluster. However, in no case did the structure of the product alkali-halide cluster change from that obtained already in the 25 ps simulations. Evidently, the internal temperatures of the product clusters are below the temperature range for transitions between the isomeric forms. Coupled with the fact that the transformations between isomeric forms involve activation barriers (e.g., estimated to be of the order of 0.5 eV for the transformation between the cubic and ring isomers of Na_4Cl_4), the reaction products remain in the isomeric structures which they assume at the early stages of the reaction process.

The origin of the behavior of the branching ratio P_C/P_R shown in Fig. 1, is related to the factors determining the probability for ejection of an energetic (translationally hot) Ar atom at the very initial stage of the collision process, since as aforementioned the reaction path (leading to products in various isomeric forms) is determined by the amount of energy transferred in that initial stage. These factors are: (i) Transfer of energy from the incident anion to the system; and (ii) ejection (escape) of the Ar atom from the cluster. Analysis of the energy dependence of these two factors [1] shows that the amount of energy transferred in a binary collision between the incident Cl^- and an Ar atom decreases with E_R, and the ejection probability is governed by geometric factors, with direct ejection of the energized Ar atom leading to ground-state (cube) isomeric products, and multiple-collisions resulting in energy redistribution yielding ring-isomers. As discussed elsewhere such considerations explain the behavior shown in Fig. 1 (including the local maxima at $E_R \sim 1$ eV).

As seen from our results [1], for the systems that we have investigated (i.e., alkali-halide clusters embedded in argon), it is possible to "tune" the branching ratio between the reaction products (i.e., modify the relative probabilities for the differnt product isomers) by varying the reaction parameters (initial vibrational temperature of the reactants and relative translational energy between the collision partners) and/or number of embedding Ar atoms. However, the nature and relative strengths of the interactions in our system (i.e., ion-ion, Ar-ion, and Ar-Ar) and the magnitude of barriers for interconversion between the reaction products (i.e., isomerization barriers) set a limit to the degree of annealing which can be achieved. Thus the fast nonstatistical cooling of the associated (compound) cluster in the initial reaction stage (via transfer of energy leading to the release of one or two hot Ar atoms) traps the reaction products in the different channels (i.e., different isomeric forms). Further cooling past the initial stage occurs at a very slow rate, because of the relatively strong Ar-ion interactions, and thus is ineffective in inducing structural transformations and further annealing of the products towards the ground state isomeric form.

Nevertheless, consideration of larger Ar environments, as well as different reactants and/or embedding cluster environments, coupled with appropriate selection of collision parameters, may enable optimal control of reaction pathways and product branching ratios.

Another class of cluster catalyzed reactions which we propose, is the collision between clusters, in which reactants are embedded, under conditions which result in the initiation of a shock wave in the intermediate compound cluster. (Similar conditions may be also achieved by colliding clusters with solid surfaces). This would lead to transient local pressure and temperature shock conditions which may induce processes such as dissociation or vibrational (and perhaps even electronic) excitations, which, in turn, may lead to subsequent reactions, with the cluster environment serving as a local heat bath. Work along these lines in our laboratory is in progress.

III. Energetics and Dynamics of Cluster Fission

To investigate energetic patterns and dynamics of fission of small Na_n^{+2} ($n \le 12$) clusters we have used our newly developed simulation method which combines classical molecular dynamics, or energy minimization, on the Born-Oppenheimer (BO) ground-state potential surface, with electronic structure calculations via the Kohn-Sham (KS) formulation of the local spin-density (LSD) functional method (for details of the BO-LSD method see Refs. 4,17). In dynamical simulations the ionic (classical) degrees of freedom evolve on the electronic BO surface which is calculated after each classical step. Minimum energy structures were obtained by a steepest-descent-like method, starting from configurations selected from finite temperature simulations. (Note that the existence of a (local) minimum-energy-configuration implies a barrier for fission.) Barrier heights and shapes were obtained by constrained energy minimization, with the center-of-mass distance, R_{cm-cm}, between the fragments specified.

From the energetics of the clusters and of the various fragementation channels, given in Table I, we observe first that in all cases the energetically favored channel (see Δ) is $Na_n^{+2} \rightarrow Na^+_{n-3} + Na_3^+$ ($n \leq 12$), i.e., asymmetric fission (except for n = 6), in contrast to results obtained from spherical jellium calculations where fragmentation via ejection of Na^+ is favored [8a] (note the shell closing in the product Na_3^+ cluster). Secondly, the first vertical and adiabatic ionization potentials (vIP and aIP) exhibit an odd-even oscillation [9] in the number of particles, as well as shell closing effects for systems containing 8 electrons. Similar effects are seen for higher ionization energies though they are sometimes complicated by structural changes upon ionization. Finally, our calculations show that for n > 6 fission involves energy barriers, in contrast to recent results [6b] obtained from adaptation of the liquid droplet model to atomic clusters. The barriers for n = 8, 10 and 12 have been determined via constrained minimization to be: 0.16 eV,

Table I

Potential energies, E_p, for Na_n^{+2} ($4 \leq n \leq 12$), in the minimum energy configurations. Energies for systems fragmenting with no barrier, calculated for the minimum energy configuration of Na_n^+, are indicated by *. Dissociation energies, $\Delta_m = E(Na^+_{n-m}) + E(Na^+_m) - E(Na_n^{+2})$, for $4 \leq n \leq 12$. vIP and aIP, are vertical and adiabatic ionization energies for $Na_n \rightarrow Na_n^+$. Energies in eV.

/n	4	5	6	7	8	9	10	12
E_p	-10.50*	-19.00*	-23.46*	-30.29	-36.95	-43.47	-50.36	-62.83
Δ_1	-2.48	0.10	-1.78	-1.07	-1.08	-0.74	-0.43	-0.26
Δ_2	-2.03	-0.24	-1.70	-1.22	-0.68	-0.82	-0.10	-0.23
Δ_3			-2.50	-1.59	-1.28	-0.87	-0.65	-0.94
Δ_4					-0.85	-0.67	0.10	-0.27
Δ_5							-0.13	-0.44
Δ_6								+0.00
vIP	4.41	4.33	4.67	4.24	4.62	3.83	4.11	3.93
aIP	4.38	4.11	4.40	4.07	4.35	3.70	4.04	

0.71 eV and 0.29 eV for the energetically favored channel and larger barriers were found for the ejection of Na^+ from these clusters (0.43 eV and 1.03 eV for n = 8 and 10, respectively). The barriers for Na_{10}^{+2} are higher because of the closed-shell structure of this parent cluster.

The potential energies along the reaction coordinates for the energetically favored channel and for Na^+ ejection, in the case of Na_{10}^{+2}

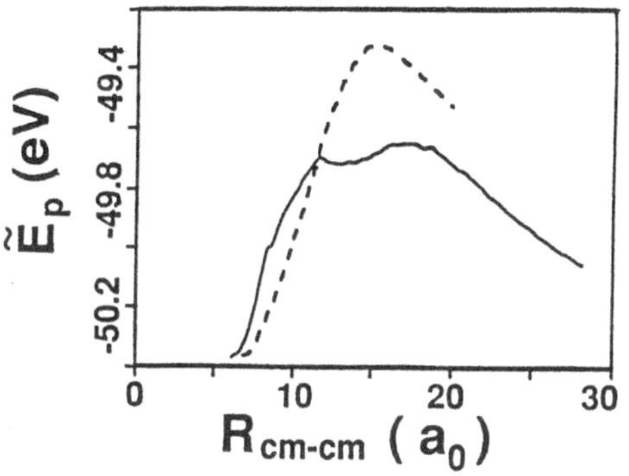

Figure 3: Potential energy vs distance between the center of mass for the fission $Na_{10}^{+2} \rightarrow Na_7^+ + Na_3^+$ (solid) and $Na_{10}^{+2} \rightarrow Na_9^+ + Na^+$ (dashed), obtained via constrained minimization of the LSD ground-state energy of the system.

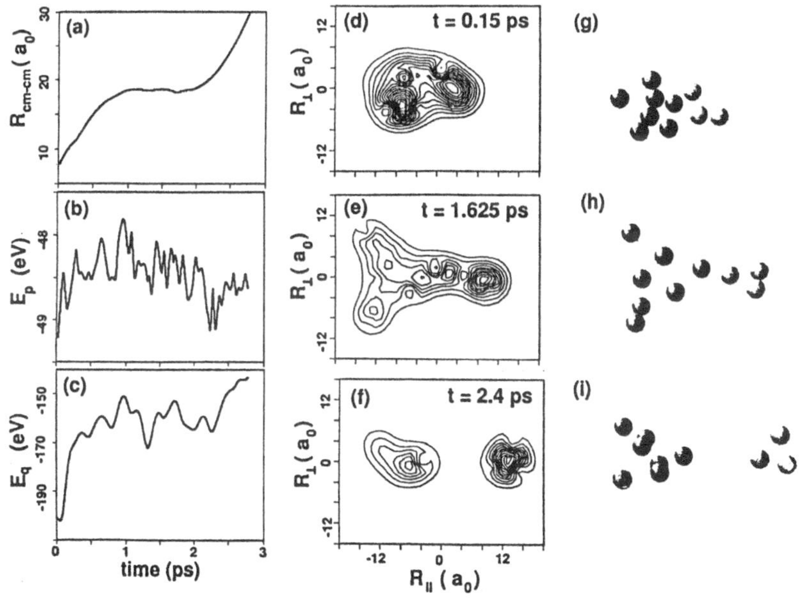

Figure 4: Fragmentation dynamics of Na_{10}^{+2}. (a-c) R_{cm-cm} between the fission products, total potential energy (E_p), and the electronic contribution (E_q), versus time. (d-f) Contours of the total electronic charge distribution, calculated in the plane containing the two centers of masses. The $R_{\parallel}$ axis is parallel to R_{cm-cm}. (g-i) Configurations for the times given in (d-f). Dark and light spheres represent ions in the large and small fragments, respectively. Energy, distance, and time in units of eV, Bohr (a_0), and ps, respectively.

are shown in Fig. 3. The most interesting feature seen from the figure is the rather unusual shape of the barrier for the favored fission channel (also found for the asymmetric fragmentation of Na_{12}^{+2}). While double-hump barriers have been long discussed in the theory of nuclear fission [18], to our knowledge this is the first time that they have been calculated in the context of asymmetric fission of charged atomic clusters. The existence of the double-humped barrier is reflected in the dynamics of the fission process of Na_{10}^{+2} displayed in Fig. 4. This simulation started from a 600K Na_{10} cluster from which two electrons were removed (requiring 11.23 eV) and 0.77 eV was added to the classical ionic kinetic energy. The variation of the center of mass distance with time (Fig. 4a) exhibits a plateau for 750 fs $\leq t \leq$ 2000 fs (see also the behavior of the electronic contribution to the potential energy of the system versus time in Fig. 4c). The contours of the electronic charge density of the system (Fig. 4(d-f)), at selected times, and the corresponding cluster configurations (Fig. 4(g-i)), reveal that the fission process involves a precursor state which undergoes a structural isomerization prior to the eventual separation of the Na_7^+ and Na_3^+ fission products. In this context we remark that examination of the contributions of individual Kohn-Sham orbitals to the total density for the intermediate stage (Fig. 4e) reveals that the lowest-energy orbital (s-like) is localized on the Na_7^+ fragment, the next s-like orbital is localized on the Na_3^+ fragment, the third is a p-like bonding orbital distributed over the two fragments, and the highest orbital is localized on the larger fragment.

Acknowledgement: Research supported by the U. S. Department of Energy Grant No. FG05-86ER15234. Calculations performed at the Florida State University Computing Center through a grant from DOE and at the Pittsburgh Supercomputer Center.

References

1. H.-P. Kaukonen, U. Landman and C. L. Cleveland, J. Chem. Phys. 95, 4997 (1991).
2. For other representative theoretical studies see (a) K. Raghavan, M. S. Stave, and A. E. DePristo, J. Chem. Phys. 91, 1904 (1989); (b) J. E. Adams, ibid. 92, 1849 (1990); (c) J. Jellinek and Z. B. Guvenc, Z. Phys. D 19, 371 (1991); (d) R. Schmidt, G. Seifert and H. O. Lutz, Phys. Lett. A 158, 231 (1991)

3. For representative experimental studies see (a) J. L. Elkind, F. D. Weiss, J. M. Alford, R. T. Laaksomen, and R. E. Smalley, J. Chem. Phys. 88, 5215 (1988); (b) W. D. Reents, Jr. and M. L. Mandich, J. Phys. Chem. 92, 2908 (1988); (c) W. F. Hoffman, E. K. Parks, and S. J. Riley, J. Chem. Phys. 90, 1526 (1989); (d) Y. M. Hamrick and M. D. Morse, J. Phys. Chem. 93, 6494 (1989); (e) S. K. Loh, L. Lian, and P. B. Armentrout, J. Chem. Phys. 91, 6148 (1989); (f) P. Fayet, A. Kaldor, and D. M. Cox, ibid. 92, 254 (1990); (g) K. M. Creegan and M. F. Jarrold, J. Am. Chem. Soc. 112, 3768 (1990); R. E. Leuchtner, A. C. Harms, and A. W. Castlemen, Jr., J. Chem. Phys. 92, 6527 (1990); 91, 2753 (1989).

4. (a) R. N. Barnett, U. Landman and G. Rajagopal, Phys. Rev. Lett. 67, 3058 (1991); (b) R. N. Barnett, U. Landman, A. Nitzan and G. Rajagopal, J. Chem. Phys. 94, 608 (1991).

5. (a) C. Brechignac, Ph. Cahuzac, F. Carlier and J. Leygnier, Phys. Rev. Lett. 63, 1368 (1989); (b) C. Brechignac, Ph. Cahuzac, F. Carlier and M. de Frutos, ibid. 64, 2893 (1990).

6. (a) W. A. Saunders, Phys. Rev. Lett. 64, 3046 (1990); (b) W. A. Saunders, ibid. 66, 840 (1991).

7. (a) Y. Ishii, S. Ohnishi and S. Sugano, Phys. Rev. B 33, 5271 (1986); (b) S. Sugano in Microclusters, edited by S. Sugano et al., (Springer-Berlin, 1987), p. 226; (c) S. Sugano, A. Tawara and Y. Ishii, Z. Phys. D12, 213 (1989); (d) M. Nakamura, Y. Ishii, A. Tamura, and S. Sugano, Phys. Rev. A 42, 2267 (1990).

8. (a) M. P. Iniguez, J. A. Alonso, M. A. Alloc, and L. C. Balbas, Phys. Rev. B 34, 2152 (1986); (b) M. P. Iniguez, J. A. Alonso, A. Rubio, M. J. Lopez and L. C. Balbas, ibid. B 41, 5595 (1990).

9. See review by W. A. de Heer et al., Solid State Phys. 40, 93 (1987), and references therein.

10. V. Bonacic-Koutecky, P. Fantucci, and J. Koutecky, Chem. Rev. (in press, 1991).

11. H. Gohlich, T. Lange, T. Bergmann and T. P. Martin, Phys. Rev. Lett. 65, 748 (1990); S. Bjonholm, J. Borggreen, O. Echt, K. Hansen, J. Pedersen, and H. Rasmussen, Phys. Rev. Lett. 65, 1627 (1990).

12. See references (6-13) in Ref. 4(b).

13. See U. Landman, R. N. Barnett, R. Rajagopal and A. Nitzan, in Nuclear and Atomic Clusters 1991, Proceedings of a European Phys. Soc. Conf. Turku, Finland, edited by M. Brenner (Springer, Berlin, 1991).

14. M. R. Flannery, in Molecular Processes in Space, edited by T. Watanabe, I. Shimamura, M. Shimizu, and Y. Itikawa (Plenum, New York, 1990), p. 145.

15. M. Born and K. Huang, Dynamical Theory of Crystal Lattices (Oxford University, London, 1954).

16. R. Ahlrichs, H. J. Bohm, S. Brode, K. T. Tang, and J. P. Toennies, J. Chem. Phys. 88, 6290 (1988).

17. (a) G. Rajagopal, R. N. Barnett, and U. Landman, Phys. Rev. Lett. 67, 727 (1991); (b) R. N. Barnett and U. Landman (to be published).

18. For a discussion of shell corrections leading to double-hump nuclear fission barriers see M. A. Preston and R. K. Bhaduri, Structure of the Nucleus (Addison-Wesley, Reading, MA, 1975), p. 589. Shell effects on symmetric fragmentations of alkali-metal clusters have been studied using an approximate expression for the energy by M. Nakamura et al., see Ref. 7d. We remark that examination of the variations in the KS orbital energies, calculated along the dynamical fission trajectory of the cluster, indicates a correlation between these variations (in particular level crossings) and the shape of the barrier which is thus related to the energetics of the deformation of the cluster (R. N. Barnett and U. Landman, unpublished).

Molecule-Cluster Collisions: Reaction of D_2 with Ni_{13}

Julius Jellinek and Ziya B. Güvenç

Chemistry Division, Argonne National Laboratory
Argonne, Illinois 60439, USA

ABSTRACT. The reactive channel (dissociative adsorption of the molecule on the cluster) of the D_2 + Ni_{13} collision system is studied via quasiclassical molecular dynamics simulations. The effects of the initial rovibrational state of the molecule and of the cluster structure are examined. Pronounced mode-selectivity of the reaction and a strong structure-reactivity correlation are found. A technique to analyze cluster-molecule complex ("resonance") formation is presented and used to characterize the direct vs indirect reaction pathways, the probability of formation of reactive resonances and their lifetimes.

I. Introduction

There is much overlap between the problems posed by atomic clusters and atomic nuclei (clusters of nucleons) considered as N-body systems, although the nature of the "bodies" and the forces acting between them are quite different in these systems. A variety of concepts and approaches, such as the liquid drop model, shell structure, stability vs. metastability, Coulomb explosion, etc., originally developed by nuclear physicists, have been adopted and used with success in the relatively young and rapidly developing field of cluster research. The different phenomena accompanying collisions involving clusters, on the one hand, and nuclei, on the other, are also similar. These include fusion, fission and inelastic scattering. The cluster analog of nuclear reactions is chemical reactions (breaking and forming bonds).

Here we present a brief discussion of a simulation study of the collision of a D_2 molecule with a Ni_{13} cluster. We focus on the reactive channel of the interaction, i.e., on the dissociative adsorption of the molecule on the cluster. Cluster-molecule reactions play an essential role in a variety of technologies, e.g., in heterogeneous catalysis, as well as in natural and environment-related phenomena. Therefore, understanding of the mechanisms of cluster-molecule interactions and of the factors affecting these interactions is of substantial importance. In section II, we outline the theoretical background. In section III the numerical results and their discussion are presented. Particular attention is paid to the role of molecularly adsorbed precursor states in dissociative adsorption. A summary is given in section IV.

II. Theoretical Background

We consider the process of collision of a D_2 molecule with a Ni_{13} cluster and characterize all the possible channels of the interaction, which include adsorption of the molecule on the cluster and its scattering from the cluster, as functions of the initial state of the collision system. The adsorption may be molecular or dissociative ("reactive"). Unless energy is dissipated from the system, the adsorbed molecule eventually either dissociates on the cluster (the reaction of D-D bond breaking and D-cluster bond formation is exothermic) or desorbs from the cluster. After dissociation the atoms may recombine again and the reborn molecule may desorb from the cluster.

The initial state of the molecule-cluster collision system is specified as follows. The initial coordinates and momenta of the D atoms are selected in accordance with a quasiclassical prescription [1] to yield an energy corresponding to a given quantal vibrational v_i and rotational

j_i state of the D_2 molecule (the subscript "i" stands for "initial"). The molecule is supplied a given amount of initial relative translational (collision) energy E^i_{tr} and sent towards the cluster with a specified impact parameter b. The center of mass of the Ni_{13} cluster, which is prepared in one of the three geometries - icosahedral (ico), cuboctahedral (cubo) or hexagonal close-packed (hcp) - is placed initially at a distance of 8.5Å ("asymptotically far") from the center of mass of the molecule. The initial momenta of the Ni atoms are chosen to yield zero total linear and angular momenta and a cluster temperature T defined as

$$T = \frac{2\langle E_k \rangle}{(3n-6)k}, \tag{1}$$

where n and E_k are the number of atoms and the internal kinetic energy of the cluster, k is the Boltzmann constant, and $\langle\,\rangle$ denotes a long-time average.

The time evolution of the system is generated by solving numerically Hamilton's equations of motion for all the degrees of freedom. The forces acting on the atoms are calculated from a combined embedded-atom (EA)-LEPS (London-Eyring-Polanyi-Sato) potential

$$V = V_{EA} + V_{LEPS}. \tag{2}$$

The embedded-atom potential V_{EA} describes the interaction between the Ni atoms in the cluster. We are using Voter and Chen's parametrization of V_{EA}[2], which in addition to properties of the bulk nickel incorporates also the binding energy and the equilibrium bond length of Ni_2; it may, thus, be expected to have a higher degree of adequacy for clusters. We have found that of the three structures considered the icosahedron is the most stable geometry of Ni_{13} described by this potential. The cubo and hcp structures are almost degenerate energetically and are by about 1.6 eV less stable (Fig. 1). The LEPS potential is representing the D-D and D-cluster interactions. We employ as V_{LEPS} the potential energy surface (PES) II of Raghavan, Stave and Deristo (RSD) [3] modified by a smoothing function used by Truong et al. [4] (details will be presented elsewhere). The reason for selecting PESII is that it is fitted to the binding characteristics of an H atom on a threefold face of an icosahedral Ni_{13} as calculated using the so-called corrected effective medium approach [3]; the parameters of PESI of RSD are chosen to mimic the dissociative chemisorption of H_2 on different faces of the (fcc) bulk nickel [5].

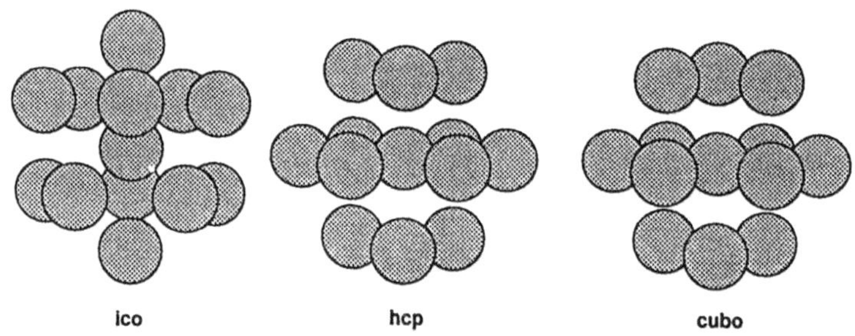

Figure 1. The icosahedral, hexagonal close-packed, and cuboctahedral structures of Ni_{13}. Their equilibrium energies, as defined by V_{EA} (see text), are -41.12 eV, -39.53 eV, and -39.51 eV, respectively.

The phase space trajectories are generated using a fourth-order variable step-size predictor-corrector propagator. With the maximal step-size of 5×10^{-16}-10^{-15} s, the total energy and total linear and angular momenta are conserved in individual runs within 0.03-0.15%. Each trajectory is propagated until the molecule either dissociates on the cluster or departs after collision with it back into the asymptotic region. The molecule is considered as dissociated if and when the D-D distance reaches the value of 2.223Å (three times the equilibrium bond length of the D_2 molecule). Since when the dissociation criterion is met the trajectory is terminated, a possible recombination and subsequent desorption of the molecule are precluded.

The different channels of the interaction are characterized quantitatively in terms of probabilities, cross sections, rate constants, etc. labeled by the initial (and, in the case of nonreactive collision, final) state of the molecule and of the cluster. Batches of at least N=500 trajectories were run for each set of initial conditions, which include the values of E^i_{tr} and b; in certain cases as many as N=4000 trajectories were run in a batch to allow for an extraction of some more delicate characteristics - see section III. The trajectories in a given batch differ by the initial phase of the D_2 oscillator and by the initial relative orientation of the molecule and of the cluster. The state-labeled probability $P_{v_i,j_i,T,...}(E^i_{tr},b)$ of a process of interest is calculated as

$$P_{v_i,j_i,T,...}(E^i_{tr},b) = \frac{\tilde{N}_{v_i,j_i,T,...}(E^i_{tr},b)}{N}, \tag{3}$$

where $\tilde{N}_{...}$ is the number of trajectories that results in that process. The cross sections $\sigma_{v_i,j_i,T,...}(E^i_{tr})$ are computed in accordance with

$$\sigma_{v_i,j_i,T,...}(E^i_{tr}) = 2\pi \int_0^{b_{max}} bP_{v_i,j_i,T,...}(E^i_{tr},b)db\ , \tag{4}$$

where b_{max} is the upper limiting value of the impact parameters contributing to the process considered. The rate constants $\kappa_{...}(T_{tr})$ are obtained using the formula

$$\kappa_{v_i,j_i,T,...}(T_{tr}) = \pi\mu\left(\frac{2}{\pi\mu kT_{tr}}\right)^{3/2} \int_0^{\infty} \sigma_{v_i,j_i,T,...}(E^i_{tr})E^i_{tr}\exp\left(-\frac{E^i_{tr}}{kT_{tr}}\right)dE^i_{tr}\ , \tag{5}$$

where μ is the reduced mass of the cluster-molecule collision system and T_{tr} is the translational (or "collisional") temperature. The activation energies $E^A_{v_i,j_i,T,...}$ are derived by fitting the low-temperature segments of the rate constants to the Arrhenius formula

$$\kappa_{v_i,j_i,T,...}(T_{tr}) = A_{v_i,j_i,T,...}\exp\left(-\frac{E^A_{v_i,j_i,T,...}}{RT_{tr}}\right), \tag{6}$$

where R is the universal gas constant.

As mentioned, the reaction (dissociative adsorption of the molecule on the cluster) may be direct or indirect. The latter involves an intermediate state of molecular adsorption; we shall refer to the transient complexes formed by the cluster and the molecule as "resonances". The resonances that eventually lead to dissociative (i.e., atomic) adsorption are the "reactive resonances". The characteristic lifetime $\tau_{...}(E^i_{tr},b)$ of the reactive resonances can be determined in the following way. Let $P^{DR}_{...}(E^i_{tr},b)$ be the probability of the direct reaction and $P^{IR}_{...}(E^i_{tr},b)$ - the probability of the indirect reaction (or, equivalently, of the formation of reactive resonances); cf. Eq. (3). The reaction probability $P^{R}_{...}(E^i_{tr},b)$ is the sum $P^{DR}_{...}(E^i_{tr},b)+P^{IR}_{...}(E^i_{tr},b)$. In order to calculate $P^{R}_{...}$ in numerical simulations correctly, one should assure that the individual trajectories in a batch are run long enough to allow for the reactive resonances to decay (i.e., for the adsorbed molecule to dissociate); the mentioned above criteria for termination of a trajectory guarantee this. If one introduces a grid t_r on the length of propagation of the individual trajectories, one can calculate a time-dependent reaction probability $P^{R}_{...}(E^{i}_{tr},b|t_r)$. Clearly, if the value of t_r is less than the " arrival time" t_{ar}, which is the time needed for the molecule to reach the cluster, $P^{R}_{...}(E^{i}_{tr},b|t_r)=0$. Assuming that the dissociation of the molecule is an instantaneous process, one defines t_{ar} as the value of t_r at which $P^{R}_{...}(E^{i}_{tr},b|t_r)$ changes its value from zero to $P^{DR}_{...}(E^{i}_{tr},b)$. If the probability of the direct reaction is zero, t_{ar} is the largest value of t_r at which $P^{R}_{...}(E^{i}_{tr},b|t_r)=0$. The dependence of t_{ar} on the initial phase of the D_2 oscillator and on the initial orientation of the molecule with respect to the cluster is neglected. Obviously, for $t_r \geq t_{ar}$, $P^{R}_{...}(E^{i}_{tr},b|t_r)$ is a monotonically increasing function of t_r: as t_r increases, the number of surviving reactive resonances decreases and more trajectories are counted are reactive. The maximal value of $P^{R}_{...}(E^{i}_{tr},b|t_r)$ is $P^{R}_{...}(E^{i}_{tr},b)$. If $\kappa^{d}_{...}(E^{i}_{tr},b)$ is the rate constant of decay of the reactive resonances and $t=t_r-t_{ar}$, then for $t \geq 0$

$$P^{R}_{...}(E^{i}_{tr},b|t) = P^{DR}_{...}(E^{i}_{tr},b) + P^{IR}_{...}(E^{i}_{tr},b)\left[1 - e^{-\kappa^{d}_{...}(E^{i}_{tr},b)t}\right]$$

$$= P^{R}_{...}(E^{i}_{tr},b) - P^{IR}_{...}(E^{i}_{tr},b)e^{-\kappa^{d}_{...}(E^{i}_{tr},b)t}. \tag{7}$$

Equation (7) can be rewritten as

$$\ln\left[P^{R}_{...}(E^{i}_{tr},b) - P^{R}_{...}(E^{i}_{tr},b|t)\right] = -\kappa^{d}_{...}(E^{i}_{tr},b)t + \ln P^{IR}_{...}(E^{i}_{tr},b)\ . \tag{8}$$

Calculating numerically the left-hand-side (lhs) of Eq. (8) on a t-grid and fitting the data by a linear function, one obtains the decay rate constant $\kappa^{d}_{...}(E^{i}_{tr},b)$ and the reactive resonance formation probability $P^{IR}_{...}(E^{i}_{tr},b)$. Note that replacing t by t_r ($=t_{ar}+t$) in Eq. (8) changes the right-hand-side of this equation by a constant but does not affect the slope of its graph. Thus, if one is interested only in $\kappa^{d}_{...}(E^{i}_{tr},b)$, it is sufficient to compute the lhs of Eq. (8) on a t_r-grid, which eliminates the necessity of determining $t_{ar}(E^{i}_{tr},b)$. The inverse of $\kappa^{d}_{...}(E^{i}_{tr},b)$ is the characteristic lifetime $\tau(E^{i}_{tr},b)$ of the reactive resonances.

One can introduce an effective, averaged over b (and, thus, b-independent), decay rate constant $\bar{\kappa}^{d}_{...}(E^{i}_{tr},b)$. This rate constant can be obtained by calculating the lhs of the equation (9) [cf. Eqs. (4), (7) and (8)],

$$\ln\left[(\sigma^{R}_{...}(E^{i}_{tr}) - \sigma^{R}_{...}(E^{i}_{tr}|t)\right] = -\bar{\kappa}^{d}_{...}(E^{i}_{tr})t + \ln\sigma^{IR}_{...}(E^{i}_{tr})\ , \tag{9}$$

on a t-grid and fitting the data by a linear function. The E^{i}_{tr}-dependent effective lifetime $\bar{\tau}_{...}(E^{i}_{tr})$ of the reactive resonances is the inverse of $\bar{\kappa}^{d}_{...}(E^{i}_{tr})$.

III. Results and Discussion

We present and discuss here results characterizing the reactive channel of the $D_2 + Ni_{13}$ collision system. Earlier accounts of our findings for this system are given in Refs. 6-8; cf. also Ref. 3. The fundamental questions pertaining to the reaction of a molecule with a cluster include the following: 1) How does the total energy of the system, in general, and the initial rotational and vibrational energy of the molecule, in particular, affect the reaction? 2) What are the effects of the structure of the cluster and of its temperature? 3) Is the dissociative adsorption of the molecule a direct process or may it also involve molecularly adsorbed precursor states (resonances)? What is the contribution of the direct process and of the resonances into the reaction? What are the characteristics (probability of formation and lifetime) of the reactive resonances?

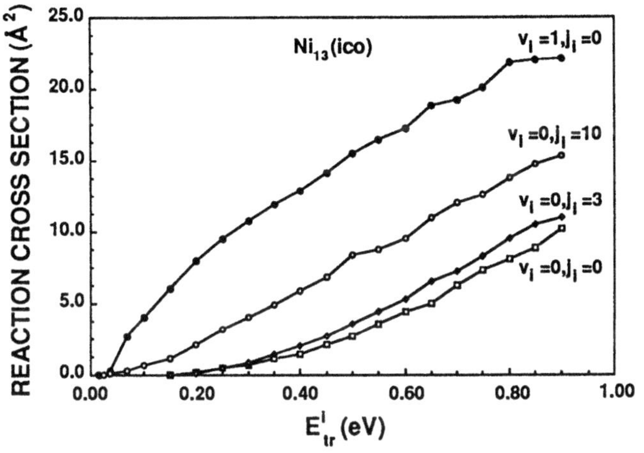

Figure 2. Reaction cross sections as functions of the collision energy for a room-temperature ico Ni_{13} and different initial rovibrational states of D_2.

In Fig. 2 the reaction cross sections for a D_2 colliding with a room-temperature (T=298 K) ico Ni_{13} are shown. The graphs clearly indicate that the larger the total energy of the system, the more readily the molecule dissociates on the cluster. The cross sections increase monotonically with E^i_{tr} for all the different initial rovibrational states of the molecule. They also display (v_i,j_i)-dependent thresholds to the reaction. In addition to the ground state of the molecule, data are presented also for (v_i=0, j_i=3), the most populated state of D_2 at room temperature, and the (v_i=1, j_i=0) and (v_i=0, j_i=10) states, which are almost degenerate energetically. The comparison of the cross sections for the latter two states points to the mode-selectivity of the reaction: a given amount of energy initially in a vibrational (v_i=1) mode only results in a considerably larger reaction cross section than the same amount of energy shared by a vibrational (v_i=0) and a rotational (j_i=10) mode. We have found that changing the temperature of the ico cluster in the range T=0-298 K does not affect its reactivity with D_2(v_i=0, j_i=0).

Typical data for the rate constants are displayed in Fig. 3. While the overall temperature-dependence is non-Arrhenius, the points corresponding to the low T_{tr}-values can be fitted by Eq. (6). We have found for all cases that the extracted from the fits (v_i,j_i)-dependent activation energies $E^A_{v_i j_i}$ are in quantitative agreement with the corresponding (v_i,j_i)-dependent threshold energies of Fig. 2. This allows one to interpret the threshold energies as measures of the (v_i,j_i)-dependent dynamical barriers for dissociative adsorption of D_2 on an ico Ni_{13}.

The reaction cross sections for the cubo and hcp Ni_{13} are shown in Fig. 4. Since these isomers are, as defined by V_{EA}, metastable, they were prepared initially at T=0 K. Analyzing the structure-reactivity correlation one notices that, on the one hand, the cross sections for the cubo and hcp structures are similar in magnitude and in their behavior as functions of E^i_{tr}. On the other hand, they are very different from the corresponding cross sections for the ico isomer. The differences are both quantitative and qualitative. The cubo and hcp Ni_{13} are noticeably

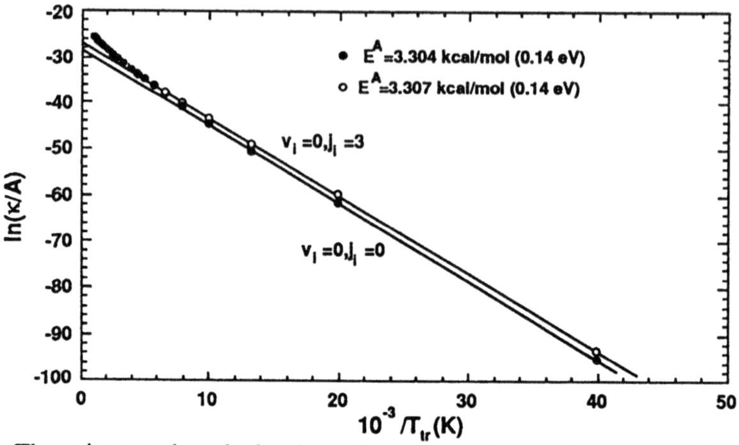

Figure 3. The points are the calculated reaction rate constants as functions of the collisional temperature T_{tr} for a room-temperature ico Ni_{13} and two different (v_i,j_i)-states of D_2 (the magnitude of A is chosen to be 1). The straight lines are the low-T_{tr} fits. The activation energies E^A are calculated from the slopes of these lines.

more reactive than the ico Ni_{13}. Although all three isomers share the feature of mode-selective reactivity, the cross sections for the cubo and hcp structures exhibit virtually no thresholds, and their change with E^i_{tr} is nonmonotonic. The main differences are in the low collision energy region, where the reaction cross sections for the cubo and hcp Ni_{13} exhibit (local) peaks. At low E^i_{tr}'s the molecule approaches the surface of the cluster slowly, spends more time in its vicinity and, thus, is more sensitive to its topology. The topologically similar surfaces of the cubo and hcp isomers are comprised of threefold and fourfold faces, while that of the ico isomer - only of threefold faces. The reaction causes the cubo and hcp Ni_{13} to convert into an ico. One can still characterize the reactivity of the different isomers of this cluster since the time scale of its possible structural rearrangements induced by the collision with the D_2 molecule is longer than the time scale of the reaction. The only possible exception is the case of very long-lived reactive resonances, when the resonance lifetime may compete with the isomerization time. But even in this case, the molecule attaches itself to a cluster with a well defined initial structure.

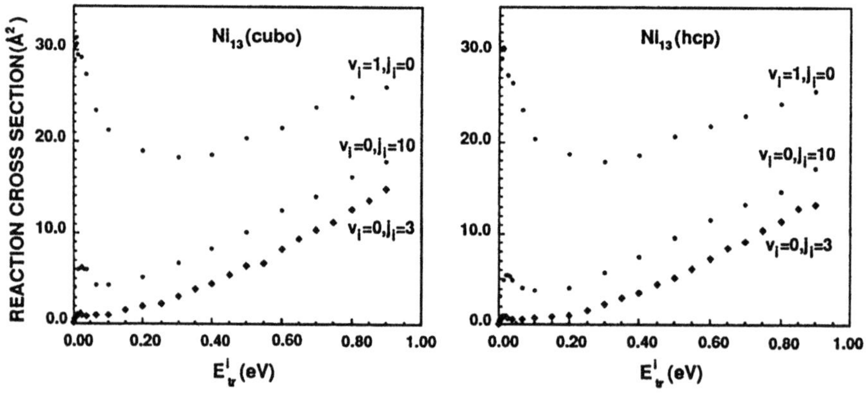

Figure 4. The same as Fig. 2 but for a zero-temperature cubo and hcp Ni_{13}.

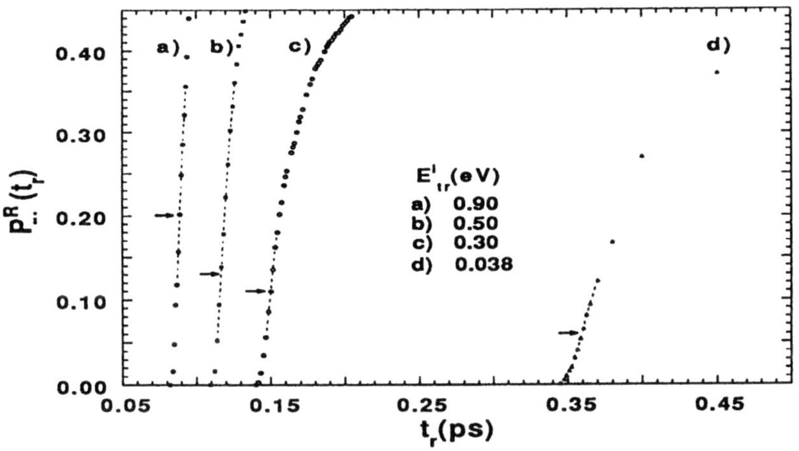

Figure 5. Segments of the time-dependent reaction probabilities as functions of the trajectory propagation time t_r for a zero-temperature cubo Ni_{13}, (v_i=1, j_i=0), b=0.25Å and different values of E^i_{tr}. The dashed lines are local polynomial fits. The arrows indicate the inflexion points that define the probabilities of the direct reaction (see text; cf. Table 1).

We turn now to the analysis of the direct vs indirect reaction pathways. In Fig. 5 segments of the time-dependent reaction probabilities $P^R_{...}(E^i_{tr},b|t_r)$ are plotted for the cubo Ni_{13}, (v_i=1, j_i=0), b=0.25Å and different values of E^i_{tr}. According to the discussion of section II, $P^R_{...}(E^i_{tr},b|t_r)$ has to change its value discontinuously from zero to $P^{DR}_{...}(E^i_{tr},b|t_r)$ at $t_r=t_{ar}$, if the probability of the direct reaction is nonzero. None of the graphs in Fig. 5, however, shows a discontinuous change. Does it mean that, independently of whether E^i_{tr} is low or high, the probability of the direct reaction is zero? The graphs of Fig. 5 display another seemingly "disturbing" feature. It follows from Eq. (7) that the rate of change of $P^R_{...}(E^i_{tr},b|t_r)$ with t_r is $\kappa^d_{...}\cdot P^{IR}_{...}\cdot\exp(-\kappa^d_{...}t_r)$, which is a decreasing function of t_r, provided $P^{IR}_{...}(E^i_{tr},b)>0$. This implies that the curves representing the t_r-dependence of $P^R_{...}(E^i_{tr},b|t_r)$ cannot be concave anywhere. The low-t_r ends of the graphs in Fig. 5 are, however, clearly concave. The reason for the "peculiar" features of these graphs is the following. The criterion for the dissociation of the molecule (see section II) makes the direct reaction a noninstantaneous event. After the molecule "hits" the cluster at $t_r=t_{ar}$ a small but finite "induction" time Δt is required for the D-D bond to reach the value defined by the dissociation criterion and, thus, for the event to be registered as a reaction. The value of Δt is different for different trajectories in a batch since it depends on the initial phase of the D_2 oscillator and on the initial orientation of the molecule with respect to the cluster; strictly speaking, so does also t_{ar}. If Δt_{max} is the largest Δt in a batch, the concave parts of the graphs in Fig. 5 correspond to the time segments $[t_{ar}, t_{ar}+\Delta t_{max}]$. If Δt_{max} is small enough, so that the contribution of the reactive resonances into $P^R_{...}(E^i_{tr},b|t_r)$ can be neglected for $t_r \leq t_{ar}+\Delta t_{max}$ (this can be achieved by a judicious choice of the dissociation criterion), then $P^{DR}_{...}(E^i_{tr},b|t_r)=P^R_{...}(E^i_{tr},b|t_{ar}+\Delta t_{max})$. Practically $P^{DR}_{...}(E^i_{tr},b)$ is found as the value of $P^R_{...}(E^i_{tr},b|t_r)$ at the inflexion points of the graphs in Fig. 5. The inflexion points are determined from local polynomial fits of the calculated $P^R_{...}(E^i_{tr},b|t_r)$ values.

Figure 6 illustrates the determination of the decay rate constants $\kappa^d_{...}(E^i_{tr},b)$ as described in section II. The points corresponding to $t_r<t_{ar}+\Delta t_{max}$ do not carry information on the reactive resonances and therefore they are excluded from the fits. The lifetimes $\tau_{...}(E^i_{tr},b)$ of the reactive

resonances for cubo Ni_{13}, (v_i=1, j_i=0), b=0.25Å and different values of E^i_{tr} are listed in Table 1.

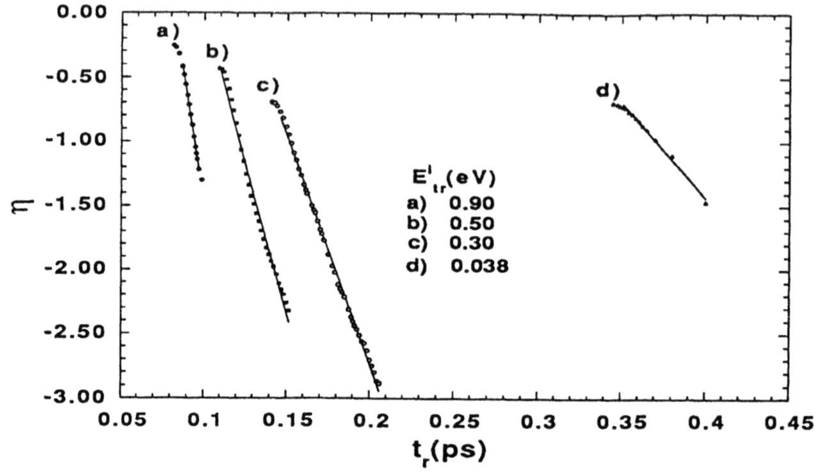

Figure 6. The points are the values of the lhs of Eq. (8) [denoted as η] calculated on a t_r-grid for a zero-temperature cubo Ni_{13}, (v_i=1, j_i=0), b=0.25Å and different values of E^i_{tr}. The slopes of the lines are the corresponding reactive resonance decay rate constants.

Also shown are the corresponding probabilities of the direct reaction $P^{DR}_{...}(E^i_{tr},b)$ and of the reactive resonance formation $P^{IR}_{...}(E^i_{tr},b)$, as well as the fractional contribution $f^{DR}_{...}= P^{DR}_{...}/P^{R}_{...}\cdot 100\%$ of the direct process to the reaction. The lifetime of the reactive resonances decreases as E^i_{tr} increases. The fractional contribution of the direct process exhibits a tendency to grow with E^i_{tr}. Noticing that for E^i_{tr}=0.038 eV the lifetime of the resonances is relatively long (on the time-scale of the zero-point vibrational period of D_2, which is ~0.01 ps) and that the value of $P^{DR}_{...}$ and $f^{DR}_{...}$ are very small, one concludes that the sharp peak in the reaction cross section at low collision energies (Fig. 4) is almost exclusively due to the reactive resonances. On the other hand, as $E^i_{tr}\rightarrow 0.9$ eV, the resonance lifetime decreases to ~0.01 ps and although f^{DR} grows only to 26%, the reaction can be viewed as almost direct. With an increase of the impact parameter, the lifetime of the reactive resonances tends to increase also [8].

Table 1. Probabilities of the direct ($P^{DR}_{...}$) and indirect ($P^{IR}_{...}$) reaction, fractional contribution of the direct reaction ($f^{DR}_{...}$) and reactive resonance lifetimes (τ...) for a zero-temperature cubo Ni_{13}, (v_i=1, j_i=0), b=0.25Å and different values of E^i_{tr}.

E^i_{tr}(eV)	0.038	0.30	0.50	0.90
$P^{DR}_{...}$	0.06	0.11	0.13	0.20
$P^{IR}_{...}$	0.44	0.39	0.52	0.57
$f^{DR}_{...}$(%)	12	22	19	26
τ(ps)	0.07	0.03	0.02	0.01

IV. Summary

We presented results on collision-induced dissociative adsorption of a D_2 molecule on a Ni_{13} cluster obtained using quasiclassical trajectory simulations. The effects of the total energy of the system as well as of the initial partitioning of the energy between the different molecular degrees of freedom were examined. The reaction is mode-selective: energizing D_2 vibrationally promotes its dissociative adsorption on Ni_{13} more efficiently than splitting the same amount of energy between a vibrational and a rotational mode. The structure of the cluster affects its reactivity both quantitatively and qualitatively. The more stable ico form of Ni_{13} is less reactive than its cubo and hcp isomers. The E^i_{tr}-dependence of the reaction cross section for the topologically similar cubo and hcp structures is very different from that for the ico geometry. An analysis is presented which allows one to differentiate between and to characterize quantitatively the direct and indirect reaction pathways and to calculate the lifetimes of the reactive resonances. Work is in progress towards obtaining the cross sections for the direct and indirect reaction and on examining the effect of heating the cluster to elevated temperatures on its reactivity.

Acknowledgments

This work is performed under the auspices of the Office of Basic Energy Sciences, Division of Chemical Science, US-DOE under contract number W-31-109-ENG-38.

References

1. R. N. Porter, L. M. Raff, and W. H. Miller, J. Chem. Phys. 63, 2214 (1975).
2. A. F. Voter and S. F. Chen, Mater. Res. Soc. Symp. 82, 175 (1987).
3. K. Raghavan, M. S. Stave, and A. E. DePristo, Chem. Phys. Lett. 149, 89 (1988); J. Chem. Phys. 91, 1904 (1989).
4. T. N. Truong, D. G. Truhlar, and B. C. Garett, J. Phys. Chem. 93, 8227 (1989).
5. C. Y .Lee and A. E. DePristo, J. Chem. Phys. 85, 4161 (1986); ibid. 87, 1401 (1987).
6. J. Jellinek and Z. B. Güvenç, Z. Phys. D, 19, 371 (1991).
7. J. Jellinek and Z. B. Güvenç, in *Mode Selective Chemistry*, J. Jortner, R. D. Levine, and B. Pullman (Eds.), Kluwer Academic Publishers, Dordrecht, 1991, p. 153.
8. J. Jellinek and Z. B. Güvenç, in *Physics and Chemistry of Finite Systems: From Clusters to Crystals*, P. Jena, S. N. Khanna, and B. K. Rao (Eds.), Kluwer Academic Publishers, Dordrecht, 1992 (in press).

Vibrational Dynamics of Large Clusters from High Resolution He Atom Scattering

U. Buck, R. Krohne and J. Siebers

Max-Planck-Institut für Strömungsforschung
Bunsenstraße 10, WD-3400 Göttingen, Germany

Introduction

The knowledge of the frequency spectrum of a cluster plays an important role both for the interpretation of static and structural properties as well as for the dynamical behaviour. This includes such quantities as the thermal free energy and the specific heat, the melting and condensation processes, and the excitation or deexcitation of the vibrational modes [1]. The latter process is of special interest, since it is very sensitive to the transition of the cluster from the discrete spectrum of vibrational modes of the molecular system to the lattice vibrations and the continuous phonon dispersion curves of the solid. The theoretical methods to study these topics reach from the classical normal mode analysis over molecular dynamics simulations to complete quantum calculations of the lattice dynamics. Especially appealing for the study of surface modes of clusters is the calculation of collective excitations in the liquid drop model of nuclear physics. The experimental methods include IR- and Raman-spectroscopy for the molecular systems and neutron, electron or He atom scattering for probing bulk and surface phonons, respectively. From all these methods the scattering of He atoms appears to be the most general process to study the vibrational spectra of clusters, since this method is mainly sensitive to surface properties and obeys nearly no selection rules. In the last 10 years it has been developed into a very successful and reliable method for measuring surface phonons of the solid and for deriving a series of very interesting surface properties [2]. In the present paper we report first measurements of He atom scattering from large Ar_n clusters in the range from n = 30 to n = 4500. There are mainly two different types of observables: (1) The angular distributions, which contain among other things information about the geometry and the size of the investigated objects through the diffraction oscillations. (2) The inelastic energy transfer, which is related to the excitation of the vibrational modes and which is measured by time-of-flight analysis of the scattered He atom.

Experimental

The experiments have been carried out in a crossed molecular beam machine which is described elsewhere [3]. Essentially it consists of a He supersonic nozzle beam and a cluster beam for the target which intersect at an angle of 90°, and a detector with an electron-bombardment ionizer and a quadrupole mass filter operating under ultrahigh vacuum conditions. The angular dependence is measured by rotating the source assembly relative to the fixed detector position. The velocity of the scattered He atoms is measured by time-of-flight analysis using the pseudorandom chopping technique with a flight path of 450 mm.

The helium atom beam is produced by expansion of the gas under high stagnation pressure (typically 30 bar) through a small orifice (diameter 30 μm) into the vacuum. Most of the measurements were carried out at a temperature of $T_0 = 77$ K. With a speed ratio of $S = 90$ an internal temperature of lower than 0.1 K and a corresponding relative width of the velocity distribution of $\Delta v/v = 0.018$ is obtained. This leads at collision energies of about 25 meV to a resolution of better than 1 meV. The cluster beam is generated by expansion from stagnation pressures of 1.2 to 5.0 bar through different nozzles of conical shape [4] which vary in diameter (60 μm to 130 μm), length (1 mm, 10 mm) and opening angle (20° to 30°). The beam contains only a distribution of cluster sizes. The maximum can be shifted to nearly any desired position by varying the different shape parameters of the nozzle and the stagnation pressure. The corresponding average cluster sizes $\bar{n}$ are taken from measurements by mass spectrometry [5,6] and electron diffraction [7] making use of special scaling laws [8]. It is interesting to note that the velocity analysis of the Ar-beam which contains a fraction of large clusters gave two contributions: a faster one at 615 m/s, which is attributed to the remaining Ar monomers and a slower one at 500 m/s, which is attributed to the clusters and is about the same for all cluster sizes.

Angular distributions

Angular distributions or, in the language of scattering experiments, the total differential cross sections, have been measured for different collision energies and cluster sizes ranging from E = 70 meV, $\bar{n} = 70$ to E = 25.0 meV, $\bar{n} = 55$, $\bar{n} = 115$, $\bar{n} = 2100$ and $\bar{n} = 4500$ for deflection angles between 3° and 80°. They can be characterized by three different regions. (1) At small angles up to 30° they are all dominated by an oscillation of the same large angular separation. (2) Then an angular range (30-60°) follows with oscillations of smaller angular separation which varies with cluster size. (3) Finally, an unregular type of intensity fluctuations is observed between 50° and 80° which also depends on cluster size. A typical example is shown in Fig. 1 for $\bar{n} = 55$ and $E = 25.0$ meV which clearly shows the oscillations of type (1) and (2), but not the intensity fluctuations of type (3). The oscillation can be attributed to diffraction which gives for the angular separation [9]

$$\Delta\Theta = \pi/(kR_0)$$

in which k is the wavenumber, and R_0 the range of the repulsive potential which is closely related to the diameter of the particle. Simple calculations of the elastic total differential cross sections show that the oscillations of type (1) can be reproduced by He-Ar scattering, that is the scattering from the monomers in the beam. The comparison with the measured data, also given in Fig. 1, shows good agreement in the position of the oscillations. In order to reproduce the oscillations of type (2), similar calculations have been carried out for different cluster sizes using the model of a sphere of homogeneously distributed Lennard-Jones (12-6) potentials [10]. The comparison is also shown in Fig. 1. To get, however, good agreement with the data of $\bar{n} = 55$, the parameter R_0 had to be lowered appreciably from 8.1 Å to 4.8 Å. A possible explanation is that the model overestimated the R_0 value and thus the size of the cluster. This suggestion, however, proved to be incorrect. A more precise evaluation of the interaction potential of $He-Ar_{55}$ based on realistic interactions

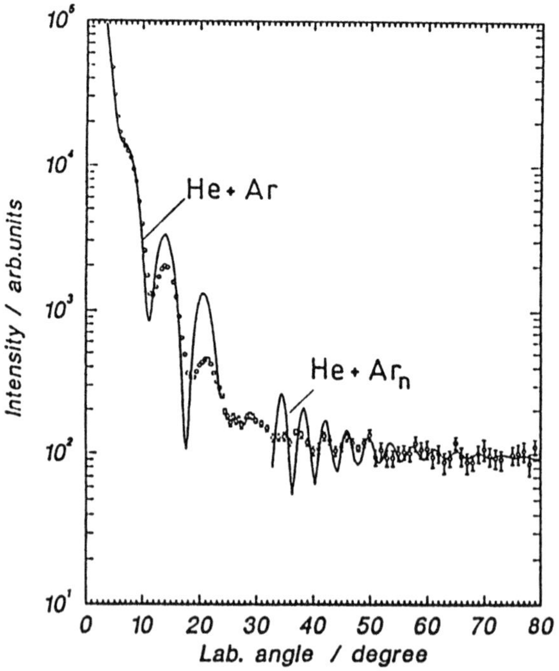

Figure 1: Measured angular distribtution for $He+Ar_{55}$ scattering. The solid lines are calculations based on the potentials for $n = 1$ and a cluster of reduced size.

and the icosahedral structure of the cluster gives about the same value $R_0 = 8.0$ Å for the most probable configuration of the approach of the He-atom to the face of the n = 55 icosahedron. Similar results were obtained for two other cluster sizes $\bar{n} = 115$, which had to be lowered from 10.3 Å to 8.1 Å which means to size $\bar{n} = 55$ and $\bar{n} = 2100$ for which R_0 decreased from 26.8 Å to 15.6 Å. Thus we conclude that the preliminary determination of cluster sizes based on diffraction oscillations is not in agreement with the commonly accepted scaling based on the source parameters. Further measurements have to be performed to clarify this problem. In the further course of the paper we keep the original notation based on the source data but keep in mind that the cluster size is probably smaller than indicated by the scaling procedure.

Vibrational dynamics

The information on the vibrational state density is obtained from time-of-flight (TOF) spectra taken at different laboratory deflection angles. Measurements were carried out at a collision energy of about 25.0 meV for $\bar{n} = 40$ at $\Theta = 5^0, 10^0$, for $\bar{n} = 55$ at $\Theta = 30^0$, for $\bar{n} = 80$ at $\Theta = 30^0$, for $\bar{n} = 115$ at $\Theta = 30^0, 36^0, 50^0$, and for $\bar{n} = 3000$ at $\Theta = 30^0$. Typical examples for $\bar{n} = 40$ at 10^0 and $\bar{n} = 80$ at 30^0 are shown in Fig. 2. From the measured beam velocities, the positions of elastically scattered monomers and clusters are known. They are used in a Monte-Carlo fitting procedure in which also the experimental resolution of the apparatus based on the measured velocity distributions and angular divergencies are properly taken into account [11]. The remaining intensities are attributed to inelastic transitions and the TOF-spectra are marked by the corresponding energy transfer ΔE in meV starting from the elastic scattering of the cluster. The monomer scattering, marked by n = 1, can be easily distinguished from the cluster scattering because of the different velocities. It is noted that for small cluster sizes, the TOF spectra give directly the vibrational modes which are excited in the collisions with He. The additional constraint for the momentum of the phonon which leads to a different procedure for obtaining bulk and surface phonon spectra of the solid [2] is probably only valid for very large clusters. The results for $\Theta = 10^0$ indicate that the energy transfer is small with a most probable value of 1 meV and a maximum value of 2 meV. The results for $\Theta = 30^0$ are quite different. Now the most probable energy transfer is found at $\Delta E = 4$ meV with an additional peak at $\Delta E = 2$ meV and the largest detectable energy transfer does not exceed $\Delta E = 8$ meV. The results for the other cluster sizes and larger angles differ in details but agree as far as the general cut off of the intensity at $\Delta E = 10$ meV is concerned.

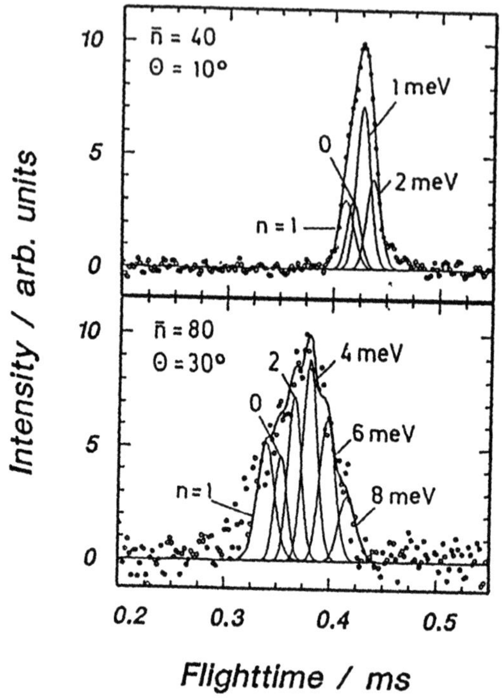

Figure 2: Measured time-of-flight distribution for different average cluster sizes and different deflection angles.

How do these experimental findings compare with what is known about the density of vibrational states from experiments or calculations? The measured bulk phonon density of solid Ar at 10 K shows a large peak at 8 meV which is caused by the longitudinal motion and a second peak around 5 meV from the transverse branch with the characteristic van Hove singularities. For clusters or particles of finite size only calculations are available. The methods include Molecular Dynamics (MD) simulations from which the frequency spectrum is obtained by the Fourier transform of the velocity autocorrelation function [13-16]. Other calculations are based on the collective motion of elastic vibrations of solid [17,18] or density fluctuations of liquid material [19] which manifests itself in the excitation of spheroidal or torsional modes [17,18] and shape or compression oscillations [19], respectively. The results of all calculations treating large clusters n = 55, 135, 419 and 14000 [14,17] or finite layers and nearly spherical blobs [13] can be summerized as follows. In addition to the peaks which already appear for the bulk material at 5 meV and 8 meV two further peaks are recognized at smaller frequencies at about 2-3 meV and at about 1.2 meV. They are attributed to surface modes with

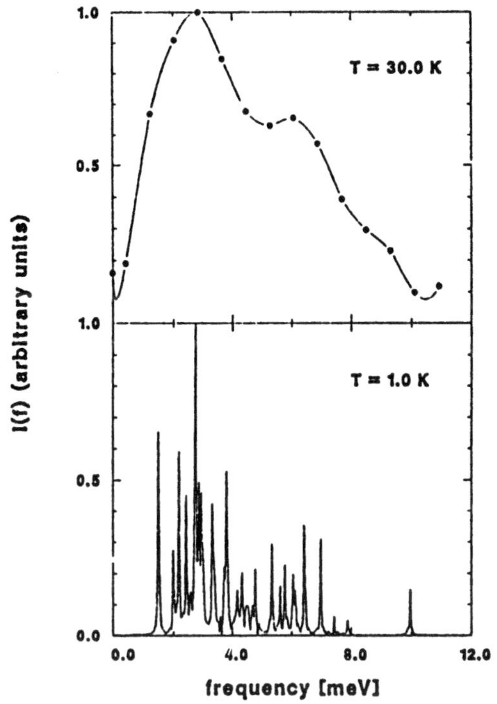

Figure 3: Calculated frequency spectra for Ar_{55} at different temperatures.

amplitudes normal to the surface and edge atoms, respectively. The overall size of the particles is reflected only in the relative amount of the intensities of the different modes to each other. Bulklike modes are less probable than surface modes in particles with a large surface-to-volume ratio and vice versa. We demonstrate this with a calculation of the vibrational modes of Ar_{55} which has icosahedral structure with one central atom, a complete inner and outer shell. Fig. 3 shows the spectrum obtained in MD-simulations from the Fourier transform of the velocity autocorrelation function for T = 1.0 K and 30 K. At the low temperature a completely resolved line spectrum is seen and with the help of a normal mode analysis each line can be identified. The single line at 10 meV is mainly attributed to the motion of the central atom, the group of lines between 5 and 8 meV is caused by the inner shell, while the frequencies around 2-3 meV are resulting from the outer shell. This picture is in good agreement with the simulations of smaller systems [15,16] as well as with the results discussed earlier. The direct comparison with the experimental results gives qualitative agreement with the experimental curve obtained for $\bar{n} = 115$ and T = 33 K. The temperature is in good agreement with the results obtained by electron diffraction [20]. The necessary reduction of the size to n = 55 corresponds nicely with what has been found in the diffraction experiment.

Summary

These first results on the measurement of the vibrational frequency spectrum of free clusters by He atom scattering look very promising and open a range of new experimental possibilities. The great advantage of this technique is the general applicability to any kind of system and any kind of vibrational spectrum. The disadvantage is, of course, the low intensity of the scattered signal and the preparation of the cluster target which has to be prepared as monosized as possible.

References

[1] M.R. Hoare and P. Pal, Adv.Phys. 24, 645 (1975)
[2] J.P. Toennies, Physica Scripta T19, 39 (1987)
[3] U. Buck, F. Huisken, J. Schleusener, and J. Schaefer, J.Chem.Phys. 72, 1512 (1980)
[4] W. Obert, Rarefied Gas Dynamics, ed. R. Campargue, CEA Paris, 11, 1181 (1979)
[5] O.F. Hagena and W. Obert, J.Chem.Phys. 56, 1793 (1972)
[6] R. Müller. Diplomarbeit, Universität Hamburg, 1990; see also J. Stapelfeldt, J. Wörmer, and T. Möller, Phys.Rev.Letters 62, 98 (1989)
[7] J. Farges, M.F. de Feraudy, B. Raoult and G. Torchet, J.Chem.Phys. 84, 3491 (1986)
[8] O.F. Hagena, Z.Phys.D 4, 291 (1987)
[9] U. Buck, in Atomic and Molecular Beam Methods, ed. G. Scoles, Oxford, New York, 1988, Chap. 20, p. 499
[10] J. Gspann and H. Vollmer, in Rarefied Gas Dynamics, ed. R. Campargue, CEA, Paris, 11, 1193 (1979)
[11] U. Buck, in Atomic and Molecular Beam Methods, ed. G. Scoles, Oxford, New York, 1988, Chap. 22, p. 525
[12] Z. Fujii, N.A. Lurie, R. Pynn, and G. Shirane, Phys.Rev.B 10, 3647 (1974)
[13] J.M. Dickey and A. Paskin, Phys.Rev.B 1, 851 (1970)
[14] W.D. Kristensen, E.J. Jensen and R.M.J. Cotterill, J.Chem.Phys. 60. 4161 (1974)
[15] T.L. Beck and T.L. Marchioro II. J.Chem.Phys. 93, 1347 (1990)
[16] J.A. Adams and R.M. Stratt, J.Chem.Phys. 93, 1358 (1990)
[17] A. Tamura and T. Ichinokawa, J.Phys.C 16, 4779 (1983)
[18] Y. Ozaki, M. Ichihashi and T. Kondow, Chem.Phys.Lett. 182, 57 (1991)
[19] A. Tamura and Ichinokawa, Surf.Science 136, 437 (1984)
[20] J. Farges, M.F. de Feraudy, B. Raoult and G. Torchet, Surf.Science 106, 95 (1981)

Collision Experiments with C_{60}^+

E.E.B. Campbell, A. Hielscher, R. Ehlich, V. Schyja and I.V. Hertel

Fakultät für Physik and Freiburger Materialforschungszentrum (FMF),
Albert-Ludwigs-Universität , Hermann-Herder-Str. 3, W-7800 Freiburg, Germany.

Introduction

Research on C_{60} and the fullerenes has been increasing at a phenomenal rate since the discovery of W. Krätschmer of a simple method to produce macroscopic amounts of the material [1]. In particular the Krätschmer discovery made possible the final confirmation of the much hypothesized and fiercely debated "soccer ball" structure of C_{60}. Due to its very high symmetry and extraordinary properties this cluster/molecule is proving to be a fascinating object of study in many branches of pure and applied research. Collision experiments with the fullerenes are only just beginning but already very interesting and unexpected results have been obtained. Here we describe two aspects of collisions between C_{60}^+ and rare gases carried out in a tandem time-of-flight mass spectrometer: Mass spectra produced on collisional fragmentation of C_{60}^+ show a marked bimodal distribution which, together with the collisional energy dependence of the fragmentation process, provides very strong evidence for the occurence of cluster fission rather than statistical evaporation of particles. In the same mass spectra additional peaks can be seen under certain conditions, first discovered by Schwarz and co-workers [2], which are due to the formation of endohedral compounds between the target atom and C_{60}^+. The collision energy dependence of this fascinating process in which a He (or Ne) atom can penetrate a C_6 ring and be trapped inside the carbon cage will also be discussed.

Preliminary results from the first mass-selected cluster-cluster collision experiments ($C_{60}^+ + C_{60}$) are reported for a collision energy of 850 eV in the laboratory frame of reference. In addition to fragmentation of C_{60}^+, cluster fusion is observed leading to products in the mass range C_{90}^+ to C_{120}^+.

Experimental setup

A schematic diagram of the apparatus is shown in Fig. 1. The C_{60}^+ beam is produced by low fluence UV laser desorption of commercially available [3] C_{60}/C_{70} deposited on an aluminium plate. Positively charged C_{60}^+ and C_{70}^+ ions are produced directly by the laser with very little fragmentation occuring [4] and are accelerated away from the surface by a constant electric field. C_{60}^+ is mass selected, by using a pulsed electric field to remove C_{70}^+ and any fragment ions from the flight path, and enters a collision chamber containing the target gas. After the collision the C_{60}^+ and product ions are mass selected in a reflectron time-of-flight mass spectrometer and detected by two channel plates operated in the Chevron configuration. The signal is read into a transient recorder and signal averager with transfer to a XT 286 PC via a CAMAC interface. The laboratory collision energy of the C_{60}^+ investigated ranged from 200 eV to 6.3 keV with an energy spread of 15±5 eV. The collision energy is scanned during an experimental run by changing the potential on the collision cell. A retarding field energy analyser placed immediately before the channel plates can be used to determine the energy of parent and product ions. The pressure in the collision cell was kept within the single-collision regime as checked by measuring the pressure dependence of the product signal. For more details of the experimental method see [5,6].

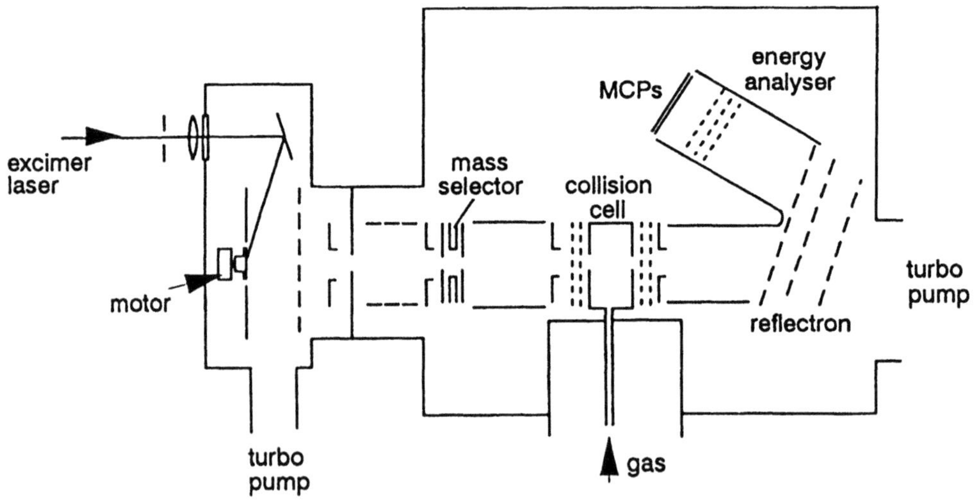

Fig. 1. Schematic diagram of the SOCCER apparatus (scattering of carbon clusters energy resolved)

Collision induced fragmentation

A typical mass spectrum produced on collisions between C_{60}^{+} and Ar in the keV laboratory energy range is shown in Fig. 2. Due to the large mass of the projectile compared to the target, a laboratory energy of 1.2 keV corresponds to a centre of mass energy of only 63 eV. Interesting is the bimodal fragment distribution with a large fragment range peaking at C_{50}^{+} and a small fragment range peaking at C_{15}^{+}. The normal fragmentation path followed by C_{60}^{+} and the other fullerenes is the loss of units of C_2 as seen in metastable decay of hot clusters [7,8] and in photofragmentation of mass-selected clusters [9]. The binding energy of a C_2 "monomer" in the fullerenes is approximately 4 eV [10,11]. If one considers the fragmentation to occur via statistical evaporation of C_2, one can quickly estimate, using standard RRK theory, that due to the very large number of degrees of freedom in C_{60}^{+} and the large binding energy, at least 50 eV internal energy is needed to

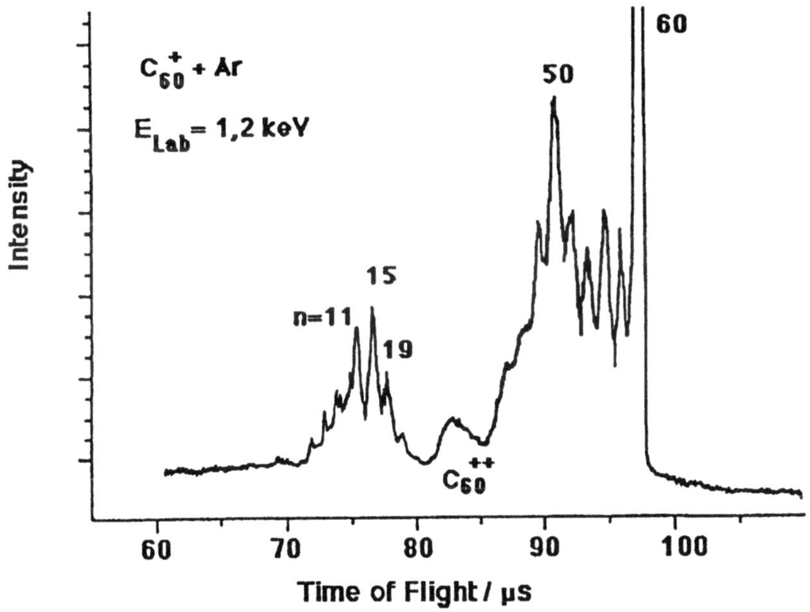

Fig. 2. Mass spectrum observed on fragmentation of C_{60}^{+} in collision with Ar at a laboratory collision energy of 1.2 keV corresponding to 63 eV in the centre of mass reference frame.

observe significant loss of C_2. In order to observe the fragmentation pattern in Fig. 2 much more than 50 eV is required for statistical fragmentation with only a maximum 63 eV available in the collision. The observed distribution can not be due to statistical evaporation but must be due to collision induced fission of the C_{60}^+. This hypothesis is supported by considering the small fragments formed in the collision. Positively charged carbon clusters with n = 11, 15, 19, 23 appear as "magic numbers" in carbon cluster spectra produced by laser vaporisation of graphite [12]. One reason for this is that the corresponding neutral clusters have exceptionally low ionisation potentials, in the range 7.2 - 7.45 eV [13], which are certainly comparable to if not less than the ionisation potentials of the fullerenes in the range n = 30 - 50 [14]. In the collision experiment we thus detect the fission product with the smallest ionisation potential.

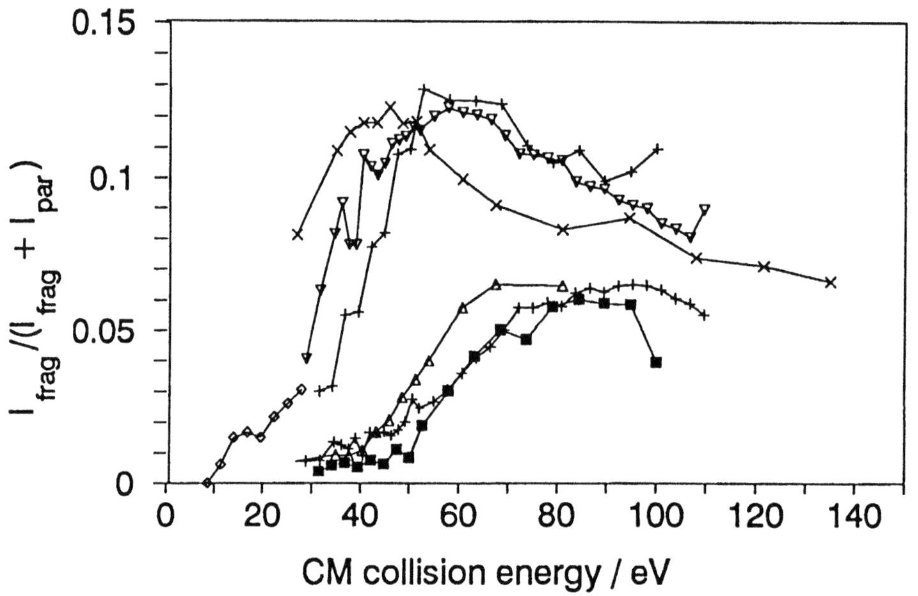

Fig.3. Ratio of fragment ion intensity to total ion intensity as a function of centre of mass collision energy for different target gases. Upper points give integrated intensity of large fragments in the range C_{32}^+ to C_{58}^+: ◊ He, x Ne, + Ar , ∇ CO_2. Lower points give integrated intensity of small fragments in the range C^+ to C_{28}^+: Δ Ne , • Ar ,+ CO_2.

Collision energy scans were carried out over a large range of energies with different target gases. In Fig. 3 the integrated intensities of the two fragment mass ranges are plotted against centre of mass energy. Within the statistical errors in the measurements the results from the different target gases agree more or less with one another. What determines the efficiency of the fission process is obviously the energy available rather than the details of the interaction potential. The threshold for production of the large fragments occurs at approximately 10 eV which is also a reasonable estimate for the energetic barrier for C_2 loss from C_{60}^+ obtained from simple bond making and breaking considerations. One can make similar estimates for other fission mechanisms. For example the energetic threshold for the process

$$C_{60}^+ + \text{Target} \rightarrow C_{50}^+ + C_{10} + \text{Target}$$

turns out to be about 15 eV if two joined hexagons are removed from C_{60}^+: One needs to break 8 single bonds in C_{60}^+ leaving 46 single bonds and 25 double bonds; the stable form of C_{50}^+ has 50 single bonds and 25 double bonds so that 4 single bonds then have to be formed leaving an energy balance of (8-4) = 4 single bonds (4 x 3.6 eV = 14.4 eV). Similarly, it is energetically possible to obtain the small fragments at centre of mass collision energies beyond about 23 eV for C_7^+ and C_{11}^+ rising to 51 eV for C_{23}^+. Although these considerations are somewhat naive and contain only the standard carbon-carbon single and double bond energies they are in surprisingly good agreement with the experimental observations. A more detailed analysis of the experimental data and possible fragmentation mechanisms will be published elsewhere [15].

He and Ne Capture

One of the fascinating properties of C_{60} is that it is hollow i.e. the delocalised electrons are concentrated on the carbon shell leaving an "empty space" inside. It has been shown to be possible to trap metal atoms inside the fullerene cage during the laser vaporisation formation process [16] which has led to numerous speculations about possible applications. Recently collision experiments have shown that under certain conditions small atoms (He and Ne) can be pushed through the carbon cage to be trapped inside [17-19]. Such endohedral compunds are distinguished from those where atoms or molecules are attached to the outside of the fullerene by using the formalism $He@C_{60}^+$ [16]. By mass selecting the $He@C_{60}^+$ formed in the collision and subjecting it to a further collision with Xe, after which the product ions retained the He but suffered the loss of C_{2x} - analogous to

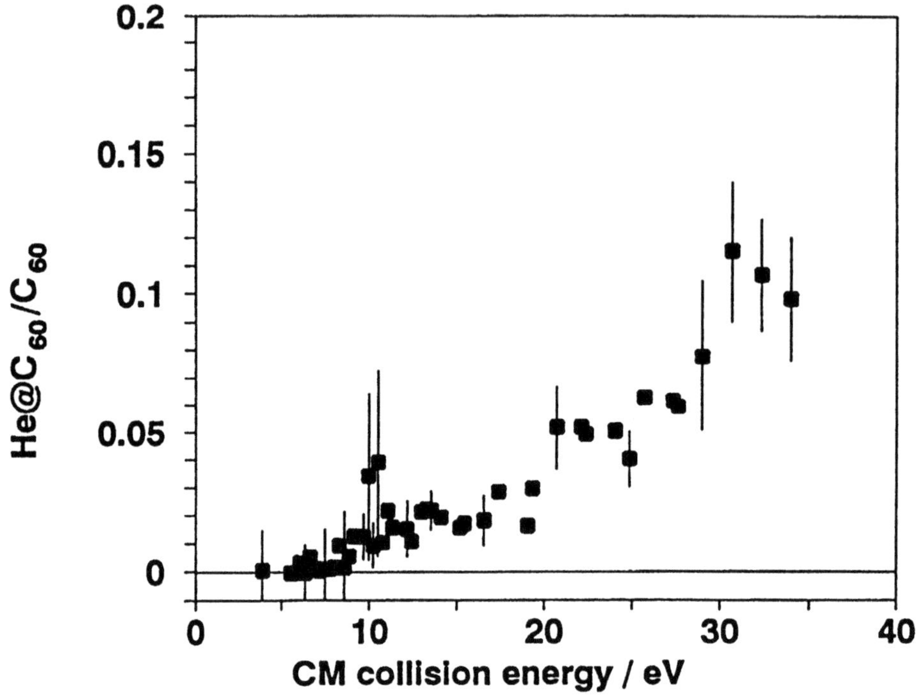

Fig.4. Centre of mass collision energy dependence of He capture by C_{60}^+.

the dissociation of C_{60}^+- Ross and Callahan were able to show that the He was indeed caught inside the carbon cage [17].

We have measured the collision energy dependence of the formation of He@C_{60}^+ and Ne@C_{60}^+ in the region of the energetic threshold [19]. The results for collisions with He are shown in Fig. 4 where the ratio of He@C_{60}^+ to C_{60}^+ is plotted as a function of centre of mass collision energy. Due to the large difference in the masses of the collision partners very low centre of mass collision energies can be easily accessed. The measured threshold lies at 6±2 eV and is consistent with ab initio molecular orbital calculations which predict a barrier of less than 10 eV for a helium atom to pass through a $C_6H_6^+$ ring [18]. The maximum capture signal occuring at about 32 eV is in very good agreement with estimates obtained from the experiments using four-sector mass spectrometers [17,18] where the maximum was seen to lie between 5 and 6 keV corresponding to a centre of mass energy range of 27.6 to 33.1 eV. The cross section for Ne capture is approximately a factor of ten smaller than that for He capture with the threshold lying at 9±1 eV. Again this is consistent with preliminary molecular orbital calculations which predict that the Ne threshold should lie more than 2 eV above that of He [18]. The maximum capture probability for Ne lies at about 50 eV.

Cluster Cluster Collisions

Simulations of collisions between small neutral sodium clusters have shown that fusion should be an important reaction channel for low impact parameter collisions and at collision energies where the energy per atom is comparable to the binding energy [20]. An experimental test of these predictions involving two beams of mass selected sodium clusters is, however, at the present time not feasible. On the other hand, the availability of macroscopic amounts of C_{60} - a neutral, mass-selected cluster with a well defined structure - makes this an ideal test case for probing the relevance of nuclear physics concepts in cluster physics.

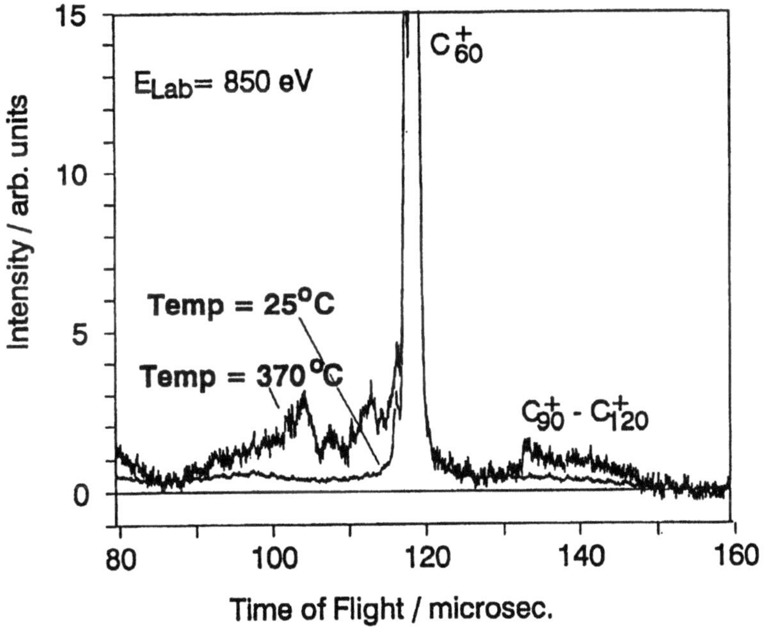

Fig.5. Mass spectrum of products of $C_{60}^+ + C_{60}$ collisions

The experimental setup used is identical to that shown in Fig. 1 with the collision cell replaced by a C_{60} oven. The C_{60}^+ ion beam was accelerated to 850 eV and the temperature of the C_{60} oven gradually increased. The resulting mass spectra, normalised to the same peak height, observed for two different oven temperatures are shown in Fig. 5. At a temperature of 25°C the pressure of C_{60} gas in the oven is negligible and the ion beam passes undisturbed through the apparatus. At 370°C the pressure of C_{60} in the oven is approximately 10^{-4} mbar and evidence of substantial fragmentation is seen (ions arriving at the detector between 90 and 115 µs). More interestingly, ions are observed with a flight time of 130 - 150 µs. Simulations of ion trajectories through the reflectron show that these species correspond to masses in the range C_{90}^+ to C_{120}^+ and where up to 400 eV kinetic energy has been converted to internal energy during the collision. This spectrum thus shows the first direct experimental evidence for the occurrence of fusion in cluster collisions.

Conclusion

In the fullerenes, cluster scientists have been presented with the means of producing intense, mass-selected beams of charged or neutral clusters with well defined structures. This opens up the possibilty of a range of detailed experimental investigations not normally feasible with standard cluster sources. The work is only just beginning but already many interesting phenomena have been discovered. One of the most interesting questions which can be tackled in the next few years with the help of the fullerenes is how far the concepts of nuclear collision physics can be applied to clusters. Investigations of mass-selected cluster-cluster collisions are now underway in our laboratory.

Financial support by the Deutsche Forschungsgemeinschaft through Sonderforschungsbereich 276 "Korrelierte Dynamik hochangeregter atomarer und molekulare Systeme" is gratefully acknowledged.

References

[1] W. Krätschmer, L.D. Lamb, K. Fostiropoulos, D.R. Huffman, Nature **347**,354 (1990)
[2] T. Weiske, D. Böhme, J. Hrusak, W. Krätschmer, H. Schwarz, Angew. Chem. Int. Ed. Engl. **30,** 884 (1991)
[3] Texas Fullerenes Corporation, 2415 Shakespeare, Suite 5, Houston, Texas
[4] G. Ulmer, E.E.B. Campbell, R. Kühnle, H.-G. Busmann, I.V. Hertel, Chem. Phys. Lett., **182**, 114 (1991)
[5] E.E.B. Campbell, A. Tittes, D. Krantz, R. Schneider, A. Hielscher, Chem. Phys. Lett., **184**, 404 (1991)
[6] E.E.B. Campbell, R.R. Schneider, A. Hielscher, A. Tittes, R. Ehlich, I.V. Hertel, Z. Phys. D., in press
[7] P.P.Radi, M.T. Hsu, J. Brodbelt-Lustig, M. Rincon, M.T. Bowers, J. Chem. Phys. **92**, 4817 (1990)
[8] E.E.B. Campbell, G. Ulmer, H.-G. Busmann, I.V. Hertel, Chem. Phys. Lett., **175,** 505 (1990)
[9] S.C. O'Brien, J.R. Heath, R.F. Curl, R.E. Smalley, J. Chem. Phys. **88**, 220 (1988)
[10] P.P. Radi, M.T. Hsu, M.E. Rincon, P.R. Kemper, M.T. Bowers, Chem. Phys. Lett., **174**, 223 (1990)
[11] C. Lifschitz, M. Iraqi, T. Peres, J.E. Fischer, Int. J. Mass Spec. Ion Phys. **107**, 565 (1991)
[12] e.g. S.W. McElvany, B.I. Dunlap, A.O'Keefe, J. Chem. Phys. **86**, 715 (1986)
[13] S.B.H. Bach, J.R. Eyler, J. Chem. Phys. **92**, 358 (1990)
[14] J.A. Zimmerman, J.R. Eyler, S.B.H. Bach, S.W. McElvany, J. Chem. Phys. **94**, 3556 (1991)
[15] E.E.B. Campbell, A. Hielscher , R. Ehlich, to be submitted to Chem. Phys. Lett.
[16] Y. Chai, T. Guo, C. Jin, R.E. Haufler, L.P.F. Chibante, J. Fure, L. Wang, J.M. Alford, R.E. Smalley, J. Phys. Chem., submitted
[17] M.M. Ross, J.H. Callahan, J. Phys. Chem., **95**, 5720 (1991)
[18] T. Weiske, J. Hrusak, D.K. Böhme, H. Schwarz, Helv. Chem. Acta, in press
[19] E.E.B. Campbell, R. Ehlich, A. Hielscher, J.M.A. Frazao, I.V. Hertel, Z. Phys. D., submitted
[20] R. Schmidt, G. Seifert, H.O. Lutz, Phys. Lett. A **158**, 231 (1991); G. Seifert, R. Schmidt, H.O. Lutz, Phys. Lett. A **158**, 237 (1991) and contributions to this volume.

Ion trap studies of ternary and radiative association processes

D. Gerlich
Fakultät für Physik, Universität Freiburg, D 7800 Freiburg, Germany

INTRODUCTION

A large fraction of the experimental and theoretical work on clusters focuses on their stability and structure and on their fragmentation and decay behavior after collisional– or photo–excitation (evaporation, fission, metastable decay, multi–fragmentation). In contrast, detailed studies of the dynamics of cluster collisions (inelastic processes, chemical reactions, cluster growth) are rather scarce. Experimentally, this is due mostly to the difficulty in preparing intense mass selected cluster beams, while theoretically, the treatment of this many–body but finite size problem requires the development of adequate techniques, combining stochastic models with molecular dynamic simulations.

Although the major aim of this contribution is to present experimental results related to the dynamics of cluster growth, we begin by making a few introductory remarks to statistical models. These models have been successfully applied to low energy bimolecular ion–molecule collisions and allow us to derive semi–empirical formulas which express association processes (i.e. cluster growth) in terms of complex formation, lifetimes, stabilization efficiencies, etc. Our experimental studies have been performed in an ion trapping apparatus which is described in the second part of this contribution. In the third part, we present and discuss results, measured for the stabilization of the $H^+ \cdot H_2$ complex and for the growth of larger hydrogen clusters from H_3^+ seed ions.

COMPOUND REACTIONS

For the treatment of collisions which proceed via compound states or strongly coupled intermediates, many statistical theories have been developed in nuclear physics, gas–phase chemistry, and other related fields. For example, Fermi [fer50] proposed a theory to calculate cross sections for pion production in which the energy of two colliding nuclei is distributed according to statistical laws. Such statistical ideas have been adapted to chemical reactions by Keck [kec58] and later by Light and coworkers [pec66] who made use of the strong coupling situation and developed the "phase space theory" for inelastic and reactive collisions. Another example is Feshbach's formal theory of resonant collisions which has found applications in nuclear theory [fes60] and has also been used to derive a generalized statistical model for molecular collisions [mil70]. More recent developments of statistical theories are reviewed in [qua81,for90].

All statistical theories for collision processes are very similar in their basic assumption: that is each individual collision process can be divided into two well-separated steps, first the complex formation and then the complex decay. Typical justifications are many, overlapping compound state resonances, strong interactions in the complex, very long complex lifetimes, chaotic trajectories, etc. Under such conditions the complex decay depends only on the conserved quantum numbers ($|c\rangle$: total energy, total angular momentum, parity, total nuclear spin, etc.), leading to a significant simplification of the theoretical treatment. Instead of performing detailed dynamical calculations for state-to-state processes $|i\rangle \rightarrow |f\rangle$, it is sufficient to determine the complex formation probability P_{ic} for *all* initial states $|i\rangle$ which can lead to $|c\rangle$. Making use of microscopic reversibility these P_{ic}'s also determine the complex decay probabilities p_{cf} into those final states $|f\rangle$ which are accessible from $|c\rangle$ ($P_{cf}=P_{fc} \rightarrow p_{cf}=P_{cf}/\Sigma P_{cf'}$, for details see [mil70]). An illustrative example of a detailed application of a dynamically biased statistical theory has been given for the prototype system $H^+ + H_2$, where the P_{ic}'s have been determined from trajectory calculations [ger80,scl83]. Results include integral, differential, state-to-state, and isotopic scrambling cross sections, complex lifetimes, and ortho-para transition probabilities [ger90].

One of the simplest dynamical criteria for the determination of complex formation probabilities is based on capture behind the centrifugal barrier of the long range interaction potential. This leads, in the case of the charge-induced dipole interaction, to the well-known polarization (or Langevin) cross section [lan05,gio58]

$$\sigma_P = 16.86\,(\alpha/E_T)^{\frac{1}{2}}\,, \tag{1}$$

where E_T is the collision energy in eV, α is the polarizability in Å^3, and σ_P results in Å^2. Multiplication of this cross section with the relative velocity, $g=(2E_T/\mu)^{\frac{1}{2}}$, leads to the rate coefficient

$$k_P = 23.42\,(\alpha/\mu)^{\frac{1}{2}}, \tag{2}$$

where μ is the reduced mass in amu and the units of k_P are 10^{-10} cm^3/s (example: $H^+ + H_2$, α=0.79Å^3, k_P=2.5x10^{-9}cm^3/s).

In order to extend these and related ideas to the description of association processes (e.g. cluster growth) one has to add additional assumptions concerning the stabilization of the intermediate complex. The overall process can be represented schematically by

complex formation	$A^+ + B \xrightarrow{k_c} (AB^+)^*$		(3)
complex decay	$(AB^+)^* \xrightarrow{\tau_c} A^+ + B$		(4)
photo emission	$(AB^+)^* \xrightarrow{\tau_r} AB^+ + h\nu$		(5)
ternary stabilization	$(AB^+)^* + B \xrightarrow{\tau_s} AB^+ + B$,		(6)

where the collision complex $(AB^+)^*$ is formed with a collisional rate coefficient k_c.

During its lifetime τ_c, this excited complex may radiate with a time constant τ_r, or it may be stabilized by a collision with a third body B with a rate

$$1/\tau_s = f'k_c'[B] \, . \tag{7}$$

Here $k_c'[B]$ is the rate for $(AB^+)^*$+B collisions, [B] the density of neutrals, and f' the stabilization efficiency factor (≤ 1). The inclusion of other possible reactive channels or an additional buffer gas is straight-forward, but not required for the results of this contribution. With these parameters, formation of stable products AB^+ can be described by an apparent second order rate coefficient, k^*, defined by

$$k^* = k_c(f'k_c'[B]+1/\tau_r)/(1/\tau_c+f'k_c'[B]+1/\tau_r) \tag{8}$$

At low densities [B], and for systems with short complex lifetimes ($\tau_c << \tau_r$), the overall stabilization rate is very small and the term $f'k_c'[B]+1/\tau_r$ in the denominator can be neglected in comparison to $1/\tau_c$. This leads to

$$k^* = k_c\tau_c f'k_c'[B]+k_c\tau_c/\tau_r \, , \tag{9}$$

and to a simple representation of the bimolecular rate coefficient for radiative association, k_r (unit cm^3/s), and of the three-body association rate coefficient, k_3 (unit cm^6/s),

$$k_3 = k_c\tau_c f'k_c' \tag{10}$$

$$k_r = k_c\tau_c/\tau_r \, . \tag{11}$$

In an actual experiment, k^* is determined as a function of the density [B] and, as illustrated in Fig.2, the results are evaluated as a linear fit to Eq. 9.

It is important to note that a quantitative treatment of the described phenomenological model requires a microcanonical description of both complex formation and stabilization by the third collision partner. In general, one has to calculate the parameters introduced in Eqs. 3-6 with extensions of the statistical models mentioned above. In most applications however, one uses simply the Langevin rate coefficient for calculating complex formation (k_c) and complex stabilisation (k_c').

EXPERIMENTAL

In our laboratory, the aggregation of hydrogen molecules onto different mass selected seed ions has been studied at different temperatures using a cooled radio-frequency (RF) ion trap. Since the fundamental principle of confining charged particles in an inhomogeneous RF field (adiabatic approximation, effective potential, stability considerations) and a variety of RF-devices (storage ion sources, ion guides, ion traps, etc.) have recently been described in detail [ger91], we give here only a very short description of our apparatus. In light of the aim of this conference it is noteworthy that our experimental method basically originated from accelerator physics (strong-focusing, alternating gradient principle) and was also inspired by different technical proposals to contain a nuclear fusion reaction.

In our apparatus, which is shown in Fig. 1, the ion trap consists of a set of ring electrodes which are mounted onto the cold head of a closed cycle helium refrigeration system. The target gas and ion temperature can be varied between 10 K and 350 K. The primary ions are formed by electron bombardment in a storage ion source, selected according to mass and energy in a quadrupole mass filter, and then injected into the ion trap via the pulsed entrance electrode. During the following storage time which can be varied between microseconds and minutes, the primary ions interact with the target and/or buffer gas. The ion cloud is then extracted with the exit electrode and analyzed with a second quadrupole mass spectrometer.

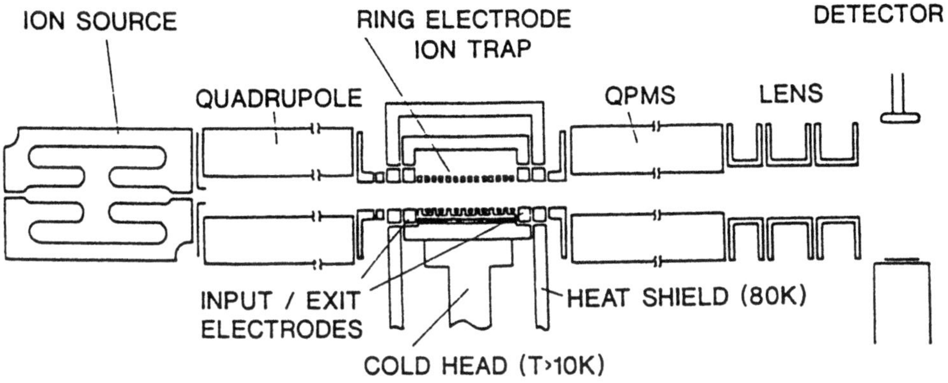

Fig. 1: Schematic diagram of the ring electrode trap apparatus. For a detailed description of the apparatus and the trapping method see [ger91].

For the energy distribution of the stored ions it is important to note that the trapping volume is distinguished by a wide, flat potential with very steep repulsive walls (the effective potential increases exponentially in comparison to the slow r^2 – dependence of a two–dimensional RF–quadrupole). As a result, nearly all collisions with the cold buffer gas occur in the field–free region, leading to an adaptation of the ion energy distribution to the thermal distribution of the neutral target gas. Only the few collisions which occur in regions of higher field strength may cause some heating resulting in an ion "temperature" which is slightly higher than that of the ambient buffer gas.

RESULTS AND DISCUSSION

One of the simplest association processes involving hydrogen molecules and hydrogen ions is the association of H^+ with H_2. It is well–known from a series of experimental and theoretical studies ([ger80],[scl83],[ger90] and references therein) that this reaction proceeds via formation of a long–lived intermediate complex. At the low energies of our ion trap, classical trajectory calculations predict a lifetime of several 10^{-10} s [scl85]. As a typical result, Fig. 2 shows the rate coefficient k^* for the formation of stable D_3^+ ions as a function of the D_2 density measured in the ring electrode trap at

350 K and 80 K. In this example the increasing number of D_3^+ product ions has been measured as a function of the storage time at several target gas densities between 10^{12} cm^{-3} and 10^{13} cm^{-3}. The slope of the linear fit (Eq.9) yields a three body rate coefficient $k_3(80\ K)=7.6\times10^{-29}$ cm^6/s and the intersection at $[D_2]=0$ can be taken as an estimate for the radiative stabilization rate coefficient $k_r(80\ K)=1\times10^{-16}$ cm^3/s. Similar results have been obtained for $H^+ + H_2$ ($k_3(80\ K)=5.4\times10^{-29}$ cm^6/s and $k_r(80\ K)=1.3\times10^{-16}$ cm^3/s).

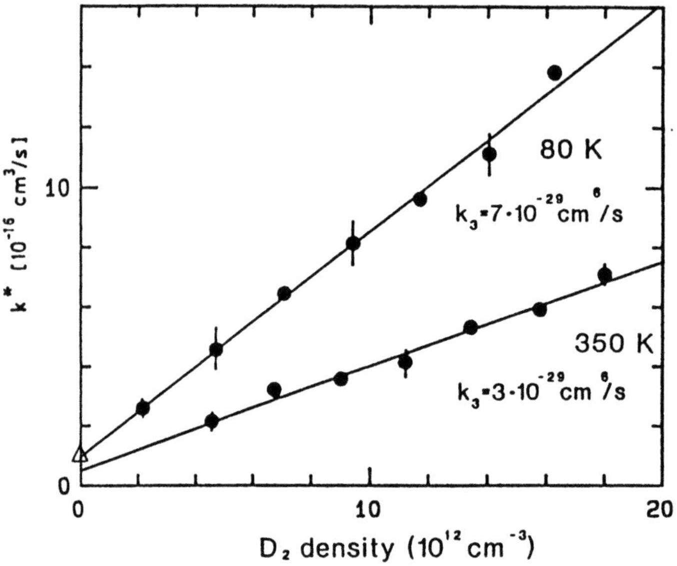

Fig. 2: Apparent binary rate coefficient for the ternary association reaction $D^+ + 2D_2 \rightarrow D_3^+ + D_2$, plotted as a function of the D_2 density. Increasing the nominal temperature of the ring electrode trap from 80 K to 350 K leads to a decrease of the ternary rate coefficient from 7×10^{-29} cm^6/s to 3×10^{-29} cm^6/s.

Based on Eqs. 10 and 11 and using Eq. 2 for k_c, one can obtain from the measured k_3 and k_r values an estimate of the complex and the radiative lifetimes. At the present stage of development of the theory, the largest uncertainty in this estimation is due to the stabilization factor f', i.e., the efficiency of deactivating the intermediate complex via collision with a third body. Assuming $f'=0.1$ for $H^+ \cdot H_2$, one obtains a complex lifetime of about 1×10^{-10} s and a radiative lifetime slightly less than 1 ms. These values are only in order-of-magnitude agreement with those determined from classical trajectory calculations (complex lifetime: 6.8×10^{-10} s [scl85], radiative lifetime: 0.14×10^{-3}s [ber88]), indicating that more detailed experiments and theoretical studies are required. Substitution of protons by deuterons augments the ternary rate coefficient by a factor of 1.4 (H_3^+ formation) and 1.5 (H_5^+). This is in accord with the statistical model since the density of complex states increases proportional to the square–root of the mass ratio. The measured isotope effect on the radiative lifetimes, $\tau(D_3^+)/\tau(H_3^+)=2.7$, also fits into this picture since the spontaneous photon emission rate is proportional to the third power of the emission frequency.

Going from H^+ to H_3^+ seed ions leads to an increase of the degrees of freedom of the complex and, from simple statistical arguments, one expects an increase of the lifetime. However, one has also to consider, that the H_2 binding energy drops from 4.35 eV for H^+ to about 0.3 eV for H_3^+ [hir87] and this leads to a reduction of the complex lifetime. A comparison of the rate coefficients measured at 80 K for H_3^+ formation (k_3=5.4×10^{-29} cm^6/s) and for H_5^+–formation (k_3=2.5×10^{-29} cm^6/s) indicates that the reduction of the binding energy more than counterweights the influence of the degrees of freedom. A more complete survey of the influence of increasing complexity and decreasing binding energies can be seen from Fig. 3. These data show the time dependence of the growth of larger hydrogen cluster ions from H_3^+ ions at a nominal temperature of 25 K and a hydrogen density of 3.7×10^{13} cm^{-3}.

The data in Fig. 3 have been fitted based on an adequate rate equation system. The resulting ternary rate coefficients which are printed in the figure show a significant trend. There is a further decrease of k_3 for the formation of H_7^+ which, again, is probably due to the smaller binding energy (0.14 eV). Going from H_7^+ to H_9^+, the binding energy remains very similar. Therefore, this is the first cluster in that series where one must anticipate an increase of k_3 due to the larger number of degrees of freedom. Indeed, the measurements show that $k_3(H_9^+)$ is 8 times larger than $k_3(H_7^+)$. Other qualitative arguments (e.g. "closed shell" in H_9^+) may also explain some of the observed features.

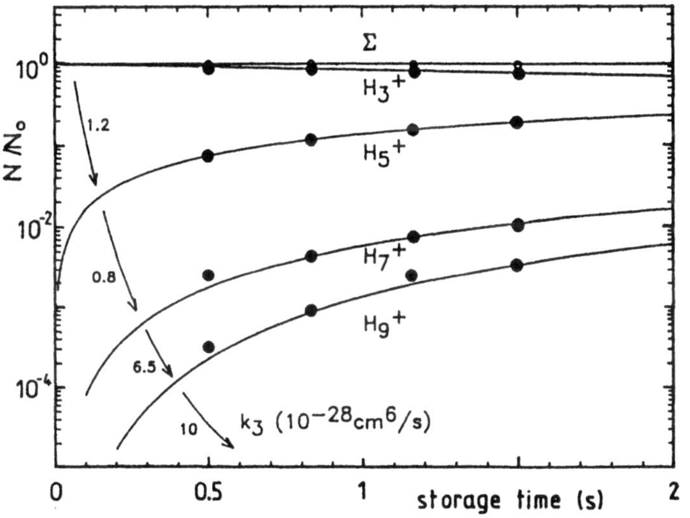

Fig. 3: Consecutive formation of hydrogen clusters injecting H_3^+ into H_2 (density: 3.7x10^{13} cm^{-13}, nominal temperature: 25 K). The lines are solutions from a rate equation system, the numbers on the arrows are the resulting ternary rate coefficients in 10^{-28} cm^6/s. It is important to note that the cluster ion concentrations are far away from the 25 K thermal equilibrium.

For a more quantitative analysis of aggregation of cold hydrogen molecules on hydrogen ions more experimental data are needed (e.g. temperature dependence of k_3, influence of ortho– and para–hydrogen) as well as a detailed statistical theory which accounts correctly for the density of states of the collision complex at the low temperatures of our experiment. For future experiments it is also planned to inject larger clusters into the ion trap in order to follow their further growth or spontaneous and collision induced decay as a function of temperature.

CONCLUSIONS

In the present work, we have presented results from an experimental method which can provide information on the dynamics of collisions between cluster ions and molecules at very low temperatures. The overall sensitivity of the RF ion trapping technique allows us to operate at such low target densities that the growth of clusters can be studied on a time scale of ms–s. The results provide information about complex lifetimes and radiative lifetimes of the highly excited intermediate "molecules". Since both positive and negative ions can be confined simultaneously in the RF field, the method can also be extended to measure cluster–cluster collisions.

Concerning the theoretical treatment, there is an interesting and open question which arises with increasing complexity of the collision systems: when is one allowed to simplify a statistical calculation by introducing a dissipative force, i.e. when can one distinguish between active modes and heat bath modes? The answer depends not only on the coupling between the different degrees of freedom, but also on the time scale of interest, for example, the time window defined by the experiment.

ACKNOWLEDGEMENTS

This work is the result of the common effort of G. Kaefer, W. Paul, and S. Horning. The author also wish to thank Prof. Ch. Schlier for many contributions and profitable discussions. Financial support of the Deutsche Forschungsgemeinschaft is gratefully acknowledged.

REFERENCES

ber88 M. Berblinger and Ch. Schlier, Mol. Phys. **63**, 779, 1988

fer50 E. Fermi, Progr. Theoret. Phys. **5**, 570, 1959

fes60 H. Feshbach, *Nuclear Spectroscopy*, F. Ajzenberg–Selove, Ed. Academic Press Inc., New York, 1960

for90 W. Forst, J. Chim. Phys. **87**, 715, 1990

ger80 D. Gerlich, U. Nowotny, Ch. Schlier and E. Teloy, Chem. Phys. **47**, 245, 1980

ger89 D. Gerlich and G. Kaefer, Ap. J. **347**, 849, 1989

ger90 D. Gerlich, J. Chem. Phys. **92**, 2377, 1990

ger91 D. Gerlich in: *State–Selected and State–to–State Ion–Molecule Reaction Dynamics*, edited by C.Y.Ng, Adv. Chem. Phys., 1991

gio58 G. Gioumousis, D.P. Stevenson, J. Chem. Phys. **29**, 294, 1958

hir87 K. Hiraoka, J. Chem. Phys. **87**, 4048, 1987

kec58 J. C. Keck, J. Chem. Phys. **29**, 410, 1958

lig64 J. C. Light, J. Chem. Phys. **40**, 3221, 1964

lan05 P. Langevin, Ann. Chim. et Phys. **5**, 245, 1905

mil70 W. H. Miller, J. Chem. Phys. **52**, 543, 1970

pec66 P. Pechukas, J.C. Light, and C. Rankin, J. Chem. Phys. **44**, 794, 1966

qua81 M. Quack and J. Troe, in *Theoretical Chemistry*, ed. by D. Henderson, Vol. 6B, p.199, Academic, New York, 1981

scl83 Ch. Schlier, in *Energy Storage and Redistribution in Molecules*, ed. by J. Hinze, Plenum, New York, 1983

CHEMICAL REACTIONS OF TRAPPED METAL CLUSTERS

Manfred P. Irion

Institut für Physikalische Chemie; Technische Hochschule Darmstadt; Petersenstraße 20; W-6100 Darmstadt/GERMANY

1. Introduction

Naked metal gas-phase clusters promise to yield interesting new information about catalysis and to enable the development of a new kind of catalyst [1, 2]. In order to study the physical and chemical behaviour of clusters, it is helpful to store them for a certain time after their production. Because of this and the greater ease in manipulation, charged clusters are preferred to neutral ones in the following experiments. Reactant ions can be carefully defined and are uniquely related to product ions. Any unintended fragmentation is clearly revealed.

During the time of their storage, the cluster ions may be selected in size, equilibrated to well-defined experimental conditions (temperature) and finally optically excited or chemically reacted with an admitted gas. Thus, absolute rate constants for their thermal reactions become available. In the special case of Fe_4^+, a catalytic cycle can be observed converting ethene into benzene. This conjecture is definitely proven by the reoccurrence of the catalyst in a complex $(MS)^5$ tandem mass spectrometric procedure.

2. Experimental

Details of the experimental arrangement reproduced in Figure 1 are published elsewhere [3, 4]. Briefly, bare metal cluster cations are continuously generated in an external chamber SC by bombarding the corresponding metal foil (T) with a primary beam of 20 keV Xe^+ ions [5, 6] from a duoplasmatron (X). They are accelerated to a kinetic energy of about 2 keV and transported by a system of steerer plates and einzel lenses toward the ICR cell of a home-built Fourier-Transform Ion Cycloton Resonance Mass Spectrometer.

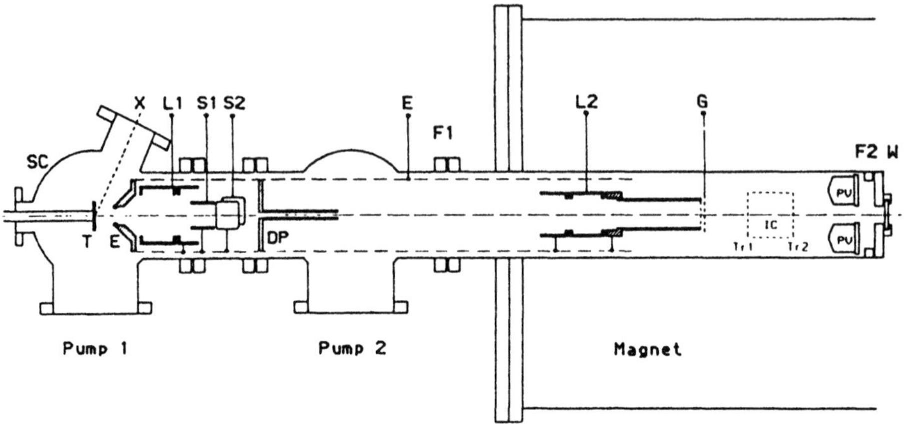

Figure 1: Experimental set-up of the FT-ICR mass spectrometer

The cubic cell of 80 mm sidelength is located in the homogeneous region of a superconducting 7 T magnet from Oxford Instruments. A grid in front of the ICR cell (G) is used to decelerate the ions electrostatically to nearly ground potential and to allow a time segment of them to enter the cell. Two piezoelectric valves are mounted near to it to pulse in inert as well as reactive gases. When neon is introduced that way, the external cluster ions lose excess kinetic and internal energy through collisions and end up trapped within the cell for time periods of up to 100 seconds. Base pressure is normally better than 10^{-9} mbar, with reactive gases mostly admitted through continuous leak valves pressure is increased up to a stationary value of $5 \times 10^{-8} ... 1 \times 10^{-7}$ mbar.

In order to isolate ions of a specific mass-to-charge ratio, radio frequency (RF) ejection pulses eliminate from the cell all unwanted ions. A special refinement of this technique, "soft ejection" [7] or "FERETS" [8], even allows the isolation of certain isotopes of metal cluster ions (close to ± 0.5 amu at a mass of 240 amu) without exciting the isolated ions [9]. DC voltages and pulses are provided by a Spectrospin data system with an ASPECT-3000 computer, which also performs the Fourier transformation.

3. Results and Discussion

3.1. Wideband FT mass spectra of metal cluster cations

Figures 2a to 2c contain overview mass spectra of cluster ions from the transition metals cobalt, nickel and iron, sputtered, transferred to the ICR cell and cooled. The typical pseudo-exponential intensity decay with increasing cluster size is quite evident. The observed distribution ends at n = 19 for Fe_n^+ ions, but somewhat higher around n = 28 and n = 25 for Co_n^+ and Ni_n^+ ions. In the cases of iron and nickel the cluster ion peak widths indicate the existence of different isotopes, whereas cobalt clusters are characterized by a single ion signal. All ion distributions have already been discussed in greater detail [10, 11].

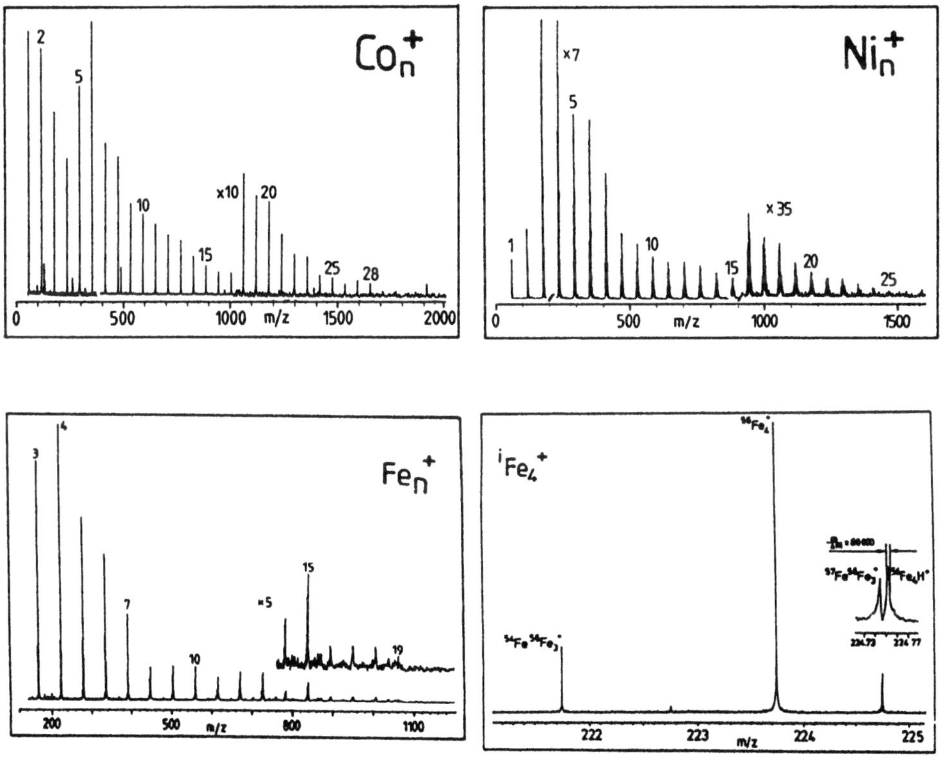

Figures 2a-2c: FT-ICR spectra of Co_n^+, Ni_n^+ and Fe_n^+ ions.

Figure 2d: FT-ICR spectrum of $^iFE_4^+$ cluster isotopes.

3.2. Narrowband FT mass spectrum of $^{i}Fe_4^+$ isotopes

The option for ultrahigh mass resolution is one of the specific advantages of the FT-ICR technique. Figure 2d demonstrates this in a very practical case. When $^{56}Fe_4^+$ was to be isolated in the ICR cell, it was discovered that the sputtering process had produced two different ions extremely close in mass: $^{57}Fe^{56}Fe_3^+$ and $^{56}Fe_4H^+$, the latter probably originating from a hydrogen contamination of the iron target. To resolve this peak doublet, it was necessary to switch detection to narrowband mode, where only a small range of masses is covered. Thus a mass resolution $m/\Delta m$ of about 80,000 was achieved [12].

3.3. Reaction of Fe_n^+ cluster ions with NH_3

When a distribution of Fe_n^+ ions with n = 2...13 (Figure 2) is exposed to ammonia at about 10^{-7} mbar, Fe_2^+ and Fe_3^+ ions are found unreactive. The larger cluster ions $Fe_{5...13}^+$ just add up to four intact NH_3 molecules. Fe_4^+, however, deviates from that pattern, in dehydrogenating the ammonia to yield $Fe_4(NH)^+$:

$$Fe_4^+ + NH_3 \dashrightarrow Fe_4(NH)^+ + H_2 \ . \qquad (1)$$

The cluster must still have a second site for dehydrogenation, as even $Fe_4(NH)_2^+$ is formed. From that step on, only intact NH_3 units are added, as in the case of cluster ions $Fe_{5...13}^+$ [9, 10, 13]. The complete reaction scheme of Fe_4^+ with ammonia is shown in Figure 3.

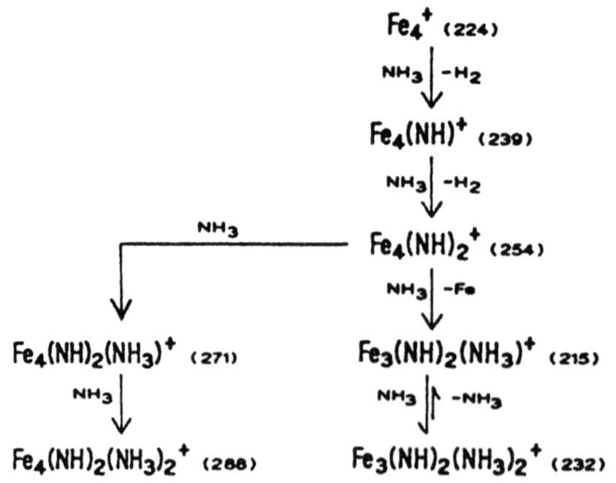

Figure 3: Reaction scheme of Fe_4^+ with NH_3 gas at 10^{-7} mbar.

3.4. Temperature of the stored and cooled cluster ions

For the study of the chemical reactions of cluster ions, it is essential to ascertain their thermodynamic state. To obtain information on the actual temperature of the metal cluster ions stored in the ICR cell, one has to modify the pulse sequence usually applied in an experiment. Generally, the procedure to isolate a single ion mass is as follows: First, the gate pulse determines for how long a certain size distribution of cluster ions may enter the cell. During that period, neon is pulsed in for cooling purposes. The gate pulse is then directly followed by the wideband RF ejection pulse that eliminates all ions of an unwanted mass-to-charge ratio out of the cell again.

In the modified pulse sequence, a variable time delay called "additional equilibration time" is inserted between gate and ejection pulses. There is a constant xenon background in the ICR cell vacuum caused by the primary ion source. The clusters that have already been partly cooled by collisions with neon atoms while they entered the cell can be finally equilibrated to the temperature of their environment (room temperature) by additional collisions with xenon atoms. The length of the above time delay directly determines the degree of equilibration.

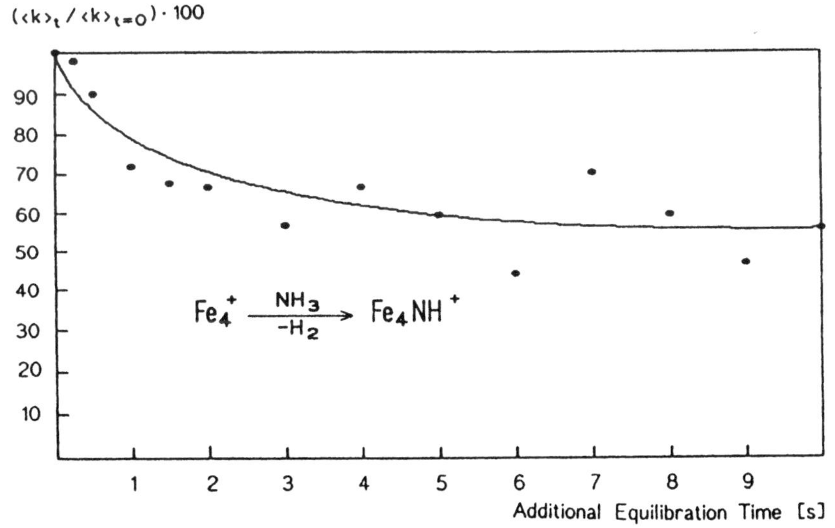

Figure 4: Initial rates of the dehydrogenation reaction of Fe_4^+ with NH_3 as a function of the additional equilibration time.

Figure 4 shows a plot of the measured initial rates of reaction (1) as a function of the additional equilibration time. We find that in the beginning, the initial rates decrease exponentially with increasing equilibration time, as is to be expected for a cluster ensemble in the process of cooling. After about five seconds, the rates approach a certain limit and cluster ions can be assumed to be totally equilibrated to their environment. Although produced through sputtering at a temperature, which is still very high a few picoseconds later they can be thermalized in a storage cell to a value of not more than 20 degrees above room temperature [14].

3.5. Comparative study of different reactivity patterns

When the chemical reactions of clusters of the three transition metals iron, cobalt and nickel are compared, one is struck by the fact that each case shows a completely different reactivity behaviour. Obviously, the way a metal cluster reacts with a gas does *not* depend just on the specific cluster alone, but on the complete system of ***both*** cluster and reactive gas [11-13].

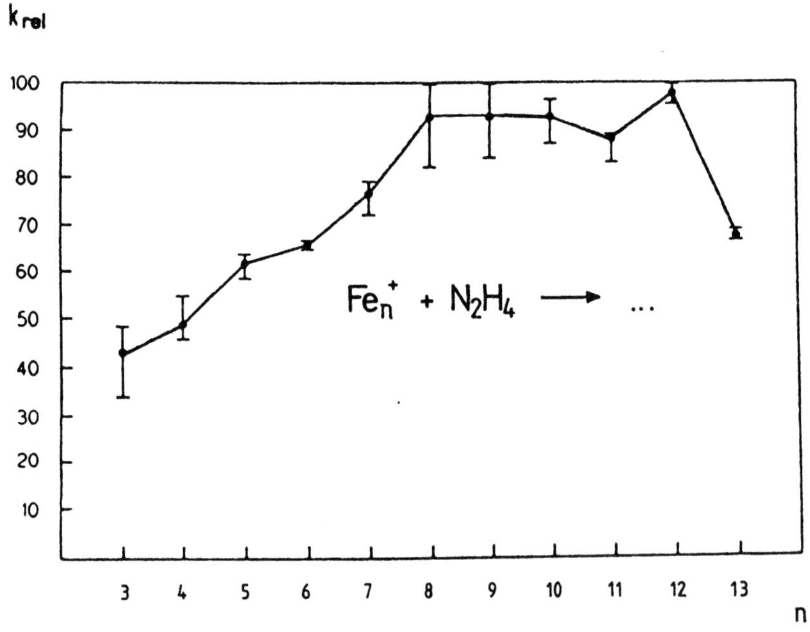

Figure 5: Rectivity of Fe_n^+ clusters toward hydrazine as a function of size.

As discussed in Section 3.3, for example, iron clusters react with ammonia in a very specific way: Out of a distribution of sizes up to Fe_{13}^+, ***only*** Fe_4^+ is able to chemically activate the NH_3 molecule, while all other ions are not [9, 10, 13]. This is a kind of abrupt reactivity change, where only a few sizes do react at all. Toward hydrazine, the same distribution of Fe_n^+ clusters behaves very differently: At about 5×10^{-8} mbar, the specific property of Fe_4^+ is lost and general reactivity varies quite smoothly with cluster size (Figure 5). The system of Co_n^+ clusters with ethene reveals a more complicated behaviour: most sizes react (n = 4, 5, 6, 10...14), but if only dehydrogenation to C_2H_2 is considered, then n = 4 (and n = 5) is distinguished in reactivity [12], as depicted in Figure 6.

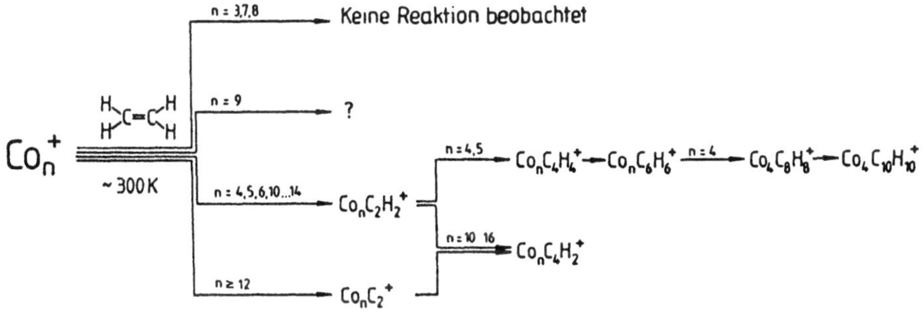

<u>Figure 6:</u> Reaction pattern of Co_n^+ ions with ethene.

Of Ni_n^+ clusters exposed to ethene at ca. 10^{-7} mbar, nearly all sizes react by binding C_2H_2 ligands in different numbers, only Ni_3^+ and Ni_4^+ are inert [11]. Ni_5^+, when isolated in the cell and stored in the presence of ethene for 40 seconds, was found to have ***nine*** C_2H_2 ligands attached to it (Figure 7). Nickel clusters behave in a very similar way toward benzene, as shown in Figure 8: Whereas Ni_3^+ and Ni_4^+ bind the molecule without change, higher clusters dehydrogenate it [12]. The reactions with both hydrocarbons belong to a class of size-specificity opposite to the one seen above: Most sizes are reactive, only a minority is not.

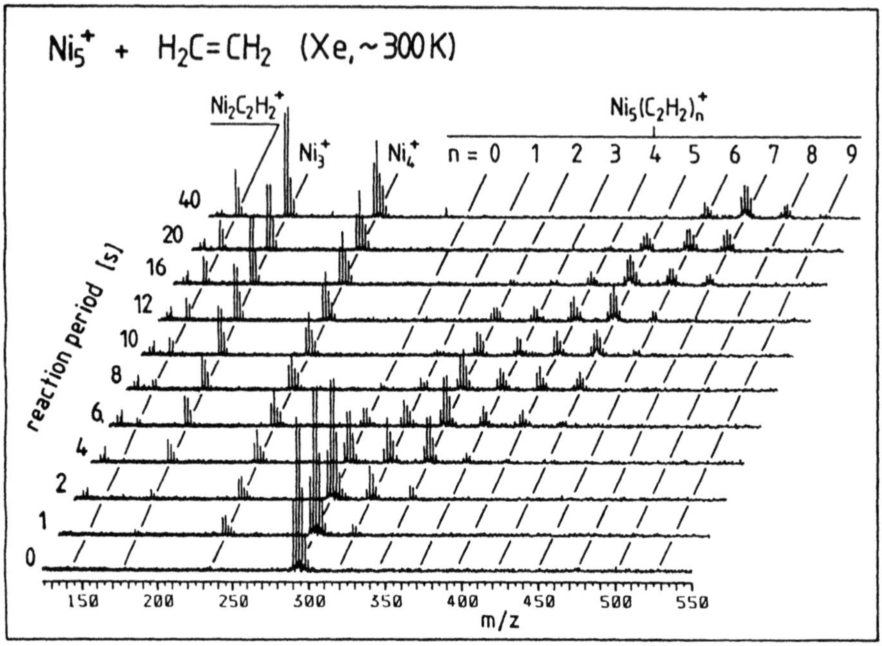

Figure 7: Reaction of Ni_5^+ ions with ethene as a function of storage time. (The inertness of Ni_3^+ and Ni_4^+ is also evident.)

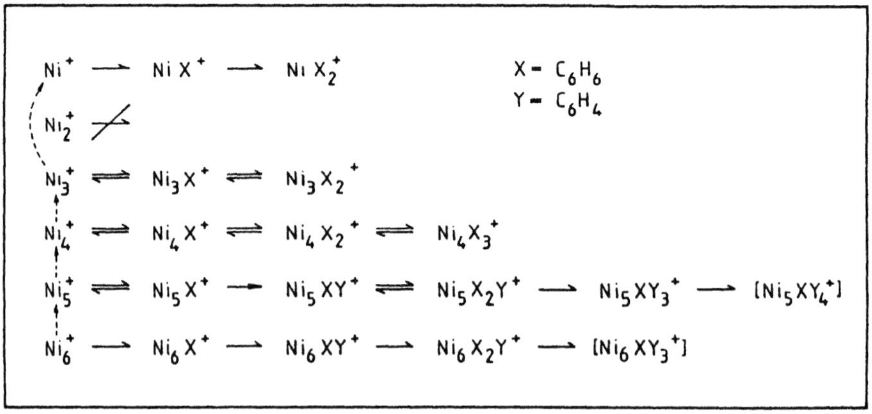

Figure 8: Reaction pattern of Ni_n^+ clusters with benzene.

3.6. Catalysis by a gas-phase naked metal cluster

In 1988 Kaldor stated in a review on the reactions of gas-phase metal clusters that "to date no example of catalytic chemistry using gas phase clusters has been reported" [15], and this was certainly true through 1991. However, in November of this year, we were able to observe the first example of this kind, namely the synthesis of benzene from ethene catalyzed by Fe_4^+ ions in the gas-phase [16 - 18].

Reminding us of the size-specificity found for cobalt clusters dehydrogenating ethene [12], we noted that of a distribution of Fe_n^+ clusters (n = 2...13) stored in the presence of ethene at about 5 x 10^{-8} mbar, only Fe_4^+ and Fe_5^+ are able to react that way. Fe_5^+ ions bind up to two C_2H_2 molecules, Fe_4^+ ions even up to four of them, as Figure 9 shows. The adduct with three C_2H_2 units already contains a ligand of C_6H_6 stoichiometry, probably a benzene precursor. To prove this hypothesis, it is necessary to study the behaviour of the adduct ions when subjected to CID (collision-induced dissociation) [16].

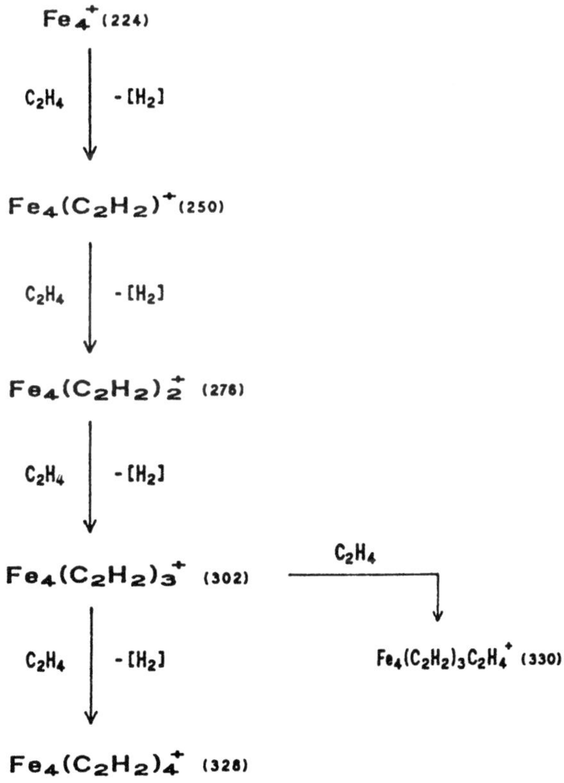

Figure 9: Reaction pattern of Fe_n^+ clusters with ethene.

For such an experiment, the ion in question is first isolated in the ICR cell by the appropriate wideband RF ejection pulse. Then it is excited to a larger ion orbit by an RF pulse at its specific resonance frequency, so that it will undergo a lot of collisions with xenon present from the primary ion gun at about 10^{-7} mbar. The fragments formed allow conclusions on the nature of binding prevalent in the parent ion. Thus, when $Fe_4(C_2H_2)_2^+$ ions are subjected to CID, they decay into carbide ions at low energy ($E_{cm} \approx 3$ eV [19]), but lose Fe atoms as it is increased ($E_{cm} > 14$ eV). In contrast, $Fe_4(C_2H_2)_3^+$ adduct ions yield naked Fe_4^+ at low energy ($E_{cm} \approx 3.4$ eV) with the simultaneous removal of all carbon and hydrogen atoms. Increasing the energy starts dehydrogenation ($E_{cm} > 6$ eV) and elimination of Fe atoms ($E_{cm} > 16$ eV) as well. $Fe_4(C_2H_2)_4^+$ ions lose at the lowest energy one C_2H_2 ligand ($E_{cm} \approx 2$ eV), at higher energies three of them ($E_{cm} \approx 4$ eV), and finally all four ($E_{cm} > 6$ eV). Figure 10 displays an overview mass spectrum of all the ions involved in this CID experiment. It is striking that *no* fragment ion with ***two*** C_2H_2 ligands appears at all!

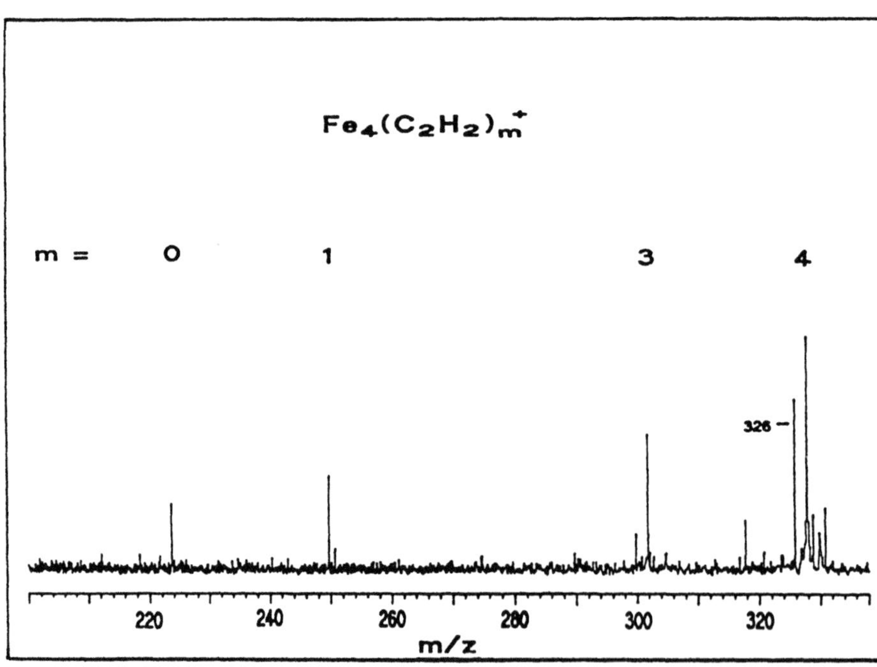

Figure 10: FT-ICR mass spectrum of all the ions partaking in the collision-induced dissociation of $Fe_4(C_2H_2)_4^+$.

All these results suggest that the metal-carbon bond is weak, whenever C_6H_6 stoichiometry is completed on the Fe_4^+ cluster. In these cases, the evidence is strong for the neutral molecule liberated being benzene. If C_6H_6 stoichiometry is not complete, the metal bonds within the cluster are easier to break than the metal-carbon bond. A measurement of the absolute rate constants for the single dehydrogenation steps of ethene nicely confirms this [17]. Normally, they should drop linearly with the number of C_2H_2 ligands already bound to the cluster. In the experiment they are quite high for the step completing C_6H_6 stoichiometry and remarkably low for any further C_2H_2 addition. With similar arguments, the group of Schwarz has recently proven catalytic oxidation of ethane by the simple ion Fe^+ [20]. The sum of our evidence reported so far is sufficient to draw a catalytic cycle where Fe_4^+ converts ethene into benzene (Figure 12).

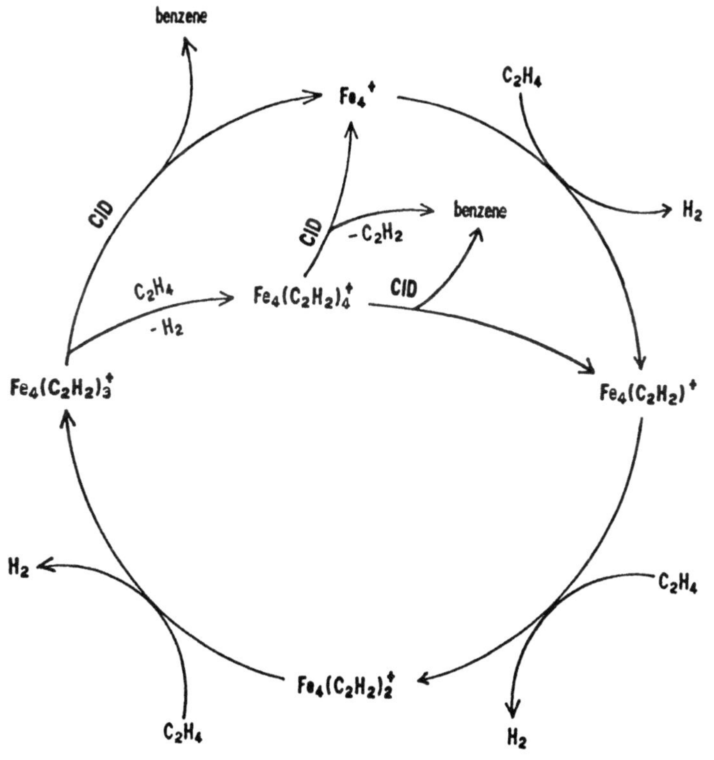

<u>Figure 11:</u> Catalytic cycle of Fe_4^+ turning ethene into benzene.

Only a high-performance tandem mass spectrometric experiment is suited to definitely prove any assertion of gas-phase catalysis by a bare metal cluster ion. By performing a complicated $(MS)^5$ experiment in our FT-ICR instrument, we are able to demonstrate that a part of the original Fe_4^+ ions will induce the synthesis of benzene ***twice!*** The five step procedure is as follows [18]:

$$Fe_n^+ \xrightarrow[(MS^1)]{\text{Isol.}} Fe_4^+ \xrightarrow{+C_2H_4/-H_2} Fe_4(C_2H_2)_m^+,\ m = 1...3$$

$$\xrightarrow[(MS^2)]{\text{Isol.}} Fe_4(C_2H_2)_3^+ \xrightarrow[(MS^3)]{\text{CID (I)}} Fe_4^+ \xrightarrow{+C_2H_4/-H_2}$$

$$Fe_4(C_2H_2)_m^+,\ m = 1...3 \xrightarrow[(MS^4)]{\text{Isol.}} Fe_4(C_2H_2)_3^+ \xrightarrow[(MS^5)]{\text{CID (II)}} Fe_4^+.$$

4. Conclusions

To summarize, a Fourier Transform Ion Cyclotron Resonance Mass Spectrometer has been described to which a secondary ion source is coupled. The instrument is especially modified for the study of bare metal cluster ions. In the exemplary case of transition metals, typical standard wideband mass spectra and in one case, a narrowband spectrum of resolution 80,000 are shown. Chemical reactions observed with admitted gases at low pressure reveal surprising size-specific effects that are not yet understood. Certainly, structure as well as electronic properties of the clusters play an important role. In that respect it would prove very helpful to have more assistance from theory. Though it is not trivial to perform model calculations on transition metal clusters a theoritician has promised to try [21]. Investigation of cluster chemistry can even provide new insights into the field of gas-phase catalysis. The first example of a catalytic synthesis by a bare metal cluster in the gas-phase is reported.

Acknowledgments

I thank Dr. Adrian Selinger and Dipl.-Ing. Patrick Schnabel for the measurements and Prof. Dr. Konrad G. Weil for enlightening discussions. Financial support is gratefully acknowledged from Deutsche Forschungsgemeinschaft, Fonds der Chemischen Industrie and Röhm-Stiftung.

REFERENCES

[1] Proceedings of Faraday Symposium 25 on Large Gas Phase Clusters, J. Chem. Soc. Far. Trans. 86 (1990) 3343-2551; see especially: A. Kaldor, D. M. Cox, pages 2459-2463.
[2] Proceedings of the Fifth International Symposium on Small Particles and Inorganic Clusters (ISSPIC-5), Z. Phys. D 19, 20 (1991).
[3] M. P. Irion, A. Selinger, Z. Phys. Chem. N. F. 161 (1989) 233.
[4] M. P. Irion, A. Selinger, R. Wendel, Int. J. Mass Spectrom. Ion Proc. 96 (1990) 27.
[5] F. M. Devienne, J.-C. Roustan, Org. Mass Spectrom. 17 (1982) 173.
[6] I. Katakuse, T. Ichihara, Y. Fujita, T. Matsuo, T. Sakurai, H. Matsuda, Int. J. Mass Spectrom. Ion Proc. 67 (1985) 229.
[7] L. J. de Koning, R. H. Fokkens, F. A. Pinske, N. M. M. Nibbering, Int. J. Mass Spectrom. Ion Proc. 77 (1987) 95.
[8] R. A Forbes, F. A. Laukien, J. Wronka, Int. J. Mass Spectrom. Ion Proc. 83 (1988) 23.
[9] M. P. Irion, A. Selinger, P. Schnabel, Z. Phys. D 19 (1991) 393.
[10] A. Selinger, P. Schnabel, W. Wiese, M. P. Irion, Ber. Bunsenges. Phys. Chem. 94 (199) 1278.
[11] M. P. Irion, A. Selinger, Ber. Bunsenges. Phys. Chem. 93 (1989) 1408.
[12] M. P. Irion, P. Schnabel, A. Selinger, Ber. Bunsenges. Phys. Chem. 94 (1990) 1291.
[13] M. P. Irion, P. Schnabel, J. Phys. Chem. (1991) in press.
[14] P. Schnabel, M. P. Irion, K. G. Weil, Ber. Bunsenges. Phys. Chem. 95 (1991) 197.
[15] A. Kaldor, D. M. Cox, M. R. Zakin, Adv. Chem. Phys. 70 (1988) 220.
[16] P. Schnabel, M. P. Irion, K. G. Weil, J. Phys. Chem. (1991) in press.
[17] P. Schnabel, M. P. Irion, K. G. Weil, Chem. Phys. Lett. (1991) submitted.
[18] P. Schnabel, K. G. Weil, M. P. Irion, Angew. Chem. (1991) submitted.
[19] The center-of-mass energy E_{cm} defines the maximum energy transfer in a collision of a molecule M with an ion I and is given by: $E_{cm} = E_{Lab}\, m_I/(m_I + m_M)$ with masses m_i.
[20] D. Schröder, H. Schwarz, Angew. Chem 102 (1990) 1466; Angew. Chem. Int. Ed. Engl. 29 (1990) 1431.
[21] V. Bonacic-Koutecky, private communication (1991).

ELECTRON SCATTERING AND ELECTROMAGNETIC RESPONSE PROPERTIES OF METAL CLUSTERS

Vitaly V. Kresin

Department of Physics, University of California
Berkeley, California 94720
and
Lawrence Livermore National Laboratory
University of California
Livermore, California 94550(*)

Abstract

The response of metal clusters to inelastic electron impact and to external electromagnetic radiation is discussed. The delocalized valence electrons play a decisive role in determining the response properties. First, a description and interpretation of an inelastic electron scattering experiment are given, and a connection to electromagnetic response studies is pointed out. Secondly, a summary of a theoretical analytical description of the response properties of metal clusters is presented. Finally, the effect of non-sphericity on the electronic properties of clusters is analyzed.

I. Introduction

A great deal of information about the electronic structure of clusters can be derived from studies of the response of cluster electrons to external probes. Two of the most fruitful tools for performing such studies are optical spectroscopy and collision experiments. In the present contribution, we discuss applications of these methods to metal cluster research.

In Section II, we summarize the results of recent experiments on inelastic electron scattering spectroscopy of sodium clusters. It is possible to formulate a consistent interpretation of the experimental data, and a connection to the studies of electromagnetic response properties is revealed.

In Section III, we discuss a general approach to the analysis of cluster response to an external field. We describe analytical results for the collective resonance frequencies and dipole oscillator strength distribution in simple metal clusters, and point out the spectral features unique to the cluster state.

Section IV considers the influence of deviations from the spherical shape on the electronic structure of metal clusters. A new technique is described, allowing to determine the effects of deformation on the electron distribution and on the resonance frequencies.

II. Inelastic Electron Scattering

Recent experiments (see [1,2] for a detailed discussion) provided the first measurement of absolute cross sections for collisions between electrons and free size-selected clusters larger than dimers. The experiments were carried out by intersecting a supersonic beam of neutral sodium clusters with a monoenergetic beam of low-energy (≅0-30eV) electrons. By measuring collision-induced depletion of the cluster beam, the total inelastic scattering cross sections were determined. The results are illustrated in Fig.1 for three spherical closed-shell clusters Na_8, Na_{20}, and Na_{40}.

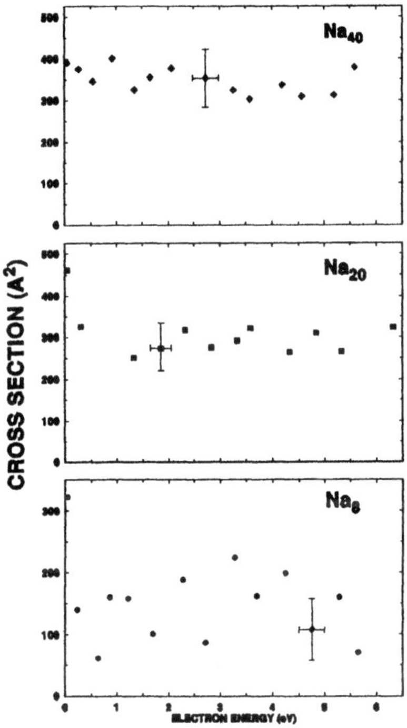

Fig.1. Total inelastic electron-impact cross sections for three spherical clusters.

Note that the cross section values increase with cluster size, and that the cross section curves display essentially two distinct regimes. Above ≅1eV the cross sections do not vary strongly with energy, while below ≅0.5eV a rise is visible [3]. In the following, an interpretation of these observations is given.

In the "rise region" (≲0.5eV), the electron energies are so low that only s-wave scattering contributes significantly to the collision process. In this case inelastic scattering is described by the "$1/v$" law [4], according to which the inelastic scattering cross section is inversely proportional to the electron velocity. This agrees qualitatively with our observations. The electron energy in this region is smaller than the cluster binding energy, therefore we propose that the dominant mechanism of cluster depletion in the s-wave scattering region is via electron attachment.

At energies above ≅1eV ("plateau region" of the cross section curves), the electrons carry sufficient energy to cause cluster fragmentation, and this is likely to be the primary inelastic interaction channel. (Plasmon excitation and ionization also occur above their respective thresholds, but they do not show up as major features on top of the overall cross section curves.) At these energies angular momenta up to $\ell = k\mathcal{A}$ will contribute to the scattering process, where $\mathcal{A}$ is the range of inelastic electron-cluster interaction and k is the wave number of the incident electron. In this case there exists an upper bound on the inelastic scattering cross section [4]:

$$\sigma_{in}^{max} = \pi\mathcal{A}^2, \qquad (1)$$

which we take to describe the sodium cluster cross sections in the plateau region. The applicability of this expression is further confirmed by the fact that the cross sections of Na_{40} at electron energies of 10, 20 and 30 eV were also observed not to differ from those at lower energies.

From the experimental data and Eq.(1), we can determine the value of the inelastic interaction range, and we find:

$$\mathcal{A} \cong (1.4\text{-}1.7)R, \qquad (2)$$

where R is the radius of the uniform positive jellium sphere representing the cluster background. Note that $\mathcal{A}$ exceeds both the extent of the spill-out of the cluster's valence electrons and the range of the potential which confines this cloud. The large value of $\mathcal{A}$ is probably due to the influence of the attractive long-range polarization interaction $\sim\alpha/r^4$, where α is the static electric polarizability of the cluster [5].

The strong effects of the polarization (sometimes referred to as "image-charge") interaction has also been observed in experiments on photoelectron spectroscopy of cluster anions [6], and in studies of collisions between noble gas cluster ions and neutral atoms [7]. This phenomenon provides a bridge between the study of electron-cluster scattering and that of electromagnetic response properties, such as electric polarizabilities and photoabsorption, which are the subject of the next section.

III. Theoretical Analysis of Cluster Response Properties

The response of the cluster valence electrons to an externally applied potential $v_0(\vec{r})exp(i\omega t)$ can be found by considering the exact linear response equation, written below both in graphical and analytical forms:

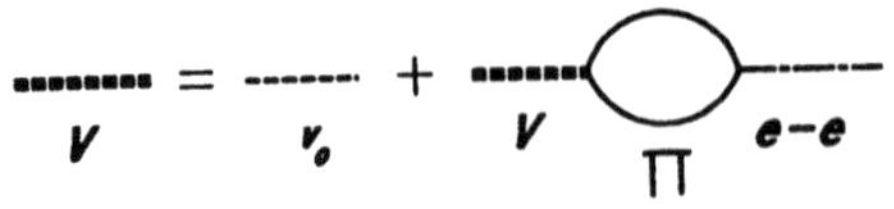

(3)

$$V(\vec{r},\omega) = v_0(\vec{r}) + e^2\iint\frac{1}{|\vec{r} - \vec{r}_1|}\Pi(\vec{r}_1,\vec{r}_2,\omega)\ V(\vec{r}_2,\omega)\ d^3r_1d^3r_2. \tag{4}$$

The second term on the right-hand side is the potential due to the induced charge; Π is the polarization operator, here used in the RPA approximation, which describes screening by the valence electrons. For spherical clusters, this integral equation can be solved analytically (see [8] and references therein) by expanding Π in powers of the ratio Δ/ω, where Δ is the single-electron level spacing in the cluster, and ω is the frequency of the dipole oscillations. In this way, we can determine the spectral distribution of the oscillator strength, and hence such properties as the resonance frequencies, static and dynamic polarizabilities, etc. Higher-order terms in the expansion account for resonance-line splittings due to interaction with single-particle levels [9]. Note that the analysis does not rely on approximations whose validity for cluster studies has not been established (such as the numerical TDLDA and GW methods, see, e.g., [10]), and is therefore more successful in accounting for the experimental observations. In the following, we summarize the main conclusions.

It was found that in a small cluster the dipole oscillator strength is concentrated in two distinct regions. In the macroscopic limit, the first corresponds to the surface plasma resonance, while the second corresponds to the volume collective mode. In a cluster, the two photoabsorption modes represent a superposition of these two types of motion, one concentrated near the cluster surface, and the other in its interior. In this way, the picture is analogous to that of the giant dipole resonance in nuclei, which is made up of both Goldhaber-Teller and Steinwedel-Jensen modes [11]. The resonance frequencies in a cluster are shifted towards lower frequencies than their macroscopic limits. This is due to the important effect of the valence electron spill-out, which reduces the average electron density in the cluster, and thus red-shifts the resonances. In fact, it turns out that the solution is uniquely determined by the ground-state density distribution of the valence electrons (the magnitude of the spill-out). It is important to point out that, in order to arrive at these conclusions, no specific assumptions are made concerning the method of calculating the ground-state density.

To obtain numerical values for the resonance frequencies and oscillator strengths of the collective modes, we made use of the electron density as given by the statistical Thomas-Fermi treatment [12]; inclusion of higher-order corrections has been shown not to alter the results by more than 1% [13]. By virtue of being analytical, our solution can be applied to a variety of cluster species, sizes, and ionization states. As shown in Table 1, the calculated resonance frequencies are in excellent agreement with the available experimental data.

Table 1. Comparison of calculated and experimental [14-16] resonance frequencies (in eV) of small metal clusters.

Cluster	Na_8	Cs_8	Na_9^+	Na_{21}^+	K_9^+	K_{21}^+
Theory	2.53	1.54	2.62	2.73	1.93	2.00
Experiment	2.51	1.55	2.62	2.66	1.93	1.98

In a small cluster, the higher-lying "volume" mode is calculated to take up approximately 20% of the valence electron dipole oscillator strength; as the cluster size increases, oscillator strength is transferred from the higher to the lower collective resonance. This reflects the fact that in bulk metals the volume plasmon does not couple to light. In small clusters, on the other hand, we just saw that the corresponding mode does acquire a significant fraction of the oscillator strength. This calculation explains quantitatively the experimentally observed fact that the surface collective resonance does not exhaust 100% of the oscillator strength [14,16]. According to the theory just outlined, the missing strength resides in the higher-lying collective excitation (located at $\cong 0.75\omega_p$, where ω_p is the bulk surface plasma resonance frequency). It would be extremely interesting to carry out an experimental search for these predicted resonances.

IV. Shape Effects in Electronic Structure of Clusters

Clusters with unfilled shells of valence electrons find it energetically favorable to acquire spheroidal or ellipsoidal shapes [17]. These shape deformations split the collective dipole resonances (see, e.g., [14]). This is an effect which is present even in a macroscopic metallic particle, since deviation from sphericity alters the restoring force acting on the electron system. However, when analyzing shape effects in a cluster, one must keep in mind special features unique to small systems. The most important one is the valence electron spill-out, already discussed in the preceding section: the amount of spill-out determines the magnitude of the frequency shift.

Thus first of all, it is necessary to determine how cluster shape affects the ground state electron density distribution. Following that, the dipole resonance frequencies have to be determined. In this section, we outline an approach (presented in detail in [18]) which takes into account both cluster shape and the electron spill-out effects in spheroidal clusters.

The method makes use of a technique [19] developed to describe particle states in spheroidal nuclei. A coordinate transformation is carried out to strain the ionic spheroid into a sphere. Clearly, this leads to a tremendous simplification of the boundary conditions. In the new coordinate system, the

Laplacian operator acquires an additional non-spherically-symmetric term. This additional term is treated in perturbation theory, and the electron distribution can, in principle, be determined to a desired degree of accuracy. In the following, we utilize the Thomas-Fermi (TF) statistical treatment. We make use of the fact that the TF equations for a spherical cluster can be solved analytically, and have been shown to successfully describe the properties of the cluster valence electrons (see preceding section and [8,12]). It is straightforward, however, to employ the general coordinate transformation approach in other methods of calculation, and to extend it to the case of triaxially deformed (ellipsoidal) clusters.

In the new ("primed") coordinate system, the aforementioned perturbation term in the Laplacian operator has the form $\varepsilon\cdot\partial^2/\partial z'^2$. Here $\varepsilon=e^2=8\delta/(2+\delta)^2$, where e is the eccentricity of the spheroid, and δ is the spheroidal deformation parameter [17]. For small deformations, ε is a small parameter, and only corrections linear in the perturbation may be considered. In this case, the electron density acquires a quadrupole term in the electron density distribution, as well as a correction in the spherical term:

$$n_e(\vec{r}') = n_0(r') + n_1(r') + n_2(r')P_2(\cos\theta'). \tag{5}$$

Here n_0 is the solution for a spherical cluster, and n_1 and n_2 are the correction terms; P_2 is the Legendre polynomial. Note that the density function is written in the primed coordinate system. The analytical expressions for these functions are somewhat long and will not be written out here; an example for the case of the oblate cluster Na_{38} with ε=-0.22 (corresponding to δ=-0.10) is shown in Fig.2.

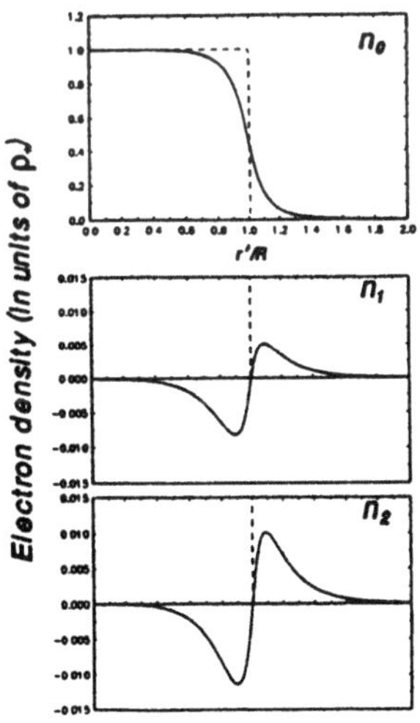

Fig.2. Electron density for Na_{38} plotted in the distorted coordinate system. The dashed line denotes the uniform ion background. The densities and the radial distance are expressed in units of the ion core density and the ionic sphere radius, respectively.

Having determined the electron density, we now need to calculate the electron response. A quantitative solution of the integral response equation (3,4) is, however, much more complicated in a non-spherical geometry. Nevertheless, an approach based on the sum-rule technique [20] can be employed to evaluate the centroid (average energy) of the photoabsorption spectrum. This method requires a knowledge of the electron density and of the electrostatic potential due to the spheroidal background. As a result of the calculation [18], we find for the average excitation energies:

$$\bar{\omega}_z^2 = \omega_C^2(T_1 - T_3), \quad \bar{\omega}_x^2 = \bar{\omega}_y^2 = \omega_C^2(T_2 + \frac{1}{2}T_3). \tag{6}$$

Here ω_C is the classical surface plasma resonance frequency of a metal sphere, and

$$T_1=\left(1-\frac{2\varepsilon}{5}\right)\left(1-\frac{\Delta N}{N_e}\right), \quad T_2=\left(1+\frac{\varepsilon}{5}\right)\left(1-\frac{\Delta N}{N_e}\right), \quad T_3=\frac{6}{5}\int_{R'}^{\infty}\frac{n_2(r')/\rho_+}{r'}dr'. \tag{7}$$

ΔN is the total amount of the valence electron spill-out outside the boundary of the spheroid, and is seen to be responsible for the red shift (cf. [21]). The term T_3 reflects the influence of the quadrupole correction to the density and contributes up to ≅10% of the total magnitude of the frequency shift. It can be deduced from the results that for a given distortion, the presence of the spill-out leads to a somewhat weaker relative splitting than would be the case in the classical limit. This is due to the fact that a large density gradient is energetically unfavorable, hence the electron cloud tends to "smooth out" the charge distribution thus reducing the effects of non-sphericity.

These results can be used to calculate the resonance splittings in a cluster of arbitrary density, size, and deformation. Conversely, it is possible to use the experimental photoabsorption data on deformed clusters to extract the deformation parameters.

V. Summary

In this paper, we have considered several aspects of the response of a metal cluster to an external probe.

1. Experimental data on inelastic electron scattering by neutral size-selected sodium clusters were described. An analysis of the scattering process was presented, and the inelastic electron-cluster interaction range was determined. This range exceeds the cluster hard-core radius by ≅50%, reflecting the influence of the long-range polarization interaction.

2. The results of an analytical theoretical treatment of the electromagnetic response properties of spherical metal clusters have been summarized. The calculated values of the resonance frequencies are in excellent agreement with experimental data. The existence of an additional resonance mode,

unique to the small size regime, is predicted. The presence of this mode is responsible for the oscillator strength missing from the measured surface resonance peaks.

3. An approach allowing to analyze the electronic structure of non-spherical clusters has been outlined. Making use of a special coordinate transformation and sum rules, analytical results are obtained for the electron density distribution and collective resonance frequencies in spheroidal clusters.

This work was carried out at Berkeley under U.S. National Science Foundation Grant No. DMR-89-13414 and at the Lawrence Livermore National Laboratory under the auspices of the U.S. Department of Energy under contract No. W-7405-ENG-48.

(*) Present address.

[1] V.V.Kresin, A.Scheidemann, and W.D.Knight, Phys.Rev.A **44**, 4106 (1991).
[2] V.V.Kresin, A.Scheidemann, and W.D.Knight, Proceeding of the International Symposium on the Physics and Chemistry of Finite Systems, Richmond, VA, 1991 (to be published).
[3] This rise is not resolved for N=40, where it occurs for energies so low that the electron gun emission was too weak to provide observable depletion.
[4] N.F.Mott and H.S.W.Massey, *The Theory of Atomic Collisions*, 3rd ed. (Oxford University Press, Oxford, 1965).
[5] E.W.McDaniel, *Atomic Collisions* (Wiley, New York, 1989).
[6] K.J.Taylor *et al.*, J. Chem Phys. **93**, 7515 (1990); L.Wang *et al.*, Chem. Phys. Lett. **182**, 5 (1991).
[7] T.Kondow *et al.*, to be published (presented at this Seminar).
[8] V.V.Kresin, Phys. Rev. B **42**, 3247 (1990); Phys.Rev.B **39**, 3042 (1989).
[9] V.V.Kresin, Z.Phys. D **19**, 105 (1991).
[10] W.Ekardt, Phys.Rev.B **31**, 6360 (1985); W.Ekardt and J.M.Pacheco, to be published (presented at this Seminar).
[11] W.D.Myers *et al.*, Phys. Rev. C **15**, 2032 (1977).
[12] V.V.Kresin, Phys.Rev.B **38**, 3741 (1988).
[13] M.Membrado, A.F.Pacheco, and J.Sañudo, Phys. Rev. B **41**, 5643 (1990).
[14] K.Selby *et al.*, Phys. Rev. B. **43**, 4565 (1991).
[15] H.Fallgren and T.P.Martin, Chem. Phys. Lett. **168**, 233 (1990).
[16] C.Bréchignac *et al.*, Chem. Phys. Lett. **164**, 433 (1989); Chem. Phys. Lett. (to be published).
[17] K. Clemenger, Phys. Rev B **32**, 1359 (1985).
[18] V.V.Kresin, to be published.
[19] S.A.Moszkowski, Phys. Rev. **99**, 803 (1955).
[20] E.Lipparini and S.Stringari, Z. Phys. D **18**, 193 (1991).
[21] M.Brack, Phys. Rev. B **39**, 3533 (1989).

Statistical Fragmentation of Clusters into Clusters - in the Example of Nuclear Fragmentation-

D.H.E. Gross

Hahn-Meitner-Institut, Bereich Kern- und Strahlenphysik,
and Fachbereich Physik der Freien Universität,
D-1000 Berlin 39, Glienickerstr.100, Germany

The technique of microcanonical Metropolis-sampling of multifragmentation is demonstrated by the example of nuclear multi-fragmentation. It is has a good chance to be sufficiently ergodic to find and cover the important part of the phase-space in a realistic CPU time. The numerical results show the peculiar physics of the breaking of a finite many-body system especially near to the liquid to gas transition. The striking differences to the behaviour of conventional macroscopic systems are emphasised.

1 Introduction

In this talk I want to give an introduction into the theory of simultanous statistical fragmentation of hot nuclei into several fragments of various size . There are two reasons for discussing this in this meeting on the dynamics of molecular clusters and their decay: First, it is the first successfull *rigorously microcanonical treatment of fragmentation* [1] and the basic techniques can also be used for a theoretical description of the decay of molecular clusters. Second, the peculiar physics of the statistical multifragmentation of finite nuclei may be representative for many other finite complex systems including our solar system.

One of the main differences between realistic finite systems and infinite systems like nuclear-matter is that realistic systems have surfaces. Because of that they can break into pieces. Also in conventional solid-state physics the usual objects are infinite systems and such a mode of inhomogenous disintegration does not exist.

There is another important difference from infinite systems: Only systems of finitesize can be subjected to long-range forces having a range larger than their size. Nuclei have a long-range Coulomb force. This is in no way an exotic but otherwise unimportant feature: We know of other systems where this is also the case, e.g. stars, planetary systems, etc are ruled by long-range gravity and centrifugal forces. This fact, and its influence on the critical behaviour of breaking, has, as we will see, far reaching consequences. I hope that the studies of disintegrating hot nuclei may show some of the interesting phenomena to be expected in such systems.

Further interest in this model comes from the fact that, although realistically complex, its number of degrees of freedom is still small enough that the microcanonical partition-sum can be calculated directly by modern powerful Metropolis-Monte-Carlo methods. The quite rich nuclear thermodynamics can thus be deduced from first principles. There are several phase-transitions which are smeared because of the finite size of the system, but nonetheless can clearly be seen. The critical fluctuations can be traced back to their origin. Even complicated quantities like the intermittency of the fluctuations can be reproduced and understood. The methods presented here may also be directly applied to derive the thermodynamics of atomic clusters as considered in this conference.

2 The Basic Picture

We assume that after a collision of a highly energetic proton or another fast small nucleus with a relatively large taget nucleus there are mainly two phases of decay: First there is an emission of fast neutrons and a few fast protons or α-particles as preequilibrium particles. Then in a second step the remaining system equilibrises and breaks into several fragments of various size. While breaking, the system may also expand chaotically while the fragments interact strongly. After reaching an overall density of something like 1/6 normal density the average distance between the surfaces of neighboring fragments is something like $2fm$. At that distance the nuclear forces between the fragments cease and the fragments separate along Coulomb trajectories. The transition configuration where the first chaotic phase of expansion turns over into a more or less isentropic Coulomb-expansion is called the freeze- out.

This picture is of course very simplified and only comparisons with experimental data can show how far it may be valid. However, it makes the theoretical description considerably easier. Furthermore, it can take the (static) effects of the Hamiltonian much more precisely into account than is possible in any other approach starting with simplified nuclear forces and working with a simplified solution of the many-body dynamics. Due to the assumption of equilibration at the freeze-out we 'only' have to sample the available phase-space at the freeze-out configuration in order to know what type of fragments are produced and with which probability. The remainig difficulty comes from the fact that at freeze-out channels with a different number of differently sized fragments are in equilibrium. This is quite unusual in statistical mechanics, as there one normaly treats systems with a fixed kind and number of particles.

3 The Method

Inspite of being based on a simple idea, the success of the model depends crucially on technical details. We will sketch some of them as far as they are relevant for a similar theory of atomic cluster disintegration.

In nuclear fragmentation the following degrees of freedom best describe the possible decay channels at the moment of emission of the fragments (freeze-out):
The number n_f of fragments, their positions $\mathbf{r}_i$, their momenta $\mathbf{p}_i$,their internal excitations ϵ_i, their masses A_i and charges Z_i, and the number of free neutrons n_ν.

Different decay-channels differ in all or some of these numbers. An equilibrium model assumes that all different channels which are consistent with the basic conservation laws of mass A, charge Z, energy E, momentum $\mathbf{p}$, and angular-momentum $\mathbf{l}$ are equally probable. The number of allowed channels is the microcanonical partition-sum $Z(E)$. Its logarithm $ln(Z(E)) = S(E)$ is the entropy. In the microcanonical statistics the temperature T is defined by

$$\frac{1}{T} = \frac{dS(E)}{dE} \tag{1}$$

and may differ from event to event.

Usually the neutrons are just counted, and their momenta and positions are integrated over. This technical point is quite essential for the success of the model. After integrating over the unobserved degrees of freedom of the neutrons, and also summing over the internal excitations of the fragments, each 'reduced channel' gets a weight W equal to the volume of the unobserved phase-space. *The weights allow one to sort out the most important reduced channels.* Before this the complete but different decay channels were equally likely.

The binding energies of the fragments are taken from experiment and therefore beyond any theoretical approximation. At typical excitation energies of a couple of MeV/A, Metropolis Monte Carlo is an effective way to find the most important parts of the available phase-space. Moreover, Metropolis sampling is ergodic. If sufficiently many channels are sampled it can be proven to cover the complete available phase-space. For molecular dynamics, for instance, this is in general not possible. In the practical situation, however, the situation is worse even for Metropolis sampling: The typical size of the available phase-space can be 10^{20} or more different channels or states. The most effective sampling program can at best calculate some few million different channels in some hundred CPU-minutes. These 'negligable few' channels will only be representative if they are selected out of the region of maximal weight.

Metropolis Monte Carlo sampling constructs a diffusive path through the phase-space that drifts into the region of maximal weight or importance, therefore its other name: importance sampling. A new reduced channel or fragmentation is constructed consistent with the conservation laws. The new channel should not differ in too many degrees of freedom from the last successfull one. This construction is called a move. It is accepted as the next successfull channel with the probability

$$P_{1\to 2} = W(2)/W(1). \tag{2}$$

If it is rejected the previous channel is repeated. Thus a chain of events is created. It can be shown that the relative probability of occurence of two events i and j along this chain becomes asymptotically $W(i)/W(j)$ [2].

One of the most difficult moves is one where the number of fragments is changed. Besides of having a rigorous microcanonical sampling, this is the other mainly new aspect in which our method differs from the conventional use of Metropolis Monte Carlo in atomic physics. For this purpose we developed an effective strategy, we either divided a randomly selected fragment into two new ones (fission), or we combined two fragments together (fusion). This is what happens in real life and should therefore be the most effective way of changing the number of fragments. In an exactly solvable model of partitioning a linear chain of integer numbers $(0-A)$ into fragments of different integer length A_i one can study the effect of exact mass- (or charge-) conservation and of the quantum-mechanical symmetries analytically [3,4]. It is interesting to note that this move by splitting a fragment or combining two fragments produces automatically the Gibbs-factors due to quantum symmetry [5,4]. Because of the high complexity of our problem, it is of vital importance to test the strategy of microcanonical Metropolis sampling by comparing its results with this non-trivial but exactly solvabel model [3].

4 Some of the most important results of the model

An interacting complicated many-body system has interesting phase transitions. Why are these interesting? In a normal macroscopic system with short range couplings of nearest neighbors, the correlation length becomes much larger than the range of the interaction near a phase transition of second order. The details of the force become unimportant and the phase transition is ruled by geometry only. The critical fluctuations, characterized by the critical exponents, carry a fingerprint of the fractal geometry of the transition.

To determine the position (T_{tr}, E_{tr}) of phase transitions conventionally one plots the canonical heat capacity $c_V = dE/dT|_V$ vs. temperature T. Anomalies or spikes in $c_V(T)$ are signals of a transition.

An excited nucleus is a microcanonical rather than a canonical system, therefore we have to look for other signals. A good method was shown [6,7] to be a scatter plot of $ln(P)$ vs.$ln(S_2')$. Here P is the size (charge) of the largest cluster and S_2' the second moment $\sum' Z_i^2$ of the charges Z_i of the fragments in a single event, the largest one excluded. This Campi scatter-plot is shown in figure 1 for nuclear microcanonical multifragmentation .

We see here, as in the case of percolation, a smooth mountain ridge going from large P and small S_2' down to small P and again small S_2'. In between the mountain ridge bends and arrives at the largest S_2' at intermediate $P = P_c$. In analogy to the phase transition of liquid to gas, in a macroscopic real gas the upper branch corresponds to the coexistence of a condensate (large P) with a vapor, whereas the lower branch is the locus of 'supercritical' events where we have a single gas phase and the largest droplet is rather small. The intermediate transition point (P_c, S_{2c}') corresponds to the 'critical point'. (This is at the moment only an analogy to the critical point of a phase transition in infinite systems. Certainly more research is neccessary here.) One can see that the fluctuations of (S_2') are also largest at the 'critical point', such that normally this point is difficult to reach, a phenomenon well known as critical opalescence.

There is, however, an important difference to percolation or a macroscopic liquid to gas transition [8]: In nuclear fragmentation there is a *second 'critical' aggregation of points* (also in phase space) at very large P and large S_2'. Events contributing to this region show that these are fission-like events, where two and only two of the fragments produced per event are large. This new aggregation in phase space (called hot fission) is naturally controlled by the interplay of the long-range Coulomb force with the surface tension and therefore totally absent in macroscopic gases or liquids as well as percolation.

Near to T_c the nuclear liquid-gas transition is also of second order. Because of the finite size of the nuclei this region is not concentrated to the 'critical point' itself. Everywhere the correlation length is of the size of the nuclear diameter, first and second order transitions are indistinguishable.

This is the first example that the study of the fluctuations (event by event analysis of the Campi-correlations) allows one to separate the two 'critical' mechanisms: hot fission and 'liquid-gas phase transition '.

It is very instructive to switch off the long-range Coulomb repulsion. Then our system behaves much more like a normal real gas at its liquid to gas phase transition . In the Campi scatter-plot discussed above this can easily be seen (c.f. figure 2). The second aggregetion by the hot fission events disappear and we have only the long mountain-ridge of the liquid to gas phase transition . Then it's behaviour is very much like that of normal water-droplets: The surface-tension is dominating and keeps the system together. There will be no real breaking. The system evaporates single nucleons and alphas. Of course, we switched-off the Coulomb interaction only to first order, we still used unchanged nuclear radii, the binding energies are the liquid-drop energies without its Coulomb part. A proper selfconsistent calculation of the ground-state without Coulomb-repulsion would lead to very different densities and radii. The correct treatment of the nuclear many-body problem under strong interactions only would require more basically new methods of analysis. This, however, is far outside the intention of this paper.

5 Acknowledgement

I thank A. R. DeAngelis for the preparation of Figure 2. To him and to H.R.Jaqaman I am gratefull for many illuminating discussions.

References

[1] D. H. E. Gross, Rep.Progr.Phys. **53** (1990) 605-658

[2] J. M. Hammersley and D. C. Handscomb, *Monte Carlo Methods* (Methuen &Co Ltd, John Wiley &Sons Inc, London,New York 1964)

[3] H. R. Jaqaman, A. R. DeAngelis, A. Ecker, and D. H. E. Gross, HMI-preprint (1991)

[4] D. H. E. Gross and A. R. DeAngelis to be published

[5] D. H. E. Gross and H. Massmann, Nucl. Phys **A 471** (1987) 339c-350c

[6] D. Stauffer, *Introduction to Percolation Theory* (Taylor and Francis, 1985)

[7] X. Campi, J. Phys. A. Math. Gen **19** (1986) 917

[8] H. R. Jaqaman and D.H. E. Gross, Nucl. Phys. **A 524** (1991) 321-343

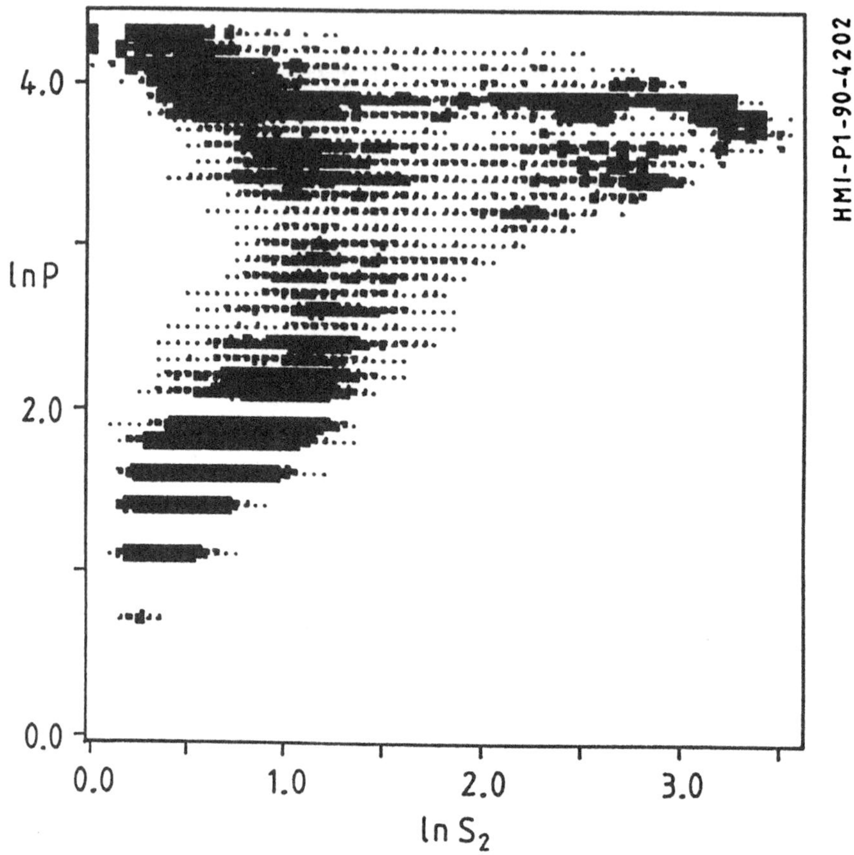

Fig.1: Fluctuation pattern in the event by event plot of the size of the largest fragment in each event, $ln(P)$, y-axis, vs. the second moment $ln(S_2') = ln[\sum' Z_i^2 / \sum' Z_i]$, x-axis, for $^{197}Au^{79}$ (from [8]).

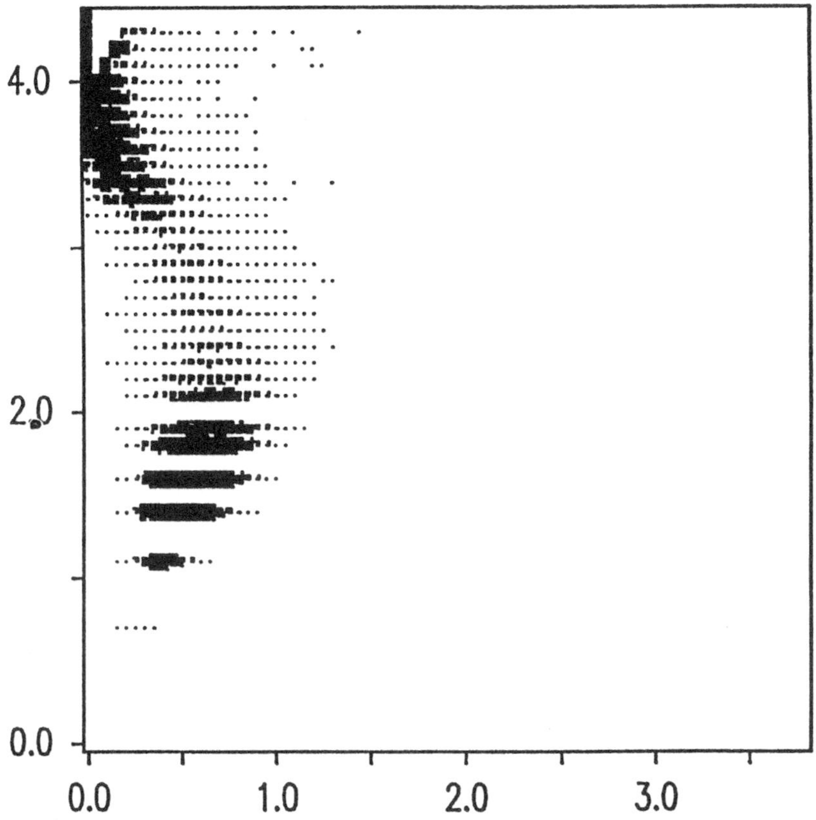

Fig.2: same as above, but the Coulomb interaction is switched off.

Shell Effects in Cluster-Cluster Collisions

O. Knospe[1,2], R. Dreizler[1], R. Schmidt[2,3] and H.O. Lutz[3]

[1] *Institut für Theoretische Physik, Universität Frankfurt, W-6000 Frankfurt a. M. 11, F.R.G.*

[2] *Institut für Theoretische Physik, TU Dresden, O-8027 Dresden, F.R.G.*

[3] *Fakultät für Physik, Universität Bielefeld, W-4800 Bielefeld 1, F.R.G.*

1. Introduction

Collisions between atomic clusters offer a new and interesting field of cluster research. First theoretical studies of cluster-cluster collisions (CCC)[1,2,3,4] have shown that close analogies to nuclear heavy-ion collisions (HIC) exist. Inspite of the different *microscopic* forces acting between the constituents, the dynamics of *collective* (macroscopic) degrees of freedom is similar in both types of systems. The reaction mechanism of HIC and CCC is characterized by *dissipation and fluctuation* phenomena of a few collective variables coupled to a large, but *finite* number of intrinsic degrees of freedom (nucleonic and atomic, respectively).[2] Therefore, CCC allow to study relaxation phenomena in finite atomic many-body systems.

One of the striking differences between HIC and CCC consists in the role of shell effects. Whereas in HIC nucleonic shell effects are "smoothed" out with increasing excitation energy of the nucleons, they survive almost completely in CCC because the excitation energy is stored mainly in chaotic *atomic* motion, while the electrons (which mediate the shell effects) remain in their ground state (at sufficiently low bombarding energies). The dominating role of shell effects in CCC shows up in the preferential (almost exclusive) occurrence of "magic" products in the exit channel.[1] The decisive influence of shell effects on the energetics[5] and dynamics[6] of cluster-fission has been demonstrated recently, too (see also the contribution of U. Landman et al., this volume).

Relaxation phenomena should be appropriately treated in terms of transport theories. As a basic entity of any transport theory, the interaction potential as function of the relevant collective degrees of freedom must be known. In the work to be reported here, a microscopic formalism for the calculation of the interaction potential between colliding clusters is presented. It is based on density functional theory in local density and a two-centre jellium approximation. The potential is calculated as function of the centre-mass distance and the size-asymmetry, being important collective degrees of freedom in collisions. First results for the $Na_{20} + Na_{20}$ system are presented. The role of shell effects and their consequences on the collisional dynamics are discussed. Relations to nucleus-nucleus potentials as well as phenomenological potentials[4] will be outlined.

2. The two-centre Kohn-Sham jellium formalism

The potential energy between colliding clusters is defined as the difference between the ground-state total energies of the two-cluster system $E_{tot}^{(1+2)}$ and the energies of the two separated clusters

$$U = E_{tot}^{(1+2)} - \left(E_{tot}^{(1)} + E_{tot}^{(2)}\right) \tag{1}$$

For the calculation of the energies we adopt the jellium approximation. The electronic ground-state configuration results from the self-consistent solution of the Kohn-Sham equation in local density approximation

$$\left(-\frac{1}{2}\Delta + V_{eff}(\vec{r})\right)\phi_i(\vec{r}) = \epsilon_i \phi_i(\vec{r}) \tag{2}$$

$$\begin{aligned} V_{eff}(\vec{r}) &= V_{eu}(\vec{r}) + V_{ee}(\vec{r}) + V_{xc}(\vec{r}) \\ &= V_{eu}(\vec{r}) + \int \frac{\rho(\vec{r}\prime)d^3r\prime}{\mid \vec{r} - \vec{r}\prime \mid} + V_{xc}(\vec{r}) \\ \rho(\vec{r}) &= \sum_i^{occ.} \mid \phi_i(\vec{r}) \mid^2 \end{aligned} \tag{3}$$

where ρ, V_{eu}, V_{ee} and V_{xc} stand for the electronic density, the electron-background potential, the electron-electron and the exchange-correlation potential, respectively. For the exchange-correlation potential and its energy density ε_{xc} we applied the formulae of Gunnarsson and Lundquist [7]. The total energy E_{tot} consists of the kinetic energy E_{kin}, the electrostatic energy E_{es} and the exchange-correlation energy E_{xc}

The positive charge density ϱ_+ as function of the centre-mass distance d betweeen the interacting clusters is treated in a simple geometrical model[5]: For $d \geq R_1 + R_2$, two spheres with radii $R_1 = r_s N_i^{1/3}$ are assumed, where r_s is the Wigner-Seitz radius ($r_s = 3.93$ a.u. for Na) and N_i the number of atoms in the cluster i. For distances $d < R_1 + R_2$, two overlapping spheres with constant charge density (i.e. volume conservation) are assumed. Thus, for $d = 0$ one approaches the fused spherical compound cluster with radius $R_0 = 2^{1/3} R_1$ in case of symmetric collisions $R_1 = R_2$. The pertinent potential V_{eu} of the positive charge is given analytically for $d \geq R_1 + R_2$. Due to the symmetry of two overlapping spheres ($d < R_1 + R_2$) of equal radii ($R_1 = R_2$), V_{eu} can be obtained using the superposition principle and solving the Poisson equation $\Delta V_{eu} = -4\pi\varrho_+(\vec{r})$ in the overlap region in spherical coordinates by a series expansion with Lengendre polynomials

$$V_{eu}(\vec{r}) = \frac{1}{2}\sum_{l=0}^{\infty}(2l+1)\{a_l r^l + b_l r^{-l-1}\}P_l(cos\vartheta) \tag{4}$$

which leads to analytical series expansions for the coefficients a_e and b_e.[9]

The two-centre Kohn-Sham problem (2,3) is treated in prolate spheroidal coordinates ξ, η, φ. The wave functions $\phi_i(\vec{r})$ are expanded in terms of the Hylleraas basis functions $\psi_{klm}(\vec{r})$ [10]

$$\begin{aligned} \phi_{i,m}(\vec{r}) &= \sum_{k,l} c^i_{klm}\psi_{klm}(\vec{r}) \\ \psi_{klm}(\xi,\eta,\varphi) &= (\xi^2-1)^{m/2}e^{-x/2}L^m_k(x)P^m_l(\eta)e^{im\varphi} \\ x &\equiv (\xi-1)/a \end{aligned} \tag{5}$$

containing generalized Laguerre and Legendre functions L^m_k and P^m_l, respectively. The expansion (5) leads to a generalized algebraic eigenvalue problem. For details and an optimal choice of the parameter a, see ref. 11.

The Poisson equation for the electrons $\Delta V_{ee} = -4\pi\varrho(\vec{r})$ is converted into a system of ordinary differential equations, using the ansatz

$$V_{ee}(\xi,\eta) = \frac{1}{2}\sum_{l=0}^{\infty}(2l+1)\chi_l(\xi)P_l(\eta) \tag{6}$$

The total energies of the individual clusters $E^{(1)}_{tot}$, $E^{(2)}_{tot}$ as well as the limiting case $E^{(1+2)}_{tot}(d=0)$ are, of course, calculated in spherical symmetry.[8]

3. Results and Discussion

3.1 Potential as function of the distance d

In Fig. 1, the calculated interaction potential $U(d)$ as function of the distance between two Na_{20} clusters is presented. The touching radius $R_{12} = R_1 + R_2$ of the jellium spheres is indicated by an arrow. Besides a small structure at large overlap ($d \approx 10$ a.u.) the potential is attractive. This behaviour corresponds to the expectation of a macroscopic liquid drop picture[4] where the gain of surface energy always induces attraction between interacting aggregates. Moreover, the maximum attractive force between both clusters is reached at the touching point R_{12}. This important property of the potential is in accord with the proximity theorem which states that the maximum cohesive force between two *curved* surfaces occurs at contact; for two spheres[4] it is given approximately by

$$F_{prox} = -4\pi\sigma\frac{R_1 \cdot R_2}{R_{12}} \tag{7}$$

with σ the surface tension. The proximity theorem has been checked to be valid for macroscopic objects [12] and has been successfully applied also for the nuclear part of the interaction potential between two heavy-ions.[12,13] In all cases we have investigated so far[14], the proximity theorem was found to be valid also for interacting clusters. This justifies the use of the phenomenological potential based on the liquid drop model and the proximity theorem to estimate the global behaviour of the fusion cross sections between metal clusters as function of cluster size, charge and energy.[4]

The absolute value of the proximity force is strongly influenced by the electronic shell structure of the clusters. For example, in case of $Na_{20} + Na_{20}$ considered here one has (with $\sigma = 200$ dyn/cm) a proximity force $F_{prox} \approx -0.23$ eV/a.u., whereas the microscopic force at contact is about -0.1 eV/a.u. and thus differs by a factor of two from the macroscopic value. This difference is due to shell effects; a similarly strong shell effect is observed for the Q-value for fusion $Q \equiv U(d = 0)$ which follows from our microscopic calculation to be ~ -1.13 eV (see Fig. 1), and differs again by a factor of two from the prediction of the liquid drop model

$$U(d = 0) = 4\pi\sigma(R_0^2 - R_1^2 - R_2^2) \tag{8}$$

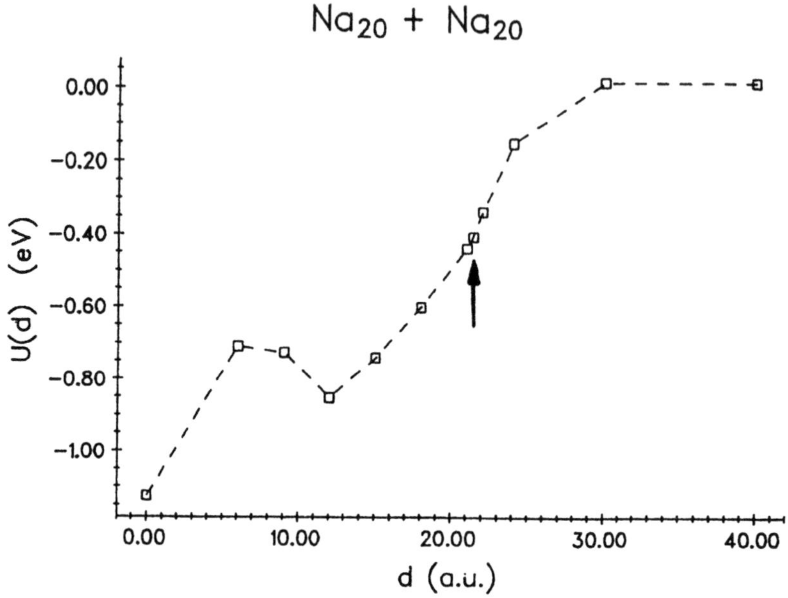

Fig. 1: Calculated potential energy U (squares) as function of the distance d for $Na_{20} + Na_{20}$. The dashed line is drawn as guide to the eyes. The arrow indicates the touching radius $d = R_1 + R_2$.

which yields -2.06 eV.

3.2 The potential as function of the size-asymmetry N_1/N

One of the most interesting relaxation processes in collisions between complex particles is the evolution of the mass-asymmetry which determines the final fragment distribution. In nuclear physics, the dynamics of the mass-asymmetry in collisions and fission has been, e.g., successfully treated using fragmentation theory based on the solution of the Schrödinger equation within the fragmentation potential[15] (see aso the contribution of W. Greiner, this volume). In the following, we will give an estimate of the fragmentation potential for atomic clusters which controls the exchange of atoms in CCC.

In quasielastic and deep inelastic collisions, only the tail of the potential energy at $d \gtrsim R_1 + R_2$ is of relevance. In this region, the interaction energy as function of the distance is always small as compared to the difference of the total cluster energies $E_{tot}^{(1)}$, $E_{tot}^{(2)}$. Therefore, to get a first insight, we neglect the former contribution and define the potential energy as function of the size-asymmetry N_1/N as

$$U(N_1/N) = E_{tot}(N_1) + E_{tot}(N - N_1) - E_{tot}(N) \tag{9}$$

where $E_{tot}(N_1)$, $E_{tot}(N - N_1)$ are the energies of the clusters containing N_1 and $N - N_1$ atoms, respectively, thereby keeping their total number $N = N_1 + N_2$ fixed. The normalization constant in eq. (9) is arbitrary; we normalize to the energy $E_{tot}(N)$ of the compound cluster having N atoms. Approximation (9) drastically simplifies the computational effort, allowing to use the spherical jellium model and thus contains the important shell effects.

In Fig. 2, the calculated fragmentation potential $U(N_1/N)$ for the Na_{40} system is presented and compared to the predictions of the liquid drop model, for which eq. (9) simply yields

$$U(N_1/N) = 4\pi\sigma r_s^2(N_1^{2/3} + (N - N_1)^{2/3} - N^{2/3}) \tag{10}$$

with the maximum value always at symmetry, $N_1/N = 0.5$. In contrast, large shell effects are clearly seen in the microscopic potential. Although, the spherical jellium model overestimates the shell energies in $U(N_1/N)$ [14], it reproduces correctly the qualitative trends; it thus allows to deduce the gross features of the final cluster-size distribution expected in experiments. For the Na_{40} system considered here in greater detail, the following points are of particular interest:

(i) In symmetric collisions $N_1/N = 0.5$ (i.e., the "double magic" system $Na_{20} + Na_{20}$) the size asymmetry remains frozen to a large extent because the potential has a pronounced minimum at this point. This prediction is in accord with molecular dynamics simulations of $Na_8 + Na_8$ collisions[2] where even in deep inelastic collisions the observed final products are identical with projectile and target.

(ii) In somewhat asymmetric collisions with $N_1/N \approx 0.4$ (i.e., $Na_{17} + Na_{23}$) one expects from Fig. 2 a strong tendency towards symmetry and thus an asymmetric final size distribution towards symmetry ($N_1 \approx N_2 \approx 20$).

(iii) For $N_1/N = 0.35$ (i.e., $Na_{13} + Na_{27}$) the potential exhibit a symmetric maximum. Hence, the size distribution of the products is expected to be very broad and symmetric.

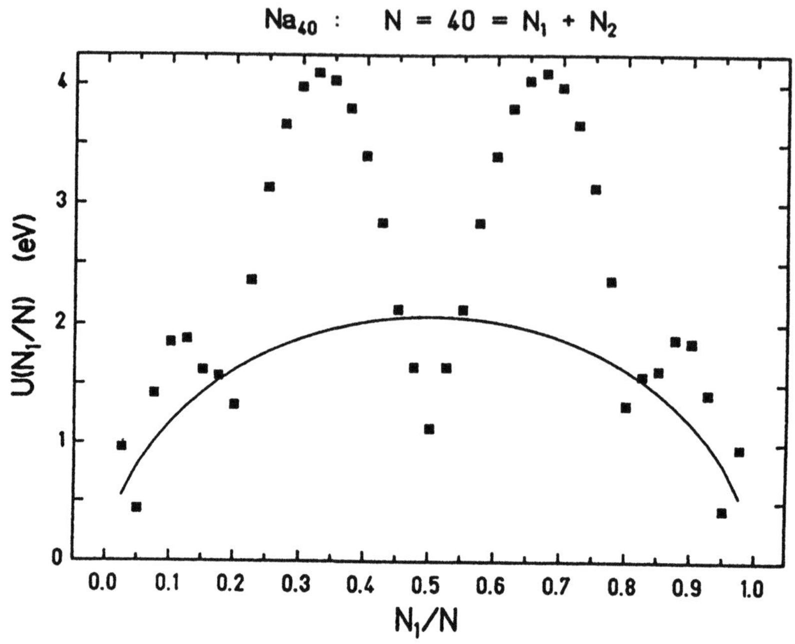

Fig. 2: Calculated fragmentation potential $U(N_1/N)$ for the Na_{40} system. Dots correspond to the quantum mechanical jellium model, the solid line is the "macroscopic" liquid drop potential.

(iv) Increasing the initial size-asymmetry further to about $N_1/N \approx 0.25$ (i.e., $Na_{10} + Na_{30}$ collisions) again an asymmetric distribution can be expected but with opposite tendency as given by (ii).

(v) Finally, for $N_1/N = 0.2$ (i.e., $Na_8 + Na_{32}$) the "magic" Na_8 produces a minimum in $U(N_1/N)$ which to a large extent should again preserve the initial asymmetry. Statistical fluctuations will lead to an asymmetric final distribution, in contrast to (i).

Altogether, the study of CCC promises a large varity of phenomena connected with the exchange of atoms, induced by the dominating influence of shell effects. For a quantitative theoretical prediction of final size distribution in CCC, the inclusion of deformation and charge of the clusters is essential.

Acknowledgement

This work has been supported by the Deutsche Forschungsgemeinschaft. One of us (O. K.) thanks the Alexander von Humboldt-Stiftung for a fellowship.

References

[1] R. Schmidt, G. Seifert and H.O. Lutz, Phys. Lett. A 158 (1991) 231; G. Seifert, R. Schmidt and H.O. Lutz, Phys. Lett. A 158 (1991) 237

[2] R. Schmidt, G. Seifert and H.O. Lutz, this volume, and to be published

[3] R. Schmidt and H.O. Lutz, Phys. Rev. A (1992), in print

[4] R. Schmidt and H.O. Lutz, Phys. Lett. A (1992), submitted

[5] S. Saito and S. Ohnishi, Phys. Rev. Lett. 59 (1987) 190; Y. Ishii, S. Saito and S. Ohnishi, Z. Phys. D 7 (1987) 289; S. Saito and M.L. Cohen, Phys. Rev. B 38 (1988) 1123 and Z. Phys. D 12 (1989) 205

[6] R.N. Barnett, U. Landman and G. Rajagopal, Phys. Rev. Lett. 67 (1991) 3058

[7] O. Gunnarson, B.I. Lundquist, Phys. Rev. B 13 (1976) 4274

[8] W. Ekardt, Phys. Rev. B 29 (1984) 1558

[9] J.D. Jackson, Classical Electrodynamics, John Wiley, New York 1962, second ed. 1975

[10] E. Hylleraas, Z. Phys. 71 (1931) 739

[11] A. Toepfer, E.K.U. Gross and R.M. Dreizler, Phys. Rev. A 20 (1979) 1808

[12] J. Randrup, Nucl. Inst. and Methods 146 (1977) 213

[13] J. Blocki, J. Randrup, W.J. Swiatecki and C.F. Tsang, Ann. Phys. 105 (1977) 427

[14] O. Knospe, R. Dreizler, R. Schmidt and H.O. Lutz, to be published

[15] H.J. Fink, W. Greiner and W. Scheid, in Heavy Ion Collisions, edited by R. Bock, North Holland, Amsterdam, Vol. 2, 1980, p. 399

Snowballs: Micro-Clusters in Liquid Helium as Tools in Nuclear and Condensed-Matter Physics

Noriaki TAKAHASHI
College of General Education,
Osaka University
Toyonaka/Osaka, 560 Japan

ABSTRACT: Snowball is a charged aggregate of helium atoms around an impurity ion in liquid helium. ^{8}Li nuclei as obtained with nuclear reaction $^{7}Li(d, p)^{8}Li$ were admitted into liquid helium and the ions were driven by an electric field to surface-barrier detectors. From the variation in number of detected alpha-particles with delay time, half-life of snowballs was obtained as 352 ± 82 ms. The observed long lifetime underlines the view of the experiment that the snowballs constitute a suitable milieu for preserving nuclear polarization throughout the lifetime of core ions ^{12}B produced in the reaction $^{232}Th(^{14}N, ^{12}B)$.

1. INTRODUCTION

Impurity ions introduced in liquid helium may behave as singly charged entities with effective mass of 50-100 He atoms, which are referred to as snowballs. A snowball is, therefore, an aggregate of helium atoms by means of electrostriction around an impurity core ion admitted in liquid helium.[1] Not only helium ions as first noticed but also any other alien ions do form snowballs.[2] The method used in the study of polarization phenomena in heavy ion reactions[3] enabled us to conceive renewed interests to approach individual snowballs through alpha- and beta-ray counting.

While the snowball has been thought of a highly permanent solid entity,[4] there are still doubt whether all the existing evidences do decisively favour the picture of a solid core.[5] The preservation of nuclear polarization provides with an important clue to this issue, enabling us to test the inherent structure of snowballs directly as well as to peripheral problems of the above experiment.

Here, in this report, preservation of nuclear polarization of core ions ^{12}B ($T_{1/2}$ = 20.4 ms, beta-radioactive) and measurement of lifetime of ^{8}Li ($T_{1/2}$ = 832 ms, beta- and alpha-radioactive) snowballs are presented. It has been proved that the snowballs are far long-lived as compared to ^{12}B, namely, half-life of snowballs was obtained as 350 ms. This observed long lifetime supports the view that the snowballs constitute a suitable milieu for maintaining nuclear polarization of short-lived core ions throughout their lifetime.

2. FORMATION OF SNOWBALLS

An impurity ion admitted in liquid helium creates a strong electric field around itself. The electrostriction exerted on the He atoms surrounding the impurity ion compels them to exhibit an extremely small but finite electric dipole moment, despite the fact that the electric polarizability of helium atom is approximately only 0.5 x 10^{-24} cm^2. To minimize the electrostatic energy, an aggregate of electrically polarized helium atoms is formed around the ion. The range of such an effect is calculated to be about 6 A, thus, we may think of a charged spherical aggregate created around the impurity ion. The internal pressure has been calculated by Atkins,[6] who attempted to simulate the aggregate by a continuous medium. We know that the pressure of at least 25 atm is needed to solidify helium at 0 K. The calculated pressure by far exceeds the value of this melting pressure at $r < 6$ A with a reasonable value for the surface tension. Such is a snowball, containing 50 - 100 helium atoms around a core ion and is singly charged.

We are thus led to consider that a snowball may well be of the solid entity, although there have been still no direct evidences for this statement.

3. EXPERIMENT

3.1. Freezing out of nuclear polarization of core ions

A 135-MeV nitrogen ion beam obtained from the cyclotron at the Research Center for Nuclear Physics, Osaka University, was shaped into pulses of 30-ms duration and 80-ms repetition and was focussed into a diameter of 4 mm on a target, a water-cooled thorium foil of thickness 20 mg/cm^2. Ions undergoing no interactions with Th nuclei were collected at a beam dump in the target chamber. Reaction products ^{12}B were taken out at 25 degrees with respect to the direction of the incident beam through a collimator into the cryogenic space and were introduced into liquid helium at a temperature of 1.7 K. The temperature was monitored by the helium vapour pressure in the superfluid chamber and by a germanium and a carbon resistors placed inside the

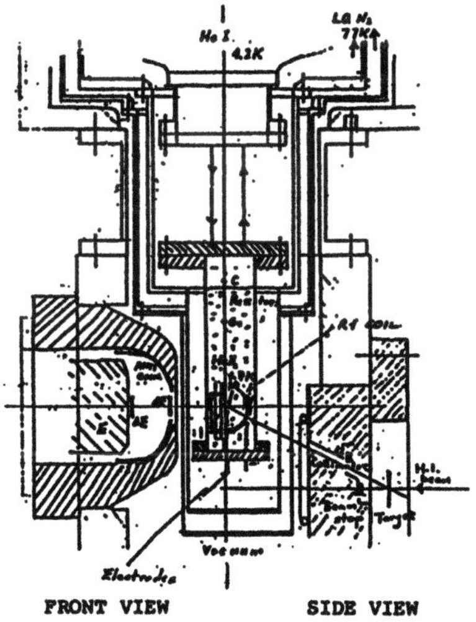

Fig. 1
Setup for polarization measurements

chamber. The chamber window was of 50-μm thick stainless steel and the reaction products with kinetic energy exceeding specified values determined from the range and the energy loss in the target and the window, penetrated into liquid helium. The residual range in the liquid helium is only a few mm at most. Most of the impinged ^{12}B ions were neutralized during the stopping process. A fraction of ions survived without being neutralized, until they formed snowballs. These snowballs were dragged to the 20-mm long central domain by a static electric field impressed in the 30-mm wide and 50-mm long superfluid chamber. The experimental setup is shown in Fig. 1.

A static magnetic field was produced all the way from the target to the superfluid chamber in the direction of the reaction normal and the strength around the domain was 0.9 kG with inhomogeniety less than 1 %. Beta rays from the unstable nuclei thus transported were detected during 40 ms of the out-beam periods by a pair of plastic counter telescopes deployed parallel to that direction.[7] Beta rays penetrated a 100 μm stainless steel window of the superfluid chamber and a 50-μm thick aluminium window of the vacuum chamber, before impinging a counter telescope which consisted of two energy-loss detectors of thickness 2 mm, one large energy detector and an anti-coincidence detector at the surface of the magnet poles. The energy threshold of the counter telescope was set at 3 MeV. Energy and time spectra were consistent with those of ^{12}B ($T_{1/2}$ = 20.3 ms, I^{π}= 1^{+} and $E_{\beta max}$ = 13.7 MeV).

Nuclear polarization of ^{12}B was reversed during every other out-beam period by applying the adiabatic-fast-passage NMR over the resonance frequency 675 kHz during 5-ms rf-periods before and after the beta-ray counting. A pair of rf-coils were installed in the chamber and were dipped completely in the liquid helium. The NMR was otherwise applied over a frequency 150 kHz apart from the resonance and the nuclear polarization was kept unaltered. By taking the average of the left-right ratios of the counting rates, the beta-ray asymmetry was obtained free of the instrumental asymmetries.

Polarization of the ^{12}B was preserved during the flight **in vacuo** from the target to the chamber window, since the ion was energetic enough to be in the fully stripped state and no hyperfine interactions in the ion were effective. The collision time is short enough whilst penetrating the solid material such as the chamber window and no deterioration of polarization took place.

The angular distribution of beta rays from the polarized ^{12}B is asymmetric, owing to the parity non-conservation in the weak interaction and reads

$$W(\theta) = 1 - P\cos\theta$$

with respect to the axis of ^{12}B polarization. Polarization P is expressed as

$$P = (1 - R^{1/2})/(1 + R^{1/2}),$$

where

$$R = (N(0)/N(\pi))_{off}/((N(0)/N(\pi))_{on}.$$

Here N's stand for the counting rates of beta rays during off- and on-resonance periods at = 0 and π.

Polarization measured was 7 ± 3 %. Though we need more counting statistics, this value agrees well with the measured average polarization of ^{12}B from the same reaction at 129 MeV and at 30 degrees, P = 10%.[7] Thus, we presume that the spin polarization of the core nucleus is essentially preserved through the lifetime.

3.2. Lifetime of Snowball

A beam of 150-MeV ^7Li ions was obtained from the cyclotron at the Research Center for Nuclear Physics and was shaped into pulses of 1.2-s duration and 3.2-s repetition. The beam was focussed into a diameter of less than 4 mm and was led to the target of CD_2 and the reaction products ^8Li were introduced into a cryogenic space at 0 degrees and into liquid helium at a temperature of 4.2 K through a window made of 50-μm thick aluminium. A part of ^8Li ions formed snowballs. These snowballs were displaced by approximately 10 mm by an electric field impressed perpendicular to the original beam direction between a grid and a surface-barrier detector in the liquid helium chamber.

The configuration near the centre of the cryogenic space is shown in Fig. 2. The electric potentials of the entrance window (A) and the detectors (E and F) were kept at +2 kV and 0 kV, respectively, and that of the grid (B) was +2 and +1 kV. The latter potential value was applied during an out-beam period after a certain delay time T_D from the end of the in-beam period and a measuring period was started. During the in-beam time and the subsequent delay time, the grid potential was kept at +2 kV. By changing the delay time, alpha-particles from $^8Li \rightarrow {}^8Be \rightarrow \alpha + \alpha$ were detected by the surface barrier detectors. This configuration enabled us to keep the produced snowballs in the vicinity of the grid when they were produced and later to transport and count them at the surface-barrier detector.

A delay time spectrum obtained with the alpha particle counting is shown in Fig. 3. A solid line is drawn for $T_{1/2}$ = 250 ms as fitted to the

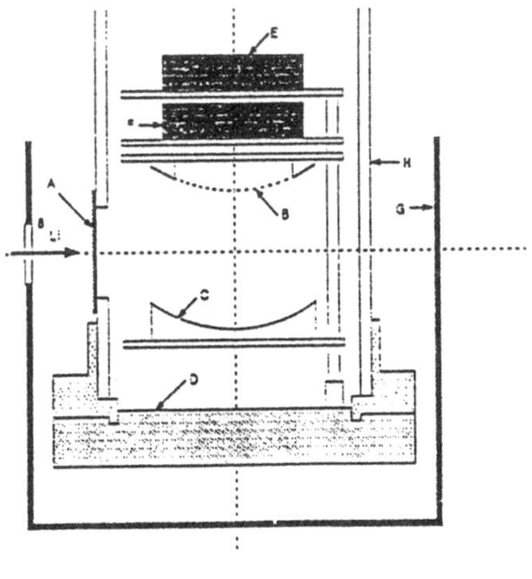

Fig. 2
Center of the cryogenic space

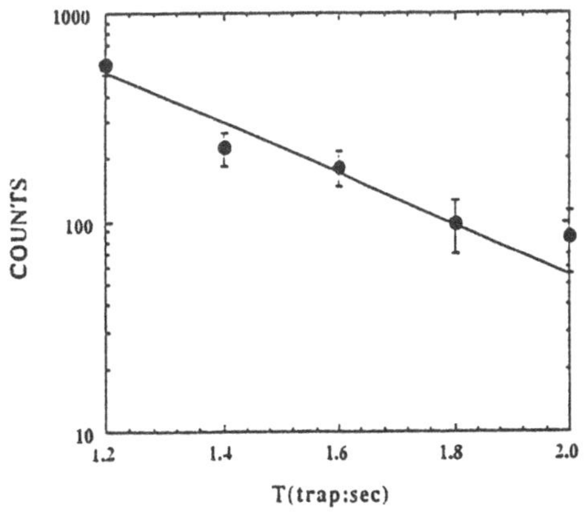

Fig. 3
Alpha-particle counting with delay time

measured points. This proves that snowballs are detected by means of alpha decay and the present alpha-particle detection method is also effective in determining the lifetime of snowball. We obtain from this observation a typical value for the half life of snowballs as 352 ± 82 ms, taking into account of the half-life of ^{8}Li nuclei. Our previous value with a conventional electric charge measurements produced a value for the half-life of about 1 s at a temperature of 1.7 K, i. e., with superfluid helium.

4. DISCUSSION

The observed lifetime strongly supports the view that the snowballs are as long-lived as many radioactive nuclei. Still the problem remains, if we are really observing snowballs. The answer is indirect but the drift velocity of the charge carriers we are dealing is in the right order of magnitude as that of snowball measured before by Meyer and Reif.[1)]

The lifetime obtained here is slightly smaller than the previous value. The reason is that the lifetime is the maximum expectation value in the measurements. The expected value depends slightly on the configuration of the apparatus used, moreover the alpha-particle is short-ranged and may be lost during transportation without reaching the surface of the detector, thus a slightly smaller value for the lifetime is quite possible to occur. In a previous measurement with a conventional electric charge method at 1.7 K, the lifetime obtained has been about 1 s. It is reported, the lower the temperature, the longer the lifetime of snowballs.[8)]

We are still far from inferring more quantitatively the internal structure of snowball, whether it is most likely of solid entity and it is highly symmetric, i. e., a picture of highly permanent cluster of helium atoms around a charged ion.[4)] The conclusion of Atkins' estimation in terms of the classical thermodynamical approach is plausible still at this stage and this description gives a clear idea; electrostriction effects increase the liquid density over a large region surrounding the ion.

5. CONCLUSION AND OUTLOOK

We have obtained a typical value for the lifetime of snowballs by use of alpha- and beta-particle detection. The lifetime was in the order of 1 s. This assures that the lifetime of snowballs are long enough so that the nuclear spin polarization of many short-lived nuclei can be maintained throughout the lifetime in them. This will supply us with an important means to explore, whether such snowball has a highly symmetric structure, most likely a permanent solid core.

The "freezing-out" of polarization characterizes a snowball as an ideal stopping material for preserving nuclear polarization for the purpose of determining the electromagnetic moments of unstable nuclei, irrespective of what atomic number they sustain and how far they are apart from the stability line. This will definitely open a way to further use of exotic heavy-ion beams[9] to expand the study of nucleus from the conventional narrow region close to the stability line to an ample unknown space in N-Z plane.

This novel detection method of snowball will enable us to observe more closely the transport phenomena of alien ions in liquid helium. Possibly the use of ^{6}He and ^{8}He beams will enable examination of new phase in the problems of liquidity and solidification of bulk helium as nucleated by naturally occurring ions.[10] For these purposes pulsed beams of beta-active nuclei will be introduced into the liquid helium as impurity ions and the snowballs formed around radioactive nuclei will be detected by the beta-rays rather than alpha-particles from the short-lived nuclei and the spatial distribution of the snowball will be measured.

6. ACKNOWLEDGEMENTS

Experiments mentioned here were carried out in a group of T. Shimoda, Y. Fujita, H. Miyatake, T. Itahashi of Osaka University and the author. The experimenters would like to thank the RCNP cyclotron staff for their help at every stage. Measurements were carried out at the RCNP under the programme numbers 26A04, 28A05, 29A08 and 30A21. This work is supported in part by the Grant in Aid of Scientific Research (Ministry of Education, Science and Culture, Tokyo) and Grants from the Mitsubishi Foundation.

REFERENCES

1) L. Meyer and F. Reif, Phys. Rev. 110 (1958) 279L and G. Careri, F. Scaramuzzi and J. O. Thomson, Nuovo Cimento 8 (1959) 1758
2) W. W. Johnson and W. I. Glaberson, Phys. Rev. Lett. 29 (1972) 214
3) See e.g., K. Sugimoto, M. Ishihara and N. Takahashi, Polarization Phenomena in Heavy-Ion Reactions in Treatise on Heavy-Ion Science, edited by D. A. Bromley, Vol. 3, Plenum Press, New York, 1985 and references therein. N. Takahashi, Hyperfine Interactions 21 (1985) 173 and K.H. Tanaka, Y. Nojiri, T. Minamisono, K. Asahi and N. Takahashi, Phys. Rev. C34 (1986) 580.
4) K. W. Schwarz, Adv. Chem. Phys. 33 (1975) 1
5) D. L. Goodstein, J. Low Temp. Phys. 33 (1978) 137
6) K. R. Atkins, Phys. Rev. 116 (1959)1339
7) N. Takahashi, Y. Miake, Y. Nojiri, T. Minamisono and K. Sugimoto, Proc. INS Intl. Symposium on Nuclear Direct Reaction Mechanism, INS, University of Tokyo, p. 635 (1979)
8) G. A. Williams and R. E. Packerd, J. Low Temp. Phys. 33 (1978) 459
9) I. Tanihata, H. Hamagaki, O. Hashimoto, S. Nagamiya, Y. Shida, N. Yoshikawa, O. Yamakawa, K. Sugimoto, T. Kobayashi, D. E. Greiner, N. Takahashi and Y. Nojiri, Phys. Lett. 160B (1985) 380, and I. Tanihata, H. Hamagaki, O. Hashimoto, Y. Shida, N. Yoshikawa, K. Sugimoto, O. Yamakawa, T. Kobayashi and N. Takahashi Phys. Rev. Lett. 55 (1985) 2676
10) M. W. Cole and T. J. Sluckin, J. Chem. Phys. 67 (1977) 746

4. Structure and Excitation

Collective Excitations in Silver Cluster Anions and Kations

J. Tiggesbäumker, L. Köller, H.O. Lutz, K.H. Meiwes-Broer

Fakultät für Physik, Universität Bielefeld, 4800 Bielefeld 1, F.R.G.

Abstract

Positive and negative silver cluster ions are produced by sputtering of solid silver by 25 keV Xe^+ bombardment. After mass selection, collinear ion beam depletion is used to measure absolute photofragmentation cross-sections in the photon energy range from $\hbar\omega = 2.0$ to 5.7 eV. Giant resonances are found which can be interpreted in terms of a collective electron oscillation. For spherical clusters (e.g., $Ag_{19}{}^-$, $Ag_{21}{}^+$) single resonances dominate the spectra, whereas deformation (e.g., $Ag_9{}^-$, $Ag_{11}{}^+$) induces a splitting into two components.

Introduction

Experimental evidence of strong photon absorption lines in metal clusters[1–6] (often termed "giant" or "plasmon" resonances due to similar phenomena in nuclei and solids) has fueled a debate on the microscopic origin and the correct description of this effect. From the solid state physics' point of view, a first estimate on the transition frequency can be obtained using the Drude-Mie theory. Its application is in principle fairly straightforward for large aggregates (R $\gtrsim$ 10nm), but necessitates increasingly severe corrections due to finite-size efffects as the cluster becomes smaller (cf. M. Vollmer and U. Kreibig, this volume). In such aggregates, the methods for the treatment of finite many-particle systems (as used, e.g., in nuclear and molecular physics) yield a rather complex behaviour of resonance structures, frequencies and widths as is expected from the decreasing density of states. In particular, for quite small clusters (say, N $\lesssim$ 10) HF-CI calculations reveal a very sensitive interplay between electronic and geometric cluster structure (cf. ref. 7); therefore, it is safe to state that in this size regime a detailed and quantitative understanding should be expected only on the basis of sophisticated *ab initio* methods. In the "transition" regime of medium cluster sizes (N $\gtrsim$ 10, where the discussion is most heated) LDA calculations on the optical response first performed by Ekardt revealed the existence of collective electron excitation.[8] Various theoretical techniques have been applied furtheron, also regarding, e.g., the influence of temperature or of the charge state.[9–12] Even here, significant differences in the excitation spectra are predicted for isoelectronic systems (for jellium-RPA calculations cf., e.g., Guet, this volume).

To shed more light on the behaviour of the "giant resonance" in the intermediate size regime, we are studying positively and negatively charged Ag clusters, in particular the isoelectronic situations ${Ag_{N+1}}^+$ and ${Ag_{N-1}}^-$. Beyond the questions touched upon above, these systems hold additional interest since they allow to study the core influence on the optical activity: It is well known that in the bulk the plasmon edge is shifted from the (Drude-)expected value of 8.8 eV to 3.9 eV due to the interband transition. Plasmon excitations in silver particles of the size 10–100 nm have been found in the near UV-region.[13–19] Very recently Harbich et al. obtained first absorption spectra of small mass selected silver clusters ($N \leq 11$) in a rare gas matrix.[20] Red- as well as blue-shifts of the measured resonance positions with varying N of these matrix deposited clusters produced some confusion in the past (for details see ref. 18). Therefore, we investigate the optical activity of mass selected silver clusters in a beam.[21] In this contribution the corresponding optical spectra of the closed-shell clusters ${Ag_{19}}^-$ and ${Ag_{21}}^+$ and the open-shell clusters ${Ag_9}^-$ and ${Ag_{11}}^+$ are presented.

Experimental set-up

The experimental set-up for production[22] and photofragmentation[21] of mass selected metal cluster ions has been described in detail elsewhere. Clusters are generated by bombarding a silver target with 25 keV Xe^+ ions. After acceleration to 1.8 keV and mass selection by a Wien filter, we obtain a monodispersed cluster kation or anion beam. Two collimators (1 mm diameter, distance 10 cm) confine the ion beam and define the interaction region. Here the clustes are irradiated collinearly with light from a pulsed excimer-pumped dye laser. In order to extend the photon energy region, second harmonic generation is achieved via frequency-doubling in BBO1 and BBO2. Additionally, fundamental and Raman-shifted excimer laser lines are used. Behind the interaction region the clusters are steered out of the laser beam axis by means of a static quadrupole ion deflector which acts as an energy selector field. Photodissociated clusters cannot reach the detector. Additionally, a grid may be used to repell cluster fragments. Care is taken to optimize the overlap of cluster and laser beam. A pyroelectrical detector serves to measure the laser intensity (typically 0.05–1.5 mJ/cm^2 per pulse) before and after each run. Depletion spectra are recorded for different photon energies $\hbar\omega = 2.0...5.7$ eV. Beer's law is used to extract absolute photofragmentation cross sections.

Results and discussion

Photofragmentation cross sections for some selected *Ag* clusters are shown in Figs. 1 and 2. Broad strong peaks dominate the spectra in the range investigated here; note that the intrinsic resolution of the experiment (mainly caused by the ion beam energy spread and the laser band width) is much narrower than the observed peaks, although to some extent the statistical error bars would allow for finer unresolved structure.

In all spectra we note an increased "background" at higher photon energies whose origin at present is not clear. Nevertheless, there is no doubt about the existence and overall shapes of the giant resonances. Their oscillator strength (OS) can be determined from the observed cross sections; normalized to the number of valence electrons, they range from approx. near 100% ($N = 8$) to 60% ($N = 20$), thus indeed being close to their maximum values. We note a general tendency of a decreasing OS with increasing number of atoms in the cluster. It cannot be excluded that this is due to the experimental method used: clusters can store a significant amount of energy for long times before they decay (cf. e.g. ref. 23); the "storage time" increases with N. This may cause a decrease of the measured apparent photofragmentation cross section within the experimental time window (roughly between 10 and 100 μs). The decreasing OS could, however, also be due to an increasing importance of photon absorption outside the covered spectral range. Corresponding further experimental work with an extended spectral range, as well as supporting calculations would be desirable.

As a first try, it might be tempting to explain the observed structures by the dipole term of the classical Mie theory.[24] In the Mie-Drude model of a free electron gas, the corresponding (N-independent) characteristic resonance energy of the surface plasmon $\hbar\omega_{Mie} = \hbar\omega_p/\sqrt{3} \approx 5.2$ eV is considerably

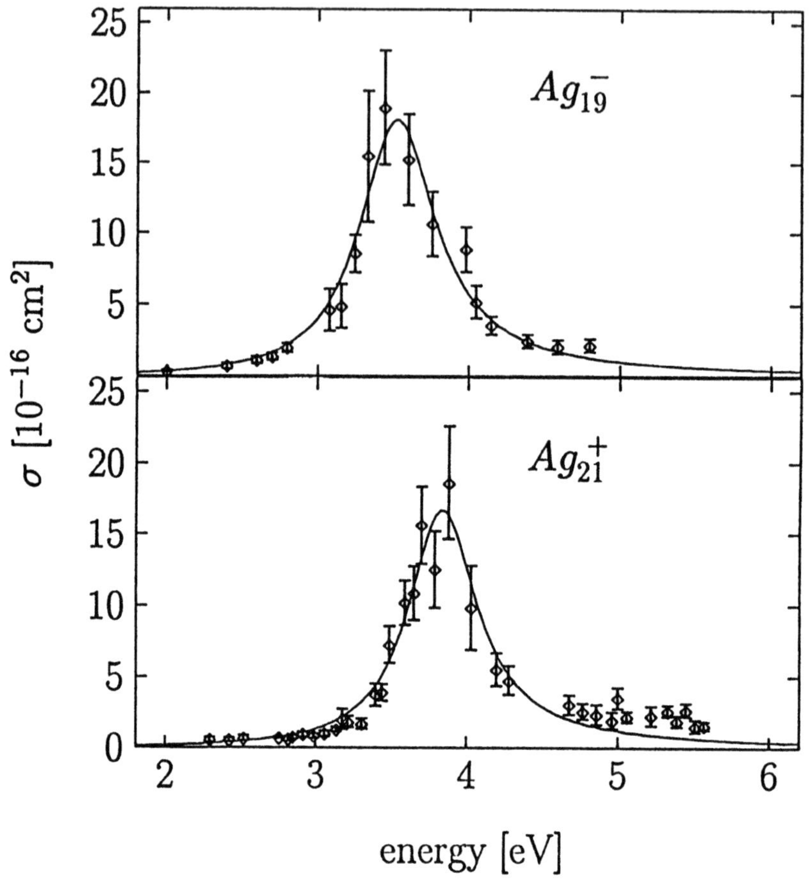

Fig. 1: Absolute photofragmentation cross sections of sputtered ${Ag_{19}}^-$ and ${Ag_{21}}^+$. According to the jellium picture these clusters are supposed to be spherical ($N = 20$). Lorentz curves as fit to the experimental values are given by the full lines. Fit parameters for:
${Ag_{19}}^-$: $\hbar\omega_0 = 3.52$ eV, $\Gamma = 0.61$ eV, $\sigma^{MAX} = 18.16$ Å^2
${Ag_{21}}^+$: $\hbar\omega_0 = 3.83$ eV, $\Gamma = 0.56$ eV, $\sigma^{MAX} = 16.69$ Å^2

higher than the observed features (between 3 and 4 eV) if one active electron per atom is assumed. Two competing effects have to be considered:

— The electron spill-out effectively reduces the restoring force on the displaced electrons, thereby increasing the cluster polarizability and thus causing a red-shift of the characteristic frequency. This effect has been discussed and tentatively been measured for alkali clusters. Assuming a similar situation in silver clusters, the additional charge of the *anions* should lead to an increased spill-out and thus to a decrease of the resonance energy. In

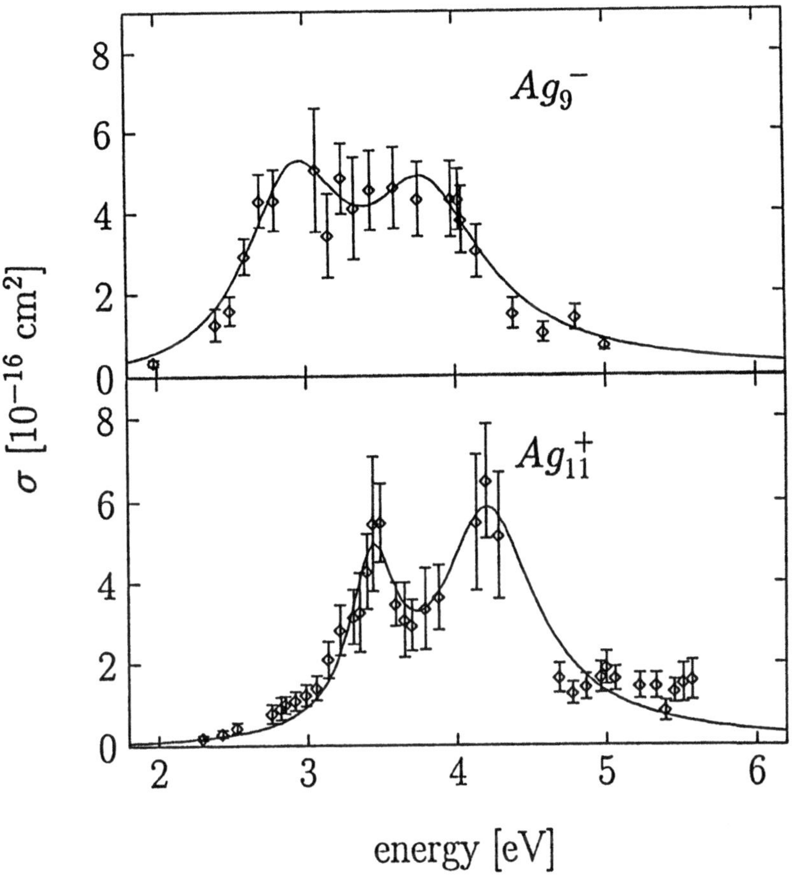

Fig. 2: Same as in Fig. 1, but now for $Ag_9{}^-$ and $Ag_{11}{}^+$ (deformed clusters $N = 10$). Fit parameters for:
$Ag_9{}^-$: $\hbar\omega_{01} = 2.94$ eV, $\Gamma_1 = 0.81$ eV, $\sigma_1^{MAX} = 4.62$ Å^2, $\hbar\omega_{02} = 3.81$ eV, $\Gamma_2 = 0.91$ eV, σ_2^{MAX} 3.87 Å^2
$Ag_{11}{}^+$: $\hbar\omega_{01} = 3.44$ eV, $\Gamma_1 = 0.39$ eV, $\sigma_1^{MAX} = 4.02$ Å^2, $\hbar\omega_{02} = 4.21$ eV, $\Gamma_2 = 0.77$ eV, $\sigma_2^{MAX} = 5.58$ Å^2

contrast, due to the missing charge in the *kations* we expect a relative contraction of the electron cloud, leading to a blue shift when compared to the negative isoelectronic clusters. Indeed, the measured resonance energies of the anions are in all cases below those of the respective positively charged clusters.

— As has been mentioned earlier, the d-electrons in silver have a considerable influence on the bulk plasmon frequency. For clusters, no quantitative calculations of their explicit influence are available; usually, this influence is empirically accounted for by the use of the material's optical constants.[25] For example, using the appropriate values for Ag in the dipole term of the Mie formula, a value of 3.5 eV is obtained, fairly close to the observed frequencies of small silver particles. For a somewhat more quantitative approach one may, e.g., apply the treatment of Apell et al.[26] who allow for a smooth variation of the dielectric function in the surface region and a non-local response of the electrons. Again using the bulk optical constants, we find a value of 3.83 eV for $N = 20$ in quite good agreement with our experimental result on $Ag_{21}{}^{+}$ (3.83 eV). A further discussion will be given elsewhere.[27]

The discussion so far rests on the implicit assumption of a collective electron response. Some shortcomings of such an approach have already been mentioned in the introduction. Another is, e.g., the eventual splitting of the resonances obtained from RPA calculations as opposed to the single frequency expected from the Mie model in case of a closed-shell situation. Further evidence for the necessity of a more sophisticated treatment can also be found in our experimental data. For example, the Mie-expectation of one strong line in closed shell cases (e.g., $N = 8, 20$) appears indeed to be realized. Also the two lines expected in case of a self-consistently deformed "non-magic" cluster may be found (c.f. Fig. 2). However, in the negative clusters the splitting is throughout less pronounced; instead the absorption profiles are relatively broad.

Admittedly, the examples discussed here refer to a size regime where a simple jellium treatment can only be a coarse approximation; it is not surprising that refinements which account for the discrete nature of the quantum mechanical electron states are necessary. For an overall representation of the "giant resonance" behaviour, however, the model of a collective valence electron excitation appears to be fairly acceptable even at cluster sizes around a dozen atoms.

In conclusion, giant resonances govern the optical response of small silver clusters ions and anions. For the resonance frequencies of the ions a semiempirical approach of Apell et al. and the use of the optical constants of silver turn out to yield values close to the measured ones. A satisfying and more detailed description based, e.g., on the jellium model, has still to be developed.

Acknowledgement

This work has been supported by the Deutsche Forschungsgemeinschaft.

References

[1] W.A. de Heer, K. Selby, V. Kresin, J. Masui, M. Vollmer, A. Châtelain, W.D. Knight, Phys. Rev. Lett. 59 (1987) 1805
[2] K. Selby, M. Vollmer, J. Masui, V. Kresin, W.A. de Heer, W.D. Knight, Phys. Rev. B 40 (1989) 5417
[3] C.R.C. Wang, S. Pollack, J. Hunter, G. Alameddin, T. Hoover, D. Cameron, S. Liu, M.M. Kappes, Z. Phys. D 19 (1991) 13
[4] H. Fallgren, T.P. Martin, Chem. Phys. Lett. 168 (1990) 233
[5] C. Bréchignac, Ph. Cahuzac, F. Carlier, J. Leygnier, Chem. Phys. Lett. 164 (1989) 433
[6] J. Blanc, M. Broyer, J. Chevaleyre, Ph. Dugourd, H. Kühling, P. Labastie, M. Ulbricht, J.P. Wolf, L. Wöste, Z. Phys. D 19 (1991) 7
[7] V. Bonačić-Koutecký, P. Fantucci, J. Koutecký, Chem. Rev. 91 (1991) 1035
[8] W. Ekardt, Phys. Rev. B 31 (1985) 6360
[9] D.E. Beck, Phys. Rev. B 35 (1987) 7325
[10] V. Kresin, Phys. Rev. B 40 (1989) 12507
[11] C. Yannouleas, J.M. Pacheco, R.A. Broglia, Phys. Rev. B 41 (1990) 6088
[12] G. Lauritsch, P.G. Reinhard, J. Meyer, M. Brack, Phys. Lett. A 160 (1991) 179
[13] J.D. Eversole, H.P. Broida, Phys. Rev. B 15 (1977) 1644
[14] Y. Borensztein, P. de Andrès, R. Monreal, T. Lopez-Rios, F. Flores, Phys. Rev. B 33 (1986) 2828
[15] L. Genzel, T.P. Martin, U. Kreibig, Z. Phys. B 21 (1975) 339
[16] K.-P. Charlé, W. Schulze, B. Winter, Z. Phys. D 12 (1989) 471
[17] J.R. Heath, Phys. Rev. B 40 (1989) 9982
[18] U. Kreibig, L. Genzel, Surf. Sci. 156 (1985) 678
[19] A. Henglein, Chem. Phys. Lett. 154 (1989) 473
[20] W. Harbich, J. Buttet, private communication
[21] J. Tiggesbäumker, L. Köller, H.O. Lutz, K.H. Meiwes-Broer, Chem. Phys. Lett. 190 (1992) 42
[22] W. Begemann, K.H. Meiwes-Broer, H.O. Lutz, Z. Phys. D 3 (1986) 183
[23] W. Begemann, K.H. Meiwes-Broer, H.O. Lutz, Phys. Rev. Lett. 56 (1986) 2248
[24] G. Mie, Ann. d. Phys. 25 (1908) 377
[25] P. Feibelmann, Prog. Surf. Sci. 12 (1982) 287
[26] P. Apell, Å. Ljungbert, Sol. St. Com. 44 (1982) 1367
[27] J. Tiggesbäumker, L. Köller, K.H. Meiwes-Broer, to be published

RPA IN NUCLEI AND METAL CLUSTERS

P.-G. Reinhard and S. Weisgerber,
Inst. f. Theor. Physik, Universität Erlangen
and
O. Genzken and M. Brack,
Inst. f. Theor. Physik, Universität Regensburg

1 Effective Mean-Field Models

The aim of this contribution is a comparison of resonance excitations in nuclei and metal clusters. Both system look quite different at first glance. In a nonrelativistic picture, nuclei consist of nucleons, protons or neutrons, and forces between them. Whereas the basic constituents of a metal cluster are atoms, split into ion-core and valence electrons. Nonetheless, there are many similarities: both systems show saturation (constant binding energy per particle, growth of the radius $\propto N^{1/3}$); both have significant shell effects which lead, e.g., to pronounced magic shells and spontaneous deformation for non-magic systems; both have a strong residual interaction and display accordingly strong collective resonances, the giant resonances in nuclei and the "plasmon" resonances in metal clusters. Of course, there are also noticeable differences: nuclei have an extraordinary strong spin-orbit force but metal clusters none; collective low-energy modes in nuclei (the surface vibrations) appear at a tenth of the energy of the giant resonances whereas the typical low-energy modes of clusters are the molecular vibrations (or phonons) which are usually much more separated from the energy of the electron resonances. The similarities and the differences make it very interesting to study the both systems in comparision. The transfer of the models and their critical test in the mutually other environment will deepen our understanding of finite Fermion systems and help to sharpen our theoretical weapons.

The preferred microscopic approach in both systems is a mean-field model using an effective Hamiltonian. An "ab initio" treatment is presently prohibitive in nuclear physics due to the enormous complications and the lack of a well defined starting point. But by far the most low-energy phenomena can be well desribed within a mean-field model using an effective energy-density funtional. The most widely used is the Skyrme energy functional [1]

$$\mathcal{E}_{Skyrme} = a\rho^2 + b\rho^(2+\alpha) + c\rho\tau + d(\nabla\rho)^2 + w\rho\nabla\mathbf{J} \quad (1)$$

$$+\text{similar with } (\rho_n - \rho_p) \quad (2)$$

Note that this functional does not only depend on the density ρ but also on the kinetic-energy density τ which is needed because the nucleon aquires significant effective mass, $m*/m \approx 0.7$ in nuclear matter. There is furthermore the dependence on the spin-orbit current $\mathbf{J}$ to account for the large spin-orbit force in nuclei. The terms in $\nabla\rho$ account for nonlocal effects to a certain extend. And finally, there is the difference between proton and neutron distribution ($\rho_n - \rho_p$ etc) as a further degree-of-freedom leading to the distiction between isoscalar (protons in line with neutrons) and isovector (protons against neutrons) vibrations. In the present stage, the functional (1) can at best be motivated from nuclear many-body theory but not derived quantitatively. One recurs to phenomenological fits of the few model parameters to nuclear ground state properties [2,3].

Ab initio calculations are performed in metal clusters [4]. Nonetheless, they are very tedious and mostly restricted to simple and small systems. A more practical approach is again to use an effective energy-density functional

$$\mathcal{E} = \mathcal{E}_{directCoulomb}(\rho) + \mathcal{E}_{ex}(\rho) + \mathcal{E}_{corr.}(\rho) \quad (3)$$

whith a local appoximation to the exchange, $\mathcal{E}_{ex}(\rho) \propto \rho^{1/3}$, and to the electron correlations, see e.g. ref. [5]. This functional is formulated completely in terms of the electron density ρ describing an electron with effective mass $m*/m = 1$ and no spin-orbit force. The functional can be derived systematically by computing a correlated homogenous electron gas at varying density. The application in finite systems implies then the Local-Density-Approximation, $\rho \rightarrow \rho(r)$. Although some precautions are in place (e.g. the Self-Interaction-Corrections), the use of effective energy-density functionals is well established in atomic and molecular systems [6].

However, there comes a further approximation into play in that only the valence electrons of metal atoms are treated explicitely. The remaining ions are smeared homogenously over a sphere of the radius $R = r_s N^{1/3}$ where r_s is the Wigner-Seitz radius of the system. This is called the jellium approximation. It is advantagous to allow for a finite surface thickness of the jellium, see the discussion around fig. 3 below.

2 Description of Resonances

The giant resonances in nuclei and metal clusters can be considered as collective modes where a large number of the particles move in phase thus ac-

Phase spaces and approximations to RPA		
Approximation	Phase space	Degree-of-freedom
Sumrule	$Q=r^L Y_{LM}$, $P=[H,Q]_{ph}$	no variation
Irrot. fluid dyn.	$Q=Q(r)$, $P=[H,Q]_{ph}$	spatial form $Q(r)$
Rotat. fluid dyn.	$P=\{v(r),\cdot\hat{j}(r)\}$, $Q=H^{-1}(P)$	velocity field $v(r)$
Standard RPA	$C^+=\sum(x_{ph}a_p^+a_h - y_{ph}a_h^+a_p)$	coefficients x_{ph}, y_{ph}
Mixed RPA	$Q,P \in \{a^+a, Q(r), \{v(r),\cdot\hat{j}(r)\}\}$	$Q(r)$, $v(r)$ and x,y

Table 1:

cumulating the observed large transition strengths. A simple and transparent description can be given in terms of fluid dynamics which employs only density- and current-distribution as the dynamical degrees-of-freedom. There are then two essentially different types of modes for finite droplets: First, a surface mode where the bulk of negative charges is displaced from the bulk of positive charges; this is the Goldhaber-Teller mode in nuclear physics. And second, the volume modes where the negative density is shifted against the positive density *within* the bounds of the droplet; this is the Steinwedel-Jensen mode in nuclear physics. The volume modes are classified by spherical Bessel functions $j_L Y_{LM}$ in a case of model with steep surfaces (L, M is the angular momentum of the mode). The surface mode, on the other hand, is generated by the simple multipole operator $r^L Y_{LM}$.

The standard microscopic description of the resonances is given by the RPA. In lowest approximation, one sees only the spectra of the pure particle-hole (ph) excitations where the transition strength is usually distributed over many modes. A strong residual interaction in RPA now recouples the ph-excitations such that one coherent superposition of states emerges which is significantly shifted in energy (e.g. shifted upwards in case of strong repulsive interaction) and with has gathered almost all collective transition strength. This is most easily understood in the schematic mode, see e.g. ref. [7].

An alternative and more flexible view of the RPA starts from the fact that RPA describes small amplitude oscillations. These can be described as a system of coupled harmonic oscillators with raw coordinates q_α and momenta p_α. The oscillators are transformed to normal coordinates Q_N and conjugate momenta P_N in the standard manner. The oscillators are quantized straightforwardly yielding the generator of an oscillator mode as $C_N^+ = \sqrt{\lambda/2}(Q_N - iP_N/\lambda)$. The normal coordinates are optimized for a given basis of raw coordinates $\{q_\alpha, p_\alpha\}$ by the variational condition

$$\delta < [C_N, [H, C_N^+]] >= 0 \tag{4}$$

with the constraint $< [C_N, C_M^+] >= \delta_{NM}$, see ref. [8]. This leaves a wide freedom in the choice of the basis which allows to cover within one and the same microscopic formalism almost all known models for resonances, from the simplemost sumrule approach, over the fluid dynamical description, up to full scale RPA. A list of raw basis sets and the corresponding level of approximation is given in table 1. It is to be noted that the irrotational fluid dynamics is equivalent to the local RPA of ref. [8], and there are also close connections to the TDLDA of ref. [9].

3 Results and discussion

We concentrate the following discussion on the modes with angular momentum L=1 which are the most prominent resonances in either system. The observation of this modes is related to the dipole operator $\propto r^1 Y_{1m}$ leading to the so called $B(E1)$ strength. It is worthwile to look first at the general spectral properties of this mode in nuclei and metal clusters. This is done in fig. 1.

The lowest resonance peak is called *surface mode* and the second strong peak *1.volume mode* indicating that there are many more volume modes at higher energies. In practice, of course, the actual modes do mix the different structures to some extend. In case of nuclei (upper part of fig. 1), we indicate only the surface mode which appears here as the dipole giant resonance although the higher modes do also exist. Note the different trends in the collective modes. The dipole resonance in nuclei decreases with increasing particle number N following nearly the trends in the *ph*-energies whereas the resonances in Na-clusters increase in energy with the increasing particle number showing a trend just opposite to the *ph*-energies. This is due to the residual interaction which is short range in nuclei but which is dominated by the long range Coulomb force in Na-clusters. As a consequence, there appears an interesting transition in the relative positions of the surface mode and the band of *ph*-excitations over three shells. The light cluster Na_8 stays clearly off any *ph*-excitation and lets us expect a unique collective peak whereas the larger clusters should exhibit some fragmentation of the strength. Similar trends are to be expected also for other metal clusters. But the relative positions (and thus the crossing point between resonance and three-shell band) change with the Wigner-Seitz radius of the system. The situation is much different in nuclei. There the spin-orbit splitting spreads the *ph*-bands significantly such that the upper end of the *ph*-band for transitions over one shell always just interferes with the resonance position which lets us expect some fragmentation in every case. It is to be noted, however, that there is still some uncertainty on the appropriate effective Skyrme force for nuclei which would cover ground-state properties as well as resonances. This means that the detailed interference effects between resonance and upper *ph*-energy can change quickly from one force to another.

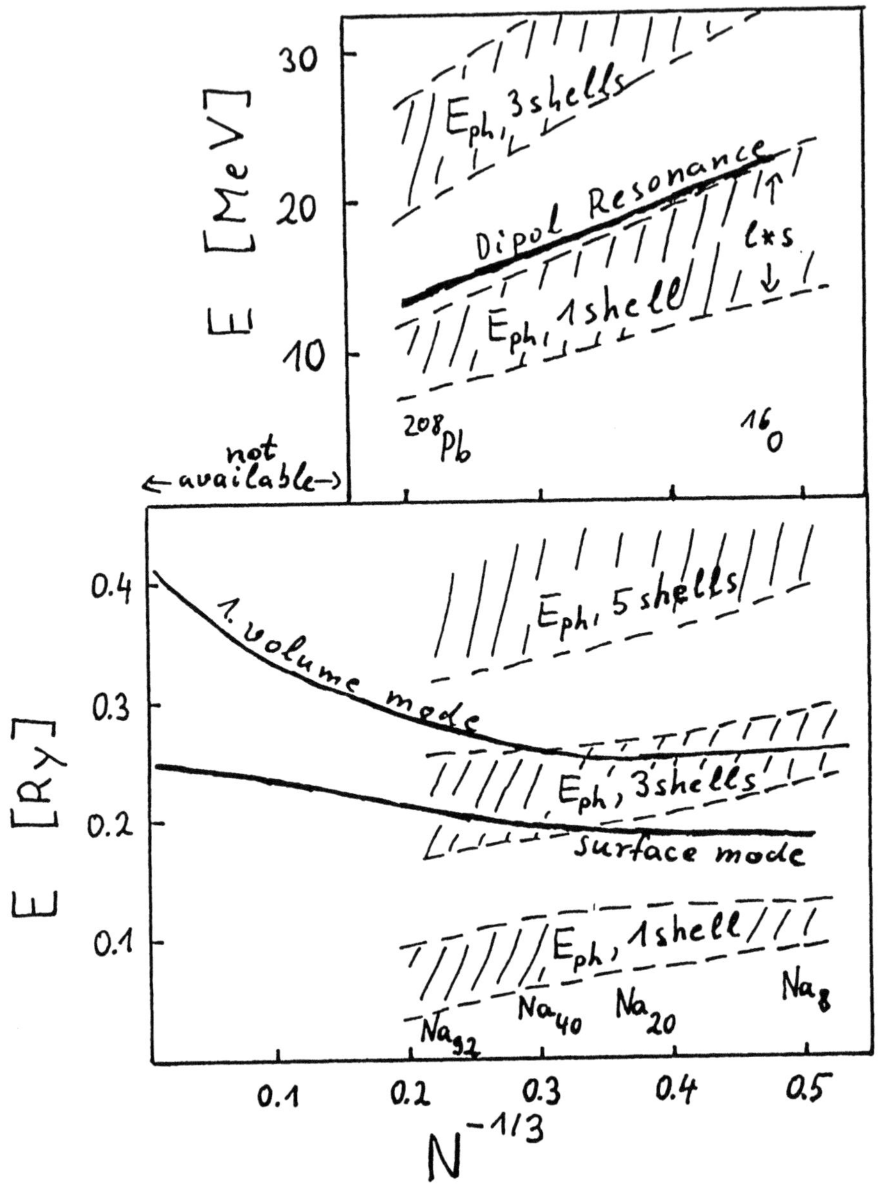

Figure 1:
The spectral properties of nuclei (upper part) and of Na-clusters (lower part) drawn versus $N^{-1/3}$. The limit $N^{-1/3} = 0$ corresponds to an infinite system. The location of special finite systems is indicated in the plot. The resonance energies are drawn in comparison to the bands of the pure ph-energies (E_{ph}).

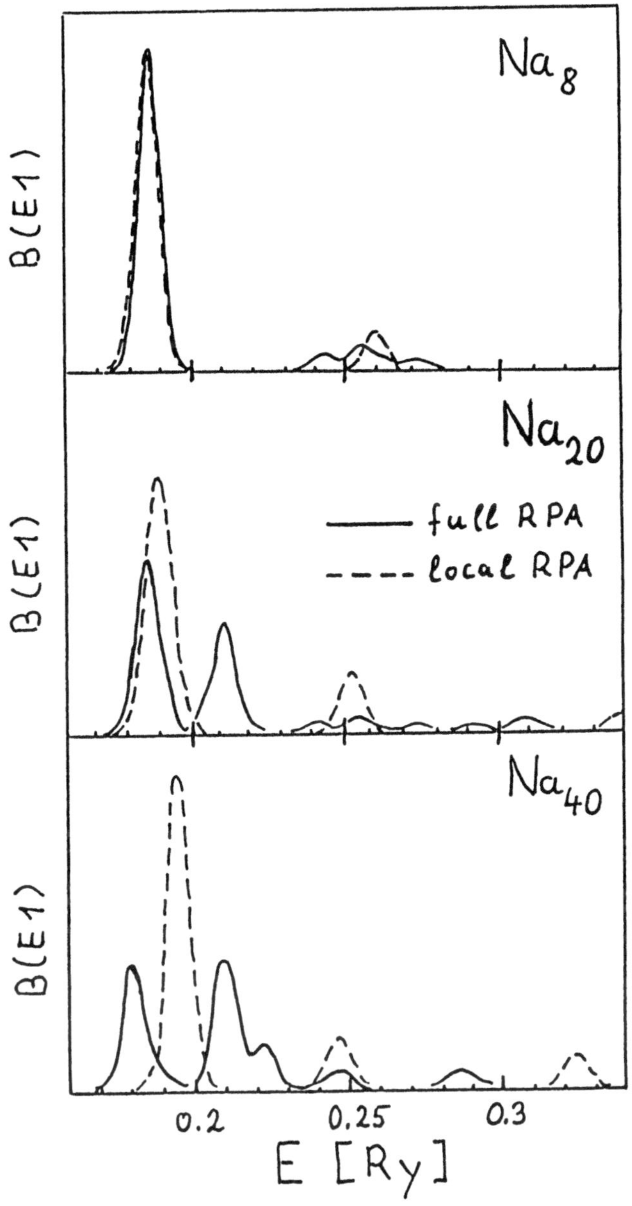

Figure 2:
The $B(E1)$ strength (in arbitrary units) versus excitation energy for the three light magic Na-clusters Na_8, Na_{20}, and Na_{40}. The discrete spectra have been folded with a Gaussian of width 0.005Ry to display a realistic resonance structure. The figures compare the full RPA (full line) with the local RPA alias irrotational fluid dynamical approach (dashed line).

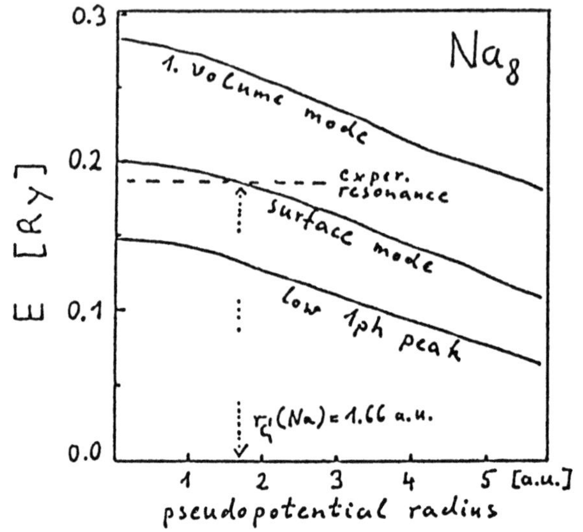

Figure 3:

The energies of the surface resonance, of the first volume mode, and of the most dominant 1ph-peak in Na_8 as function of the core radius r_C of the pseudopotential.

The transition of the fragmentation structure of the spectrum in *Na*-clusters is shown in fig. 2.

The local RPA (≡ fluid dynamical approach) shows clear collective pattern where already some dipole strength is distributed amongst the higher resonances (dominantly volume modes). The full RPA reproduces very well the unique resonance peak in Na_8. There is clearly an interference with one *ph*-state visible in Na_{20} leading to the splitting in just two peaks. That is the point where the resonance joins the band of three-shell transitions. The resonance has penetrated deeper in the *ph*-band for Na_{40} thus leading to a sizeable fragmentation (Landau damping) of the peak. This damping will hold on for a long series of higher clusters as can be read-off from the schematic view in fig. 1. We can also read-off from fig. 1 that the situation in nuclei will always look similar to the case of Na_{20} in fig. 2. This is indeed qualitatively true. We have to skip the details for reasons of space.

All the above calculations in *Na*-clusters employ the jellium approximation for the ionic background. The straightforward assumption of a steep jellium surface produces usually somewhat to high resonance energies and accordingly to low polarizabilities. It has been demonstrated in ref. [10] that a finite surface width can cure the problem. We have found an explanation for such a finite width: in the jellium approach, only the valence electrons are treated explicitly with the effective density functional. The core electrons are completely neglected. But the valence electrons should feel at least the

Pauli-repulsion from the core electrons. This can be taken into account fairly well by the so called pseudopotentials [11]. We thus have to fold the steep jellium background with a pseudopotential of some core radius r_C. The effect is drawn in fig. 3.

One sees that all energies react sensitively and similar with a decrease to an increasing core radius. It is an interesting coincidence that the experimental energy is reached just at the appropriate core radius for *Na* [11]. One would very much like to take this as the simple solution to the mismatch in resonance energy. However, there is an alternative explanation from the SIC effect [12] and we now probably overcorrected the energy. One needs to check carefully both explanations and their possible interference. Finally, there may very well be further contributions to the total polarizability which need to be looked at.

References

[1] P. Quentin, H. Flocard: Ann.Rev.Nucl.Part.Sci. **28** (1978) 523

[2] J. Bartel, P. Quentin, M. Brack, C. Guet, H.B. Hakansson, Nucl.Phys. **A386** (1982) 79

[3] J. Friedrich, P.-G. Reinhard, Phys.Rev. **C33** (1986) 335

[4] V. Bonacic-Koutecky, *see these proceedings*

[5] O. Gunnarson, B.I. Lundquist, Phys.Rev. **B13** (1976) 4274

[6] R.M. Dreizler, E.K.U. Gross, *Density Functional Theory*, Springer verlag, Berlin 1990

[7] P. Ring, P. Schuck, *The Nuclear Many Body Problem*, Springer Verlag, Berlin 1980

[8] P.-G. Reinhard, M. Brack, O. Genzken, Phys.Rev. **A41** (1990) 5568

[9] W. Ekardt, Phys.Rev. **B31** (1985) 6360; **B32** (1985) 1961

[10] A. Rubio, L.C. Balbas, J.A. Alonso, in *5th Int. Symposium on Small Particles and Inorganic Clusters*; Z.Phys. **D** in print

[11] N.W. Ashcroft, Phys.Lett. **23** (1966) 48

[12] J.M. Pacheco, W. Ekardt, *see these proceedings*

Optical Response of Doped S^1-Electron Metal Clusters

Tina A. Dahlseid, Geoffrey M. Koretsky, Stuart Pollack, C. R. Chris Wang, Joanna Hunter, George Alameddin, Douglas Cameron, Shengzhong Liu and Manfred M. Kappes
Department of Chemistry, Northwestern University, Evanston, IL 60208, USA

Introduction

Since the first photodepletion measurements of neutral alkali cluster electronic absorption [1], much effort has been devoted to obtaining more comprehensive and better resolved (~1Å) spectra for a number of pure and mixed lithium and sodium clusters [2],[3]. This effort has been complemented by extensive spectroscopic work on size selected s^1-electron metal cluster cations [4] and anions [5]. Comparison of the resulting data sets with the predictions of various computational models [1-3,6] has led to several conceptual advances including: (i) new insights into the evolution of collective electronic response as particle size (and electronic count) is varied and (ii) the realization that geometry (i.e. ion core positions) plays an important role in determining even the global absorption response of small clusters. While our understanding of neutral cluster optical response is rapidly developing, it is still far from complete. Three experimental avenues for further work have recently arisen: (a) dephasing dynamics, (b) temperature effects and (c) heteroatom doping [7]. We have recently been involved in the latter.

Impurity Doping:

Here the idea is to incorporate into an s^1-electron metal cluster (M_x) one heteroatom (A), which for large enough x perturbs only slightly the electronic properties of the heterocluster relative to those of a pure metal species with the same number of "delocalized" valence electrons. We are then interested in resulting changes to cluster geometry and electron localization and whether we can describe them qualitatively in terms of available models. There are two obvious variants: doping with either an electron withdrawing or electron donating impurity atom. We chose for the former a halogen atom (X = Cl, Br, I) and for the latter a group IIB element (Zn).

The global size dependence of electronic structure in Na_xX and Na_xZn has been previously accessed experimentally via abundance measurements in continuous beams [8],[9]. From these it is known that local stability islands occur for Na_9X, $Na_{21}X$, as well as for Na_8Zn, $Na_{18}Zn$. These observations have been rationalized in terms of a perturbed spherical jellium model in which the uniform potential assumed for homogeneous alkali clusters is modified to include a central heteroatom corresponding to a region of either lower or higher background positive charge density [10]. Within this paradigm Na_9Cl represents an 8-electron closed shell species containing a central Cl^- while Na_8Zn is a 10-electron closed shell due to a level filling order inversion associated with stabilization of all low angular momentum jellium states (relative to higher l's) by the centrally located Zn atom. The resulting photodepletion spectra are shown in figure 1. Compared to the homonuclear closed shell at Na_8, the Na_9Cl spectrum appears slightly red shifted and broadened. In contrast, the centroid of the Na_8Zn absorption spectrum (which now contains two peaks) is blue shifted relative to Na_8. Spectral shifts relative to Na_8 can be rationalized to first order in terms of lower and higher "metallic" electron densities, respectively. A more detailed analysis at the level of jellium on jellium RPA [11] and pseudopotential molecular orbital [12] calculations is ongoing.

Acknowledgements: This work was sponsored by the National Science Foundation under Grant No. CHE 9011341. Partial support was also obtained from the National Science Foundation through the Northwestern University Materials Research Center (Grant No. DMR 8821571). TAD acknowledges support through the ONR-NDSEG fellowship program. MMK acknowledges support through the Sloan Foundation.

References

1. W. de Heer, K. Selby, V. Kresin, J. Masui, M. Vollmer, A. Chatelain, and W. Knight, Phys. Rev. Lett, 59, 1805 (1987); K. Selby, M. Vollmer, J. Masui, V. Kresin, W. A. de Heer, and W. D. Knight, Z. Physik D12, 477 (1989); K. Selby, V. Kresin, J. Masui, W. A. de Heer, W. D. Knight, Phys. Rev., B40, 5417 (1989).
2. M. Broyer, G. Delacretaz, P. Labastie, J. Wolf and L. Wöste, Phys. Rev.

Lett., 57, 1851 (1986); M. Broyer, J. Chevaleyre, P. Dugourd, J. Wolf, and L. Wöste, Phys. Rev. A, 42, 6954 (1990); J. Blanc, V. Bonačić-Koutecký, M. Broyer, J. Chevaleyre, P. Dugourd, J. Koutecký, C. Scheuch, J. Wolf, and L. Wöste, submitted for publication.

3. C. R. C. Wang, S. Pollack, D. Cameron, M. M. Kappes, J. Chem. Phys., 93, 3787 (1990); S. Pollack, C. R. C. Wang, M. M. Kappes, J. Chem. Phys., 94, 2496 (1991); S. Pollack, C. R. C. Wang, T. A. Dahlseid, M. M. Kappes, J. Chem. Phys., to be published.
4. M. Jarrold and K. Creegan, Chem. Phys. Lett., 166, 116 (1990).
5. C. Petiette, S. Yang, M. Craycraft, J. Conceicao, R. Laaksonen, O. Cheshnovsky, and R. Smalley, J. Chem. Phys. 88, 5377 (1988); O. Cheshnovsky, K. Taylor, J. conceicao, and R. Smalley, Phys. Rev. Lett. 64, 1785 (1990); G. Ganteför, M. Gausa, K.-H. Meiwes-Broer, H. Lutz, J. Chem. Soc. Faraday Trans. 86, 2483 (1990).
6. V. Bonačić-Koutecký, P. Fantucci, J. Koutecký, Chem. Rev. 91, 1035 (1991).
7. H. Fallgren, T. P. Martin, Chem. Phys. Lett. 168, 233 (1990).
8. U. Heiz, U. Röthlisberger, A. Vayloyan, E. Schumacher, Israel J. Chem. 30, 147 (1990).
9. S. Pollack, C. R. C. Wang, M. M. Kappes, Z. Phys. D., 12, 241 (1989).
10. S. B. Zhang, M. L. Cohen, M. Y. Chou, Phys. Rev. B. 36, 3455 (1987); C. Baladrón, J. A. Alonso, Phys. Lett. A 140, 67 (1989).
11. S. Pollack, C. R. C. Wang, T. A. Dahlseid, G. M. Koretsky, M. M. Kappes, to be published.
12. V. Bonačić-Koutecký, to be published.

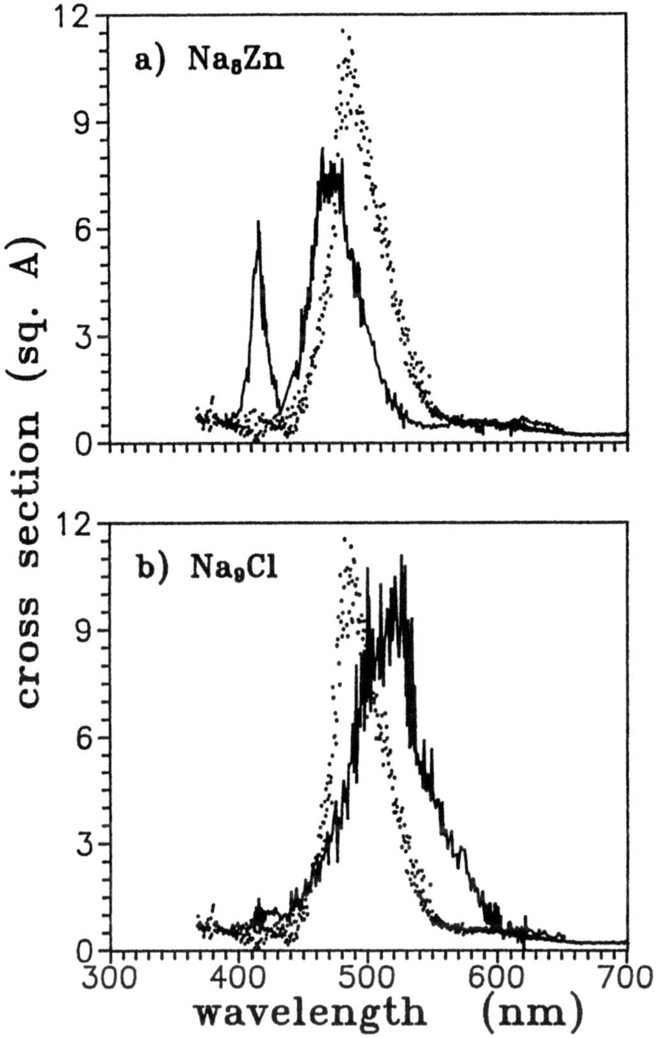

Figure Caption: Photodepletion spectra obtained for Na_9Cl and Na_8Zn. In each case the measurements are compared to Na_8(...).

Collective Excitations in Large Metal Clusters

M. Vollmer[1] and U. Kreibig[2]

[1] Fachbereich Physik der Universität Kassel,
Heinrich Plett Str. 40, 3500 Kassel, FRG
[2] 1. Physikalisches Institut der RWTH Aachen,
Sommerfeldstraße, Turm 28, 5100 Aachen, FRG

Abstract:
Studies of the evolution of the optical properties of metal clusters as a function of size have gained considerable attention in the last few years. One theoretical approach starts from large metal clusters, which can be described by classical electrodynamics, provided the dielectric functions of the clusters are known. The resulting resonant features in the absorption spectra are commonly called surface plasmons and are collective excitations of the electron system.The present paper discusses the electrodynamic *(Mie)* theory for large clusters of different metals, also considering the range of validity of this approach towards smaller cluster sizes. More details can be found in an extended review on this topic which is to be published soon [1].

Optical Response of Metal Spheres

The interaction of spheres of arbitrary material with electromagnetic waves was treated already by Gustav Mie in 1908 [2] by solving Maxwell's equations with appropriate boundary conditions using a multipole expansion of the incoming electromagnetic field (see also [3, 4]). Input parameters are the refractive indices or the dielectric functions of the particle and of the surrounding medium. The boundary conditions are defined by the electron density, which is assumed to have a sharp discontinuity at the surface of the cluster with radius R. Following e.g. Bohren and Huffman [4] the extinction, scattering, and absorption cross sections can be calculated analytically to yield

$$\begin{aligned} \sigma_{ext} &= \frac{2\pi}{k^2} \sum_{L=1}^{\infty} (2L+1)\operatorname{Re}\{a_L + b_L\} \\ \sigma_{sca} &= \frac{2\pi}{k^2} \sum_{L=1}^{\infty} (2L+1)\left(|a_L|^2 + |b_L|^2\right) \\ \sigma_{abs} &= \sigma_{ext} - \sigma_{sca} \end{aligned} \tag{1}$$

where k is the wavevector and a_L and b_L are coefficients, containing Bessel and Hankel functions which depend on the complex index of refraction of the particle, the real index of refraction of the surrounding medium and the size parameter x=kR. The general problem consists in the calculation of the coefficients a_L and b_L with recurrence relations for the Bessel functions or series expansions.

The sum index L gives the order of the vector spherical harmonic functions which enter the expressions for the electric and magnetic fields and thus describes the order of spherical multipole excitations in the clusters. L=1 corresponds to dipole fields, L=2 to quadrupole and L=3 to octupole fields. Far away from the cluster the waves are identical to waves coming from equivalent multipoles. Each multipole is split into electric and magnetic partial waves due to plasmon polariton and eddy current excitations, respectively [5,6]. In the following the magnetic contributions will be omitted since they are usually small in the investigated spectral region.

Quite often plasmon polaritons are briefly called plasmons. Whenever used the latter notation should not to be mixed with free plasmons in clusters which are excited e.g. by fast electrons. Plasmon polaritons are restricted to excitations by the electromagnetic field. Only in the *quasistatic approximation* for very small spheres (R«λ) there is no difference between both.

In the *quasi static regime*, phase retardation and effects of higher multipoles are neglected and the problem is simplified considerably. The extinction cross section is then solely due to the electric dipole, i.e. to the lowest order term a_1:

$$\sigma_{ext}(\omega) = 18\pi \frac{\varepsilon_m^{3/2}}{\lambda} V_0 \frac{\varepsilon_2(\omega)}{\left(\varepsilon_1(\omega)+2\varepsilon_m\right)^2+\varepsilon_2(\omega)^2} \qquad (2)$$

V_0 denotes the particle volume, ε_m and $\varepsilon(\omega) = \varepsilon_1(\omega) + i\,\varepsilon_2(\omega)$ denote the dielectric functions of the medium and particle, respectively. Extinction is only due to the dipole absorption being proportional to the particle volume whereas scattering is negligible in this case.

The cross section has a resonance whose position and shape are governed completely by the dielectric functions. The dipole resonance is determined by the condition $\varepsilon_1(\omega) = -2\,\varepsilon_m$ provided $\varepsilon_2(\omega)$ is not too large and does not vary much in the vicinity of the resonance. Steep $\varepsilon_1(\omega)$ spectra yield narrow resonances whereas low $d\varepsilon_1(\omega)/d\omega$ and large $\varepsilon_2(\omega)$ tend to smear out the resonances, sometimes past recognition. The resonance condition also implies that it is possible within certain limits to choose a matrix material for embedded spheres such that the plasma frequency shifts to a desired wavelength range.

In the Drude/Lorentz/Sommerfeld model the dielectric function for free electron metals (e.g. the alkali metals) is governed by transitions within the conduction band. It is calculated from the equation of motion of a free electron of mass m_e and charge e subject to an external electric field. Using a phenomenological damping constant Γ for a system of free electrons the condition $\omega \gg \Gamma$ gives the dielectric function $\varepsilon(\omega) = \varepsilon_1(\omega) + i\,\varepsilon_2(\omega)$:

$$\varepsilon_1(\omega) \approx 1 - \frac{\omega_p^2}{\omega^2}, \qquad \varepsilon_2(\omega) \approx \frac{\omega_p^2}{\omega^3}\Gamma \qquad (3)$$

Under the restriction $\omega \gg \Gamma$ the resonance position and shape can be determined by inserting Eq. 3 in Eq. 2. In the vicinity of the resonance the lineshape is then described by a Lorentzian

$$\sigma_{ext}(\omega) = \sigma_0 \frac{1}{(\omega - \omega_1)^2 + \left(\frac{\Gamma}{2}\right)^2} \qquad (4)$$

In other metals a substantial amount of interband transitions from lower lying bands into the conduction band or from the conduction band into higher unoccupied levels is possible, which alters the simple form of Eq. 3 for $\varepsilon(\omega)$ [7]. For the alkali metals the interband threshold given by excitations of conduction band electrons to higher levels lies at about 0.64 E_F (e.g.[8]), however, their contributions to the dielectric function are small. In contrast, the relevant interband transitions for the noble metals are due to excitation of d-band electrons into the conduction band. They give a positive contributiom $d\varepsilon_1$ to ε_1. Consequently the dipole resonance frequency given by $\varepsilon_1 = -2\,\varepsilon_m$ is shifted to lower frequencies, in the case of noble metals beyond the low frequency interband transition edge. For other metals these effects complicate the optical spectra appreciably and in fact, only a few materials like the alkali and the noble metals as well as aluminum exhibit sharp resonances.

With Eqs. 1, 2 the optical response of metal spheres to incident electromagnetic waves can be calculated. As input parameters the Mie theory uses the phenomenologically introduced dielectric function $\varepsilon(\omega)$ for the clusters. It should be stressed that Mie theory gives no insight whatsoever to the microscopic excitation mechanisms in the particle material. These are exclusively contained in the applied $\varepsilon(\omega)$. Any strong deviations from free electron behavior make it essentially impossible or at least very difficult to derive $\varepsilon(\omega)$ from a microscopic theory.

Fortunately a huge amount of experimental optical material functions for bulk solids which already incorporate all electronic effects is available (e.g. [9]).

Optical properties of clusters show two different kinds of size effects which can be readily attributed to different parts of the theoretical description:

1) Extrinsic size effects

For clusters larger than about 10 nm diameter, the optical material functions $\varepsilon(\omega)$ are size independent, having the values of bulk material (see below). The change of the spectra with size is then dominated by retardation effects of the electric field across the dimension of the particle which can cause huge shifts and broadening of the resonances. Consequently the size dependence of the optical spectra of large clusters is an *extrinsic cluster size effect* due to electrodynamics of the excitation which is governed only by the dimension of the particle with respect to the wavelength of the light.

2) Intrinsic size effects

For small clusters the optical material functions do no longer have the values of bulk material, but vary as a function of particle size. This is an *intrinsic cluster size effect*, as the material properties give rise to a change of the optical response. Still, the Mie theory formula Eq. 2 can be a good description, if a proper dielectric function is used. In fact, various size dependences of the optical material functions have been investigated by studying the optical response of cluster samples. The values of $\varepsilon(\omega)$ ascribed to some cluster are therefore those which - when being introduced into Mie theory - yield the very spectral response which is observed experimentally.

Roughly speaking extrinsic size effects dominate for clusters with $R \geq 10$nm whereas intrinsic size effects become important for smaller ones.

Extrinsic Size Effects

In the following we present some examples of Mie theory calculations for sodium, potassium, and silver clusters. Fig. 1 shows the absorption cross section as function of wavelength for sodium cluster spheres in vacuum ($\varepsilon_m=1$) with fixed mean cluster radii R_0 = 20 nm, 40 nm, 60 nm, 80 nm, and 100 nm, and FWHM of the cluster size distribution of 5% and 100%, respectively. The wavelength range was chosen from 330nm to 800 nm, using the optical functions from inagaki et al [10]. Several important features can be illustrated with Fig. 1.

1) For the nearly monodisperse particles (5% FWHM) of 20 nm mean radius the absorption cross section is dominated by the dipole resonance centered around λ = 400 nm. Contributions of higher multipoles, though present, are negligible. The position of the resonance at a wavelength λ_{res}= 400 nm already indicates the importance of a complete electrodynamic calculation. The electrostatic approximation without phase retardation effects would lead to a size independent resonance at λ_{res}= 384 nm for the dipolar mode. Although the particle size is only 10% of the wavelength of the light, the shift that is introduced by phase retardation amounts to about 16 nm. A further decrease of cluster size in our calculation (not indicated in Fig. 1) leads to dipole resonances whose peak positions move towards the lower limit of 384 nm. The difference between λ_{res}= 384 nm derived from the optical functions and the prediction of λ_{res}= 362 nm from the simple Drude theory is due to the neglect of ion core effects for the polarizability. The width of the resonance for small clusters in the quasistatic regime is determined by the dielectric functions [11] and reduces to the phenomenologically introduced damping constant Γ (Eq. 3) in the case of free electrons (but see also below). For larger clusters the width is increased by phase retardation effects.
2) With increasing cluster radius, higher order resonances come into play in the absorption cross section. For R = 40 nm the quadrupole oscillation is clearly visible. It even overcomes the dipole contribution for R = 60 nm. Further increase in particle size leads to octupole resonances and for R = 100 nm the dipole resonance is completely negligible as compared to

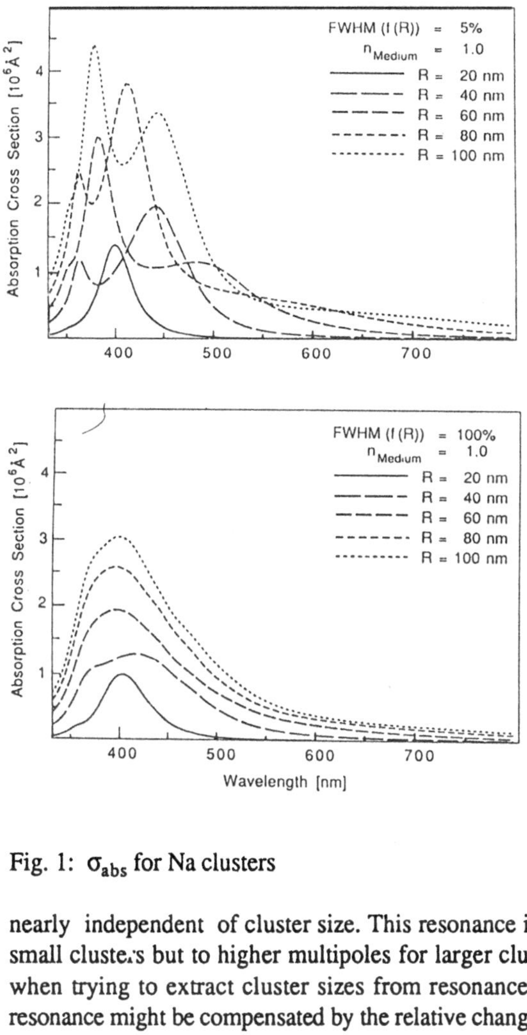

Fig. 1: σ_{abs} for Na clusters

quadrupole and octupole terms. The situation is different for the extinction cross section: due to dominating dipolar scattering the dipole contributions still determine these spectra. Fig. 1 illustrates the limitations for applicability of the *quasistatic limit* of Mie theory. In all cases where particles are not small compared to the wavelength of the light higher multipole resonances and phase retardation effects are important leading to redshifts of the resonances with increasing cluster size.

3) Finally Fig.1 also illustrates the pronounced changes of the absorption cross section which are due to a non-zero width of the cluster size distribution. In the lower part of Fig. 1 the FWHM of the size distributions is increased from 5% which corresponds to nearly monodisperse clusters to 100%. A FWHM of 100% of a cluster size distribution with mean radius R = 100 nm means e.g. that the distribution contains mainly clusters from about 50 to 150 nm size. Clearly the structures due to the various multipoles are completely washed out and it seems that only one single resonance dominates the spectrum whose position λ_{max} is nearly independent of cluster size. This resonance is, in fact, due to dipole contributions for small clusters but to higher multipoles for larger clusters. Consequently caution is necessary when trying to extract cluster sizes from resonance positions, as the shift of one multipole resonance might be compensated by the relative change of other multipole contributions.

In most experiments a directly accessible quantity is the extinction cross section, which is the sum of absorption and scattering processes. Figure 2 depicts a plot of the theoretical extinction cross section of nearly monodisperse K clusters (FWHM = 5%) versus wavelength and radius. As can be seen by the change in the absolute scale, scattering dominates the extinction spectra for large particles (absorption cross sections of Na and K are similar). With increasing particle size scattering, i.e. radiation damping, causes a drastic shift, broadening, and increase of the magnitude of the scattering cross section. It is this effect which mainly dis-

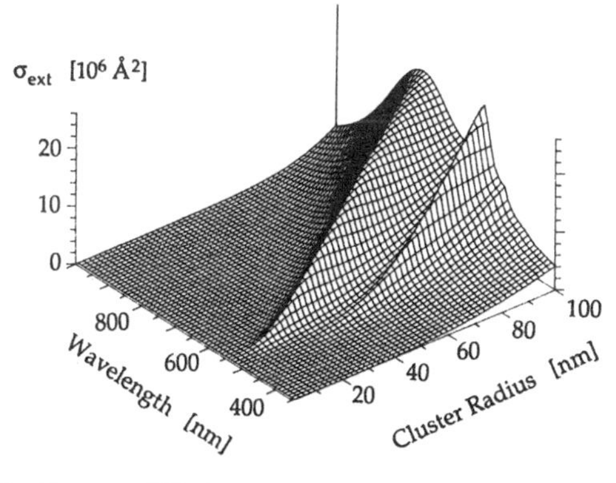

Fig. 2: σ_{ext} of K clusters

tinguishes plasmon polariton from free plasmon excitations. For 20 nm and 40 nm particles the shape of the extinction spectra is still similar to absorption but an increase of cluster size leads to a considerable broadening of the extinction spectra. Increasing also the width of the cluster size distribution leads to an additional broadening of the resonances. Any kind of structure is washed out if the FWHM of the cluster size distribution exceeds about 50%. In this case extinction spectra are thus not very sensitive on mean cluster size.

In experiments with clusters in matrices the value of ε_m has to be increased above 1. The same is possible for clusters on surfaces. The assumption that the effect of the substrate may be taken into account by simply replacing the spherical particle on a substrate by a spherical particle in a matrix with an averaged effective dielectric constant serves as a first approximation of the problem [1] and gives reasonable results. The influence of an effective index of refraction on the absorption cross section is a redshift which for Na amounts to about 50 nm when increasing n_m from 1.0 to 1.15 [12]. Neglecting a slight broadening due to an increased imaginary part of the dielectric function, the shape of the resonances remains more or less the same, as the boundary conditions are only slightly changed.

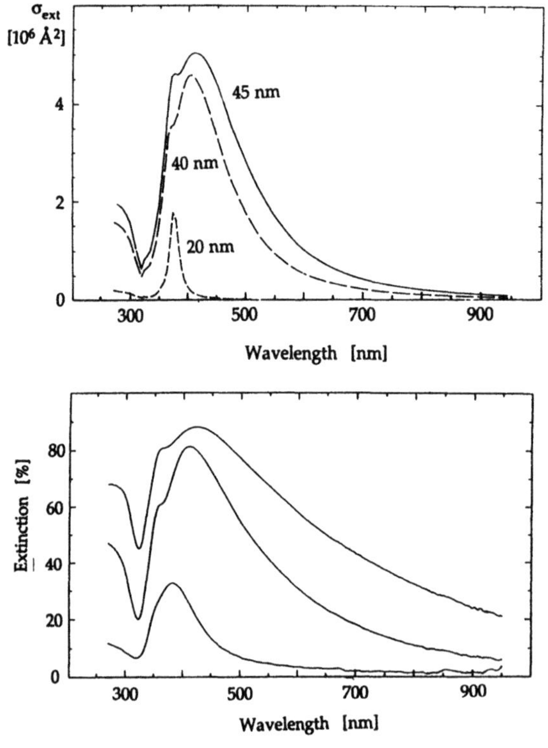

Fig. 3: Theoretical (top) and experimental (bottom) extinction spectra for silver clusters.

Fig. 3 (top) gives another example for theoretical extinction cross sections, this time for spherical silver clusters with FWHM of 50%. For sizes above R ≈ 40 nm, a shoulder on the small wavelength side develops which reflects the growing importance of the quadrupole resonance (L=2). Fig. 3 (bottom) shows corresponding experimental extinction spectra of annealed Ag-clusters on LiF surfaces which are in good agreement with theory [13]. The results of Mie theory calculations for other large free electron metal clusters are essentially similar. Without doubt, experimental evidence proves that this classical electrodynamic theory provides a very satisfactory description of the optical response of metal clusters with sizes above 10 nm. For smaller sizes, intrinsic size effects have to be taken into account.

Intrinsic Size Effects

The physics underlying intrinsic size effects may be clarified by considering one effect in more detail, namely the contribution of damping of the small clusters (R≤10nm). In the classical theory of free electron metals, the damping is due to scattering of the electrons with phonons, lattice defects or impurities. Hence Γ follows from the average of the respective collision frequencies of the electrons. Simple free electron theory gives for the inverse of the relaxation time τ

$$\Gamma_{bulk} = \Gamma_\infty = \frac{v_F}{\ell_\infty} \tag{5}$$

v_F is the Fermi velocity and ℓ_∞ the mean free path of the electrons in bulk material, given by $\ell_\infty = v_F \tau$. For Na and Ag one finds e.g. ℓ_∞ (Na) = 34 nm and ℓ_∞(Ag) = 52 nm at room temperature. If the particle sizes become comparable or smaller than ℓ_∞ the collisions of the conduction electrons with the particle surfaces become important as an additional collision process [14] resulting in a reduced effective mean free path ℓ. A number of theoretical approaches has been developed to compute ℓ. The common feature of all theories is that $\Delta\Gamma \sim 1/R$. Therefore Eq. 5 is usually replaced by

$$\Gamma = \frac{v_F}{\ell_\infty} + A\frac{v_F}{R} \tag{6}$$

where A is a theory dependent constant of the order of 1. For small particles surface scattering obviously becomes the dominant contribution to the damping. As an example of the variation of the dielectric function due to the particle size effect of Eq.6, Fig.4 shows experimental size dependent functions $\varepsilon_1(\omega)$ and $\varepsilon_2(\omega)$ (from [15]) which are in full agreement with calculated ones according to Eq. 6 for $A \approx 1$.

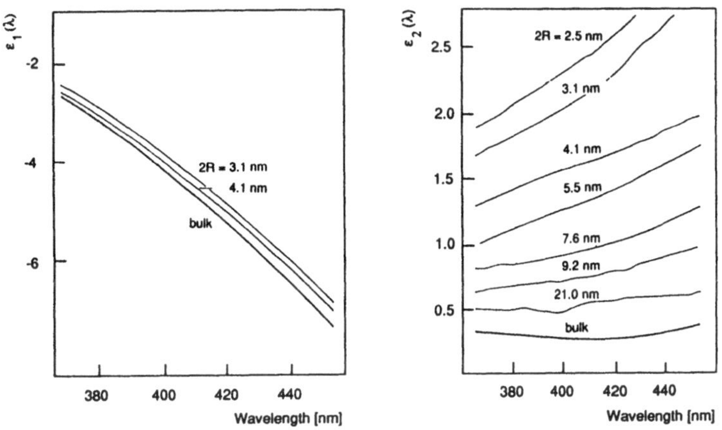

Fig. 4: Experimental dielectric function of Ag- clusters as a function of size

A listing of the various theories which deal with plasma peak positions and plasma resonance halfwidths are given in the reviews [1, 16].

Summary: Size Dependences of Plasma Resonances

The overall size dependence of the dipole resonance wavelength of metal clusters is schematically illustrated in Fig.5. In the quasistatic regime the resonance position is size independent, given by $\varepsilon_1(\omega) = -2\,\varepsilon_m$, if bulk optical functions are used. Larger particles will suffer a red shift due to phase retardation and the influence of higher multipoles. Smaller clusters of simple metals usually exhibit shifts due to intrinsic size effects. One effect is that the dielectric function is changed from the bulk values due to the electron spill out beyond the edge of the ion core [17]. Therefrom an enhanced polarizibility results which for free electron metals gives rise to a red-shift (e.g. [18]). On the other hand noble metals like Ag and Au can show blue shifts due to the influence of d-electron interband transitions [19,20]. This scheme is also

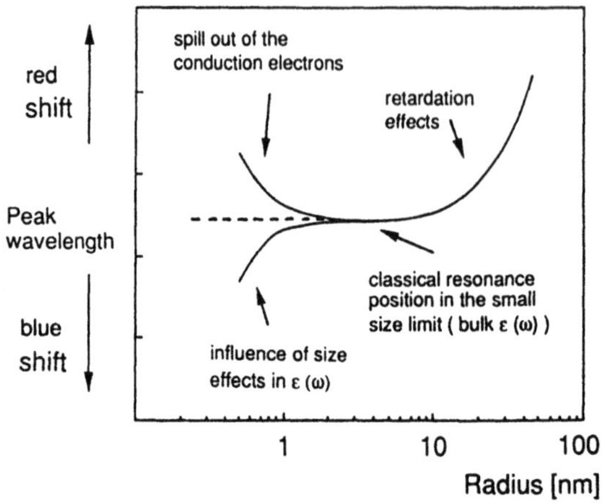

Fig. 5: Schematical size dependence of resonance positions

valid for clusters in matrices. The dielectric surrounding leads to additional shifts with regard to free clusters.

A similar dependence exists for the width of the resonances as function of size. For large clusters radiation damping due to retardation causes the broadening with increasing size. For small clusters the Mie theory would give constant width, however additional damping effects show up due to the above discussed size effects. For some small clusters like Au_{55} the damping effects are not yet fully understood [21].

The size scale in Fig. 5 must not be considered strict, it rather should give an idea of the order of magnitude. In general the size dependencies of the resonance wavelengths and widths for spherical metal clusters can be schematically described as follows:

Cluster radius R	$R \leq 10$ nm	$R \geq 10$ nm
Mie theory	$\neq f(R)$	$= f(R)$
optical material functions	$\varepsilon = \varepsilon(R)$	$\varepsilon \neq \varepsilon(R)$
size effects	intrinsic	extrinsic

Mie´s theory was developed in 1908. Only a few extensions of the theory were performed later. Briefly they can be classified as dealing with other geometrical particle shapes, shell particles, shape dependent substrate effects, diffuse electron density boundaries, non local optical effects, and cluster-cluster interactions in cluster matter (for more details see [1]).

Applicability of Electrodynamic Calculations to Small and Very Small Clusters

Recently experiments using the new technique of beam depletion spectroscopy for studying spectra of free small clusters suggested the excitation of surface plasmons already in sodium clusters of only 8 atoms [22]. The technique was subsequently applied by other groups [23-27] and nowadays detailed absorption spectra for several size selected alkali metal neutral and ionic clusters are available, as well as for some silver cluster ions and - using direct absorption spectroscopy for Hg-cluster distributions in beams [28]. These results were originally interpreted in terms of surface plasmons or collective resonances, i.e. Mie resonances. This was supported by jellium calculations [17] which regarding the collective excitations can be considered as the quantum mechanical extension of the Mie theory taking into account the electron spill out effects and discrete electron energy levels in the ion core potential. Later it was successfully shown that the spectra of very small alkali clusters with up to 8 atoms can also be understood in terms of a quantum chemical molecular ab initio approach (see review [29]). In the latter calculations the term collective excitation is not used at all, rather correlated electron excitations are stated to give rise to the observed experimental features. At present, the theoretical situation is still unclear insofar the various ranges of validity of the theoretical concepts and the use of terminology are subject of intense and controversial discussion. Unfortunately, the resonance positions of the collective and the single particle spectral features are often in adjacent regions, which makes an unequivocal identification of the resonances difficult. In the following an attempt is made to clarify at least the conceptual notations like plasmon, collective excitation and single electron excitation. This might also help to understand the limits of applicability of the Mie theory towards small clusters. In addition, suggestions for further experiments are given, which might help to interpret the spectra more easily.

The following aspects which are mostly interconnected have to be considered when discussing collective or plasmon excitations versus single electron excitations:

- degree of coherence of the electronic excitations
- strength of electron-electron-interactions
- occurence of nonmetal-metal transition
- energy gain per electron during excitation compared to thermal energies
- oscillator strength and sum rules
- resonance positions and widths

From the solid state physics point of view a plasmon is the elementary excitation of plasma oscillations or waves. One may visualize them as the oscillation of the Fermi sphere as a whole. Two properties of this oscillation are essential:

1) The energy gain per electron during this oscillation is minute due to the macroscopic number of electrons in the metal, so the single electron contribution is almost elastic.

2) The motion of all electrons is in phase, i.e. coherent.

In a Drude model like description of the electron system the coherent character of the excitation is due to the assumption of independently oscillating electrons which are all in phase. This corresponds as well to very strong inherent electron-electron-interactions. Hence Drude theories are a priori models involving collective excitations.

As a matter of fact the strength of the electron-electron-interactions governs the degree of coherence of the electronic excitations and hence also the degree of collective character. This is also directly correlated to the nonmetal-metal-transition of clusters as a function of size, as has been demonstrated recently for Hg clusters [30].

We suggest the following use of notations. If the cluster size is getting so small that the concept of the oscillating Fermi sphere with only small energy gains per electron loses sense, the notation plasmon for the collective escitation is useless. Using a rule of thumb this limit occurs if the energy gain per electron starts to exceed thermal energies. Hence, $h\nu/N \geq kT$ i.e. $N \leq \approx o(100)$, or $R \approx 9$Å for Na and photons of 2.5 eV. For example the single electron energy gain in a Na_8 cluster for such photon energies amounts to ≈ 0.3eV. In this case details of higher lying unoccupied single electron states may become important.

Nevertheless the motion of the electrons may still be coherent both for the case that the total electron energy is purely kinetic and if strong electron-electron interactions occur. The strength of the interaction manifests itself e.g. in the nonmetal metal transition. In these cases one may still describe the excitation as resembling a collective electronic behavior, depending e.g. on the degree of coherence. The ab initio concepts make no distinction between single and collective excitation as the change of the all electron wavefunction is computed.

The above arguments suggest the following classification of optical excitations in metal clusters in four size regimes, defined by the number N of atoms in the cluster:

Cluster sizes:	very small $N \leq 20$	small $20 \leq N \leq 500$	large $500 \leq N \leq 10^7$	bulk $10^7 \leq N$
Molecular orbital calculations: ab initio theories	<————————> ? excitations of the all-electron wave-functions (no distinction between single and collective excitation)			
Solid state theories	? <——————————————————————————————I collective electron excitations		<———————————I plasmon polariton excitations	

Concerning the experimental situation for very small clusters, it is difficult to unequivocally interpret the resonant features as being due to either collective or molecular like spectral features. This is due to the fact, that for many of the investigated metal clusters, atomic or molecular resonance lines lie in the vicinity of the predicted collective resonances. Two clear exceptions were reported in the literature so far, although both dealt with cluster size distributions rather than single sized clusters. First, the Hg experiments [28] showed the evolution of the collective resonance as a function of cluster size, being strongly correlated to the nonmetal-metal-transition. The atomic and dimer spectral features are well separated and disappear as the collective resonance builds up with increasing cluster size. This already occurs for sizes of ≈20 atoms in the cluster. Second, early experiments on Ag clusters in photosensitve glass show a marked size dependence of the optical absorption (see Fig. 6) [19]. The development of Ag clusters from the single atom to sizes of about 10^5 atoms showed that the atomic absorption at about 330 nm is followed by complex, obviously molecular absorption bands between 300 and 350 nm. These structures are then cut down, until the minimum of absorption typical of large clusters remains at 320 nm. Simultaneously, the plasmon polariton peak develops at 400 nm, i.e. at an energy lower by about 0.5 eV. As shown in Fig.6, the differences between both kinds of spectral structures are far beyond experimental limits of accuracy. The according cluster sizes where molecular structures still dominate the spectra (curve 1) are estimated to range between 10 and 50 atoms. Plasmon, i.e. collective excitations, which are also present increase with size and become the prominent features (curve 12) for sizes of about 300 to 500 atoms. The energetic differences between both spec-

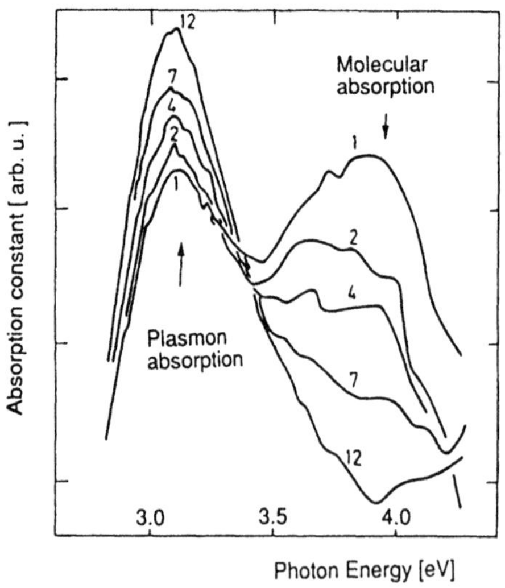

Fig. 6: Absorption spectra of Ag clusters in glass

tral features are particularly large in these experiments since the considered Ag-clusters were embedded in a matrix of glass, the dielectric constant of which shifted the plasmon peak for about 1/4 eV from the position for clusters in vacuum. This is due to the localization of the excitation close to the cluster interface and is caused by changes of the electrodynamic boundary conditions. Less well separated were these spectra in rare gas matrix isolated Ag clusters. This simultaneous existence of both spectral structures in Fig. 6 - though with varying magnitudes - in each spectrum most probably is an artifact due to the distribution of cluster sizes in the investigated samples which in the transition region included both small molecular and somewhat larger metallic Ag-clusters. During the cluster growth process, the number of the former decreases while the number of the latter increases. There is no hint of a continuous shift of structures from the molecular to the plasmon band. Instead, the plasmon peak position is almost independent of cluster size once it is observed. This supports the appearance of the metallic properties to be due to a well defined structural "phase" transformation, postulated earlier [31].

This experiment may offer interesting perspectives for future experiments dealing with the interplay of atomic/molecular and collective spectral features of metal clusters. The embedding of metal clusters in a medium with given dielectric constant will shift the collective resonance positions in a predictable way according to Eq. 2. Hence, the embedding in matrices offers a separation of the two features provided that the molecular features are not shifted to the same amount. The same holds for clusters on surfaces (e.g. [32]).

The energy difference between the atomic resonance lines (-250 nm) and the Mie absorption (500 nm) is even more pronounced in Au compared to Ag, amounting to about 2.5 eV. The Au_2 molecular bands measured in vapor are located at 480/650 nm and 380/410 nm; bands of more complex molecules appear not to have been identified. The fact that atomic and molecular lines almost coincide with the plasmon band in the case of Na and Ag-particles, but not in Au should allow a crucial experiment to verify the model evaluated above: in extremely small Ag clusters there should be oscillator strength in the environment of the resonance, in Au clusters, however, not. It would be interesting to look upon recent experiments by Meiwes-Broer [27] from this point of view, also keeping in mind that in chemically stabilized Au_{55} clusters no hints of a plasma resonance could be detected experimentally [21].

Conclusions:

Optical properties of metal spheres (R≥10nm) can be described by classical electrodynamics, i.e. Mie theory. The corresponding excitations are commonly called plasmon polaritons. Application of this theory to large metal clusters is successsful, although for clusters with less than 100 atoms it makes more sense to use the notation collective rather than plasmon excitations. The transition from molecular like to collective features apparently lies in the size region of about 8 to 20 atoms for free electron metals, however, unequivocal identification is difficult. Experiments are proposed with clusters in dielectric media, which may separate molecular like and collective features and thus help to interpret the spectra. Finally it should be mentioned that plasmon excitations in metal clusters have interesting properties like e.g. surface plasmon induced desorption processes [12,33] which similar to photoelectron spectroscopy or Surface Enhanced Raman Scattering are strongly enhanced in the vicinity of the resonance frequency since they sensitively depend on the electric near field intensities.

Acknowledgements: We want to thank W. Hoheisel for letting us use still unpublished results. This work was supported by the Deutsche Forschungsgemeinschaft.

References:

[1] M. Vollmer, U. Kreibig, *Optical Properties of Metal Clusters*, Springer Ser. Mat. Sci., to be published
[2] G. Mie, Ann. Phys. **25**, 377 (1908)
[3] M. Kerker, *The Scattering of Light*, Academic Press (1969)
[4] C.F. Bohren, D.R. Huffman, *Absorption and Scattering of Light by Small Particles* Wiley 1983
[5] U. Kreibig, P. Zacharias, Z. Phys. **231**, 128 (1970)
[6] U. Kreibig, B. Schmitz, H.D. Breuer, Phys. Rev. B **36**, 5027 (1987)
[7] H. Ehrenreich, H.R. Philipp, Phys. Rev. **128**, 1622 (1962)
[8] P.O. Nilsson, Solid State Physics **29**, 139 (1974)
[9] J.H. Weaver, C. Krafka, D.W. Lynch, E.E. Koch, Physics Data: Optical Properties of Metals, Parts 1,2, Fachinformationszentrum Karlsruhe, FRG (1981)
[10] T. Inagaki, L.C. Emerson, E.T. Arakawa, M.W. Wiliams, Phys. Rev. B **13**, 2305 (1976)
[11] U. Kreibig, Appl. Phys. **10**, 255 (1976)
[12] W. Hoheisel, U. Schulte, M. Vollmer, F. Träger, Appl. Phys. A **51**, 271 (1990)
[13] Dissertation W. Hoheisel, Heidelberg (1991), also : W. Hoheisel, T. Götz, F. Träger, M. Vollmer, to be published
[14] U. Kreibig, C.v. Fragstein, Z. Phys. **224**, 307 (1969)
[15] U. Kreibig, Z. Phys. **234**, 307 (1970)
[16] U. Kreibig, L. Genzel, Surf. Sci. **156**, 678 (1985)
[17] W. Ekardt, Phys. Rev. Lett. **52**, 1925 (1984); Phys. Rev. B **31**, 6360 (1985)
[18] M.A. Smithard, M.Q. Tran, Helv. Phys. Acta **46**, 869 (1974)
[19] L. Genzel, T.P. Martin, U. Kreibig, Z. Phys. B **21**, 339 (1975)
[20] H. Abe, W. Schulze, B. Tesche, Chem. Phys. **47**, 95 (1980)
[21] U. Kreibig, K. Fauth, C.-G. Granqvist, G. Schmid, Z. Phys. Chem. NF **169**, 11 (1990)
[22] W.A. de Heer, K. Selby, V. Kresin, J. Masui, M. Vollmer, A. Châtelain, W. D. Knight, Phys. Rev. Lett. **59**, 1805 (1987); also Phys. Rev. B **40**, 5417 (1989); **43**, 4565 (1991)
[23] C.R. Wang, S. Pollack, M.M. Kappes, Chem. Phys. Lett. **166**, 26 (1990); also J. Chem. Phys. **93**, 3787 (1990); J. Chem. Phys. **94**, 2496 (1991)
[24] C. Bréchignac, P. Cahuzac, F. Charlier, J. Leygnier, Chem. Phys. Lett. **164**, 433 (1989); also Phys. Rev. Lett., submitted
[25] H. Fallgren, T.P. Martin, Chem. Phys. Lett. **168**, 233 (1990); Z. Phys. D **19**, 81 (1991)
[26] J. Blanc, M. Broyer, J. Chevaleyre, Ph. Dugourd, H. Kühling, P. Labastie, M. Ulbricht, J.P. Wolf, L. Wöste, Z. Phys. D**19**, 7 (1991); also Phys. Rev. Lett. **67**, 2638 (1991)
[27] J. Tiggesbäumker, L. Köller, H.O. Lutz, K.H. Meiwes-Broer, Chem. Phys. Lett., in press; also contribution in *The Physics and Chemistry of Finite Systems: From Clusters to Crystals*, Eds.: P. Jena, S. Khanna, B. Rao, Nato Asi Series B, in press
[28] K. Rademann, submitted to Phys. Rev. Lett.; also: p. 45 in *Symposium on Atomic and Surface Physics*, T.D. Märk, F. Howorka (Eds.), Studia, Innsbruck (1990)
[29] V. Bonacic-Koutecky, P. Fantucci, J. Koutecky, Chem. Rev. **91**, 1035 (1991)
[30] K. Rademann, Ber. Bunsenges. Phys. Chem. **93**, 653 (1989)
[31] U. Kreibig, in *Growth and Properties of Metal Clusters*, Ed.: J. Bourdon, Elsevier (1980)
[32] J.H. Parks, S.A. McDonald, Phys. Rev. Lett. **62**, 2301 (1989)
[33] W. Hoheisel, K. Jungmann, M. Vollmer, R. Weidenauer, F. Träger, Phys. Rev. Lett. **60**, 1649 (1988); see also Appl. Phys. A **52**, 445 (1991)

Quantum Molecular Interpretation of Optical Response Properties of Simple Metal Clusters

V. Bonačić–Koutecký[1], P. Fantucci[2] and J. Koutecký[1]

[1]Freie Universität Berlin, Institut für Physikalische und Theoretische Chemie, Takustr. 3, W–1000 Berlin 33, Germany

[2]Dipartimento di Chimica Inorganica e Metallorganica, Centro CNR, Universita di Milano, Via Venezian 21, I–20133 Milano, Italy.

Abstract: The ab–initio configuration–interaction study of excited states of small s^1 metal clusters accounts fully for the spectroscopic patterns recorded by depletion spectroscopy and permits geometrical assignments. Specific structural and electronic properties responsible for excitations in small alkali metal clusters allow for the interpretation of theoretical and experimental findings in the framework of molecular many–electron description. Similarities and differences between absorption spectra of Li_n and Na_n clusters will be discussed. The reliability of simplified methods such as random phase approximation for description of optically allowed transitions in metal clusters have been examined.

I. Introduction:

Theoretical investigation of excited states of simple s^1 clusters using relatively accurate ab initio quantum chemical methods offers an excellent opportunity to gain an understanding about the reasons for appearance of the specific electronic and structural features as a function of the cluster size [1]. In recent years the interest in cluster research has been rapidly increased, very often emphasizing the analogies with the phenomena found in other fields such as nuclear and solid state physics. In this contribution two aspects will be pointed out: 1) Small clusters do not represent only a bridge between atoms and solids but they possess several of their own specific features as well; 2) Conditions under which physical laws act in small alkali metal clusters might give rise to characteristic features similar to those found for atomic nuclei (giant resonances) in spite of the fact that the constituent particles and forces acting among them are basically different. In order to avoid the erroneous conclusions about the physical nature of these parallels, a careful analysis of the reasons for these similarities is needed.

For this purpose the ab–initio methods which account adequately for the electronic correlation effects (such as multireference configuration interaction (MR–CI)) are well suited since they are free from apriori assumptions. In addition to their predictive power these types of methods, although limited to relatively small number of electrons, can serve as a guidance to find out the essentials which must be included in the simplified procedures for description of ground and excited states of simple metal clusters. This can be in particulary useful when comparing the results with those obtained using more approximate methods which are well established and justified in other fields such as nuclear and solid state physics.

There are three such model approaches presently used for discussion of absorption spectra of clusters: a) Mie–Drude classical theory employed for an estimate of the surface plasmon frequencies in spherical metal droplets [2]. b) time–dependent local density approximation (TDLDA) based on jellium model [3] and iii) random–phase approximation (RPA) also in connection with the jellium model [4]. All these models ignore more or less structural aspects of small alkali metal clusters and interprete the measured spectra in terms of 'surface plasmon' [2], inhanced plasmon [5] or frag–mented plasmon [4] if more than one intense transitions are present. These simple models seemed to be very appealing for two reasons: a) The exact temperature of depletion experiments is not known and has been assumed to be relatively high at least for large clusters; b) The first photodissociation experiments on Na_n clusters larger than tetramers with low resolution and for selected visible wavelengths yielded small number (one to three) of intense relatively broad bands [2]. Also for cationic clusters K_n^+ (n=9,21) a single intense transition resembling a giant resonances found for nuclei has been observed [6].

Nevertheless, the presence of quantum effects became apparent from the high resolution spectra of Na_n (n=3–8,20) [7–9]; Li_n (n=2–4,6–8) [10–12] and Li_yNa_x [13] which cover the spectral region up to 3.2 eV and the needs for quantum molecular inter–pretation evident [14–19]. Moreover, increasing experimental evidence on different cluster sizes called for revision of the oversimplified picture of 'surface plasmon' interpretation of intense bands of absorption spectra of small alkali metal clusters. For example Na_{20} [8], and Na_{21}^{+} [20] do not exhibit single bands as expected. Also recent experiments on optical response of large clusters K_{500}^{+} and K_{900}^{+} [21] show that the expected transition to the bulk properties does not yet take place. The blue shifts of the maxima of the dominant bands are very small with respect to K_{21}^{+}. In contrast, the classical approach and quantum mechanical random phase approximation (RPA) based on jellium model predict considerably larger blue shifts towards bulk type of behaviour [21].

In this contribution it will be shown on selected examples of Na_n and Li_n that the ab–initio MRD CI transition energies and oscillator strengths calculated for the structures with the lowest energies determined in the earlier work on the ground state porperties [22,23] are in good agreement with the recorded absorption spectra [9,12] demonstrating clearly that the structural properties cannot be ignored. The successful structural assignment of topologies determined at T=0 to the recorded depletion spectra suggests that 1) the atoms in clusters are not so mobile as assumed although the experiments are not carried out at T=0 and that 2) the experimental temperature does not seem to be considerably higher than 300K for small clusters. This is also supported by the quantum molecular dynamics work on Na_n at the T≠0 [24]. A comparison of the CI predictions with those obtained by the ab–initio RPA based on the HF–SCF procedure serves to investigate the importance of treating single and double excitations at the same footing which is not the case in the RPA [25,1]. The results obtained from the ab initio RPA calculations are also useful for comparison with the predictions of the RPA based on jellium model [4] in order to find out the influence of exact potential and the position of the nuclei in within equivalent approximate treatments of the correlation effects. This will be illustrated on example of Na_{20} [26]. Finally, the analysis of the electronic excitation will serve for interpretation of observed spectroscopic patterns.

II. Absorption Spectra of Li_n and Na_n Clusters

In alkali metal clusters the bonding is characterized by multicenter delocalized bonds and very small clusters do not assume highly symmetrical shapes due to Jahn–Teller–deformation (cf. also [1]). Tetramers with four valence electrons have rhombic 2D geometries but octamers assume 3D compact symmetrical T_d structures with tetrahedral subunits since they have sufficient number of valence electrons to fill the degenerate one–electron levels [22,23]. Trapezoidal planar 2D structures of Li_5 and

Na_5 have considerably lower energies than 3D–trigonal bipyramids (~ 0.2 eV). In the case of hexamers there is a competition between a planar structure built from triangles, a flat pentagonal pyramid and a 3D C_{2v} structure with tetrahedral subunits [22,23]. For Li_6 the 3D C_{2v} structure has the lowest energy [22b] in contrast to Na_6 for which the planar form and the flat pentagonal pyramid have the lowest and almost degenerate CI energies [23]. The 3D pentagonal bipyramids represent equilibrium geometries for both Li_7 and Na_7 [22,23].

The calculation of the optically allowed states and the oscillator strengths in the energy interval up to 3.2 eV for the most stable ground state structures and a comparison with the depletion measurements offers a good opportunity to carry out the structural assignments. For details of calculations and the estimate of their accuracy compare references [1,16,17 and 11]. For both Li_4 and Na_4 rhombic tetramers three intense transitions located at ~ 1.7–1.8 eV and in the 2.5–3.0 eV energy interval as well as a large number of weak bands [14,16,17,11] are in a good agreement with the recorded spectra [7,10] and the assignment is straightforward. The 3D T_d structures of Na_8 and Li_8 give rise to two dominant features located at ~ 2.5– 2.7 eV and the red shifted weak transitions [7,15,16,11] coinciding with measured features. The questions can be raised at which cluster size the transition from the planar to the three dimensional structures occurs. The predicted spectrum for Na_5 planar structure [19] is in substantially better agreement with the recorded spectrum [9] than the one calculated for the trigonal bipyramid. The broad dominant band is located in the energy interval 2.0–2.2 eV and a tentative assignment has been made to the 2D planar structure. Theoretical and experimental work on Li_5 is still in progress.

From a comparison of calculated and recorded spectra for Li_6 and Na_6 given in Figures 1 and 2 a structural difference between hexamers is apparent. The 3D C_{2v} structure of Li_6 gives rise to dominant features in the energy interval 2.5–2.7 eV and to fine structure with considerably less intensity located at ~ 1.8 eV in the complete agreement with the experimental finding [12]. The predicted spectra of other two structures with the dominant features at ~ 2.1 eV do not correspond to any of the measured features. In contrast, the calculated transition energies and oscillator strengths for the planar 2D geometry of Na_6 [19] agree well with the recorded spectrum which is characterized by a dominant transition at ~ 2.1 eV and the considerably weaker shoulder at ~ 2.8–3.0 eV (Fig. 2). The flat pentagonal pyramid for Na_6 gives rise to very similar spectrum as the planar structure. Two topologies differ only in the position of one atom. Although the pentagonal pyramid does not represent the local minimum on the HF energy surface, its contribution to the recorded spectrum cannot be completely ruled out presently, since the energy difference between this and the

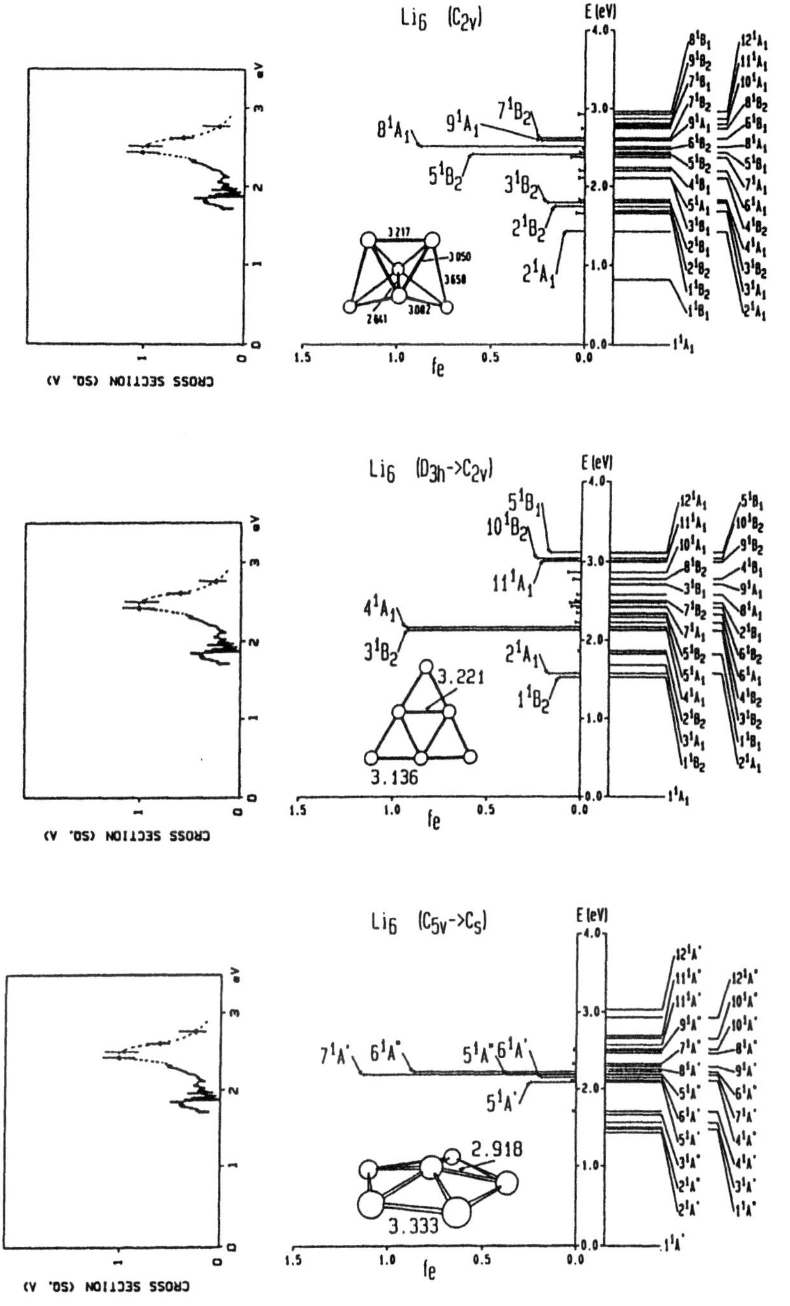

Fig.1: Comparison of photodepletion spectrum (on the left) and CI predicted optically allowed transitions and oscillator strengths f_e for the three Li_6 structure [12].

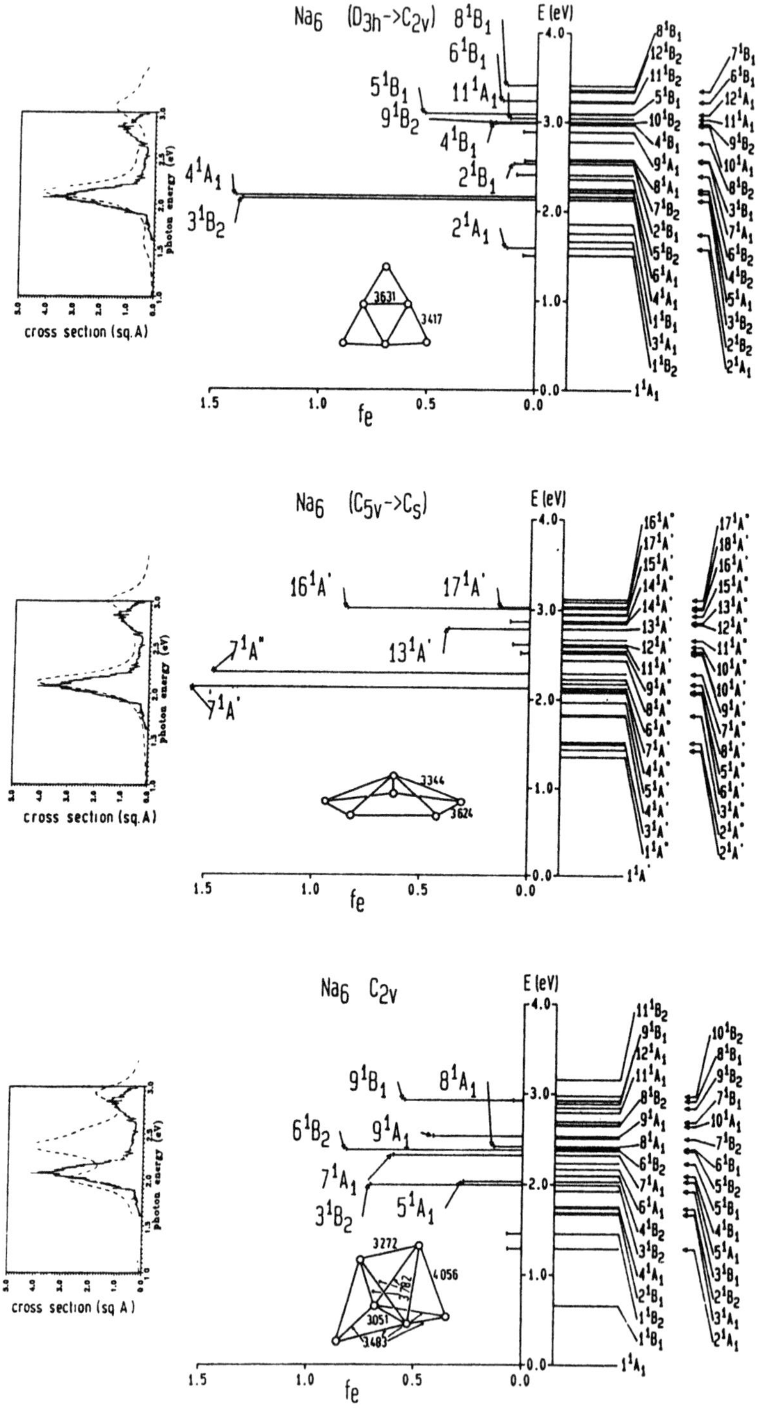

Fig.2: Comparison of photodepletion spectrum [9] and CI predictions for three Na_6 structures [19]. On the left of the figure the experimental finding is compared with the Lorentzian simulation based on the CI results (dashed lines).

planar structure is very small, the calculated spectroscopic patterns for both structures are very similar and the experiment is not carried out T=0. In constrast, the 3D C_{2v} structure of Na_6 with substantially higher energy of 0.25 eV gives rise to transition energies and oscillator strengths which do not agree with the experimental finding. Consequently the contribution from the 3D C_{2v} structure of Na_6 to the recorded spectrum can be excluded. Figures 1 and 2 illustrate clearly that Li_6 seems to be three dimensional and Na_6 is still planar or close to planarity. It is likely that the sp hybridization is stronger for Li than for Na. The findings of Figures 1 and 2 demonstrate that the position of nuclei is important and therefore the structural aspects should not be neglected because of possible experimental conditions (e.g. unknown exact temperature).

If the average contributions of all three structures with comparable weights would be responsible for the recorded spectrum, three dominant features located at ~ 2.2, 2.5 and 2.8 eV would be present. Since this is not the case, it seems that the temperature of both experiments for Li_6 and Na_6 is suficiently low so that the vibrations are not able to bring cluster from one class of related geometries to the other one.

Pentagonal bipyramids for Li_7 and Na_7 give rise to the spectra with dominant transitions at 2.5 eV which are in agreement with experimental findings. It seems that the occurance of intense transitions at ~ 2.5 eV is characteristic for the three dimensional structures of Li_6, Li_7, Na_7, Li_8 and Na_8.

The ab–initio RPA calculations based on the SCF Hartree–Fock ground states have been carried out for all three geometries of Na_6 (cf. Fig. 3). From comparison of the CI and RPA results (Fig. 4), the role of explicit consideration of double excitations has been studied. Moreover, it is of interest to investigate whether the dominance of single excitations might be connected with the symmetry of the cluster geometries and in this connection with the degeneracy of the one–electron levels among which the excitations take place. For the planar structure the CI wavefunctions of the 4^1A_1 and 3^1B_2 (to which the most intense transitions have been calculated) are dominated by single excitations. Therefore, the RPA results for these transitions do not differ substantially, they are just slightly blue shifted with respect to the CI values. For all other characteristic spectroscopic features, differences between the results obtained from the RPA and the CI procedures are considerably larger. For example, the leading role of single and double excitations in 3^1B_1state gives rise to the transition energy T_e=2.58 eV and to low value of oscillator strength f_e=0.0298. The RPA yields considerably higher values: T_e=2.85 eV and f_e=1.04, due to defficient treatment of double excitations. Since the location of the 3^1B_1 state obtained from RPA coincides with the

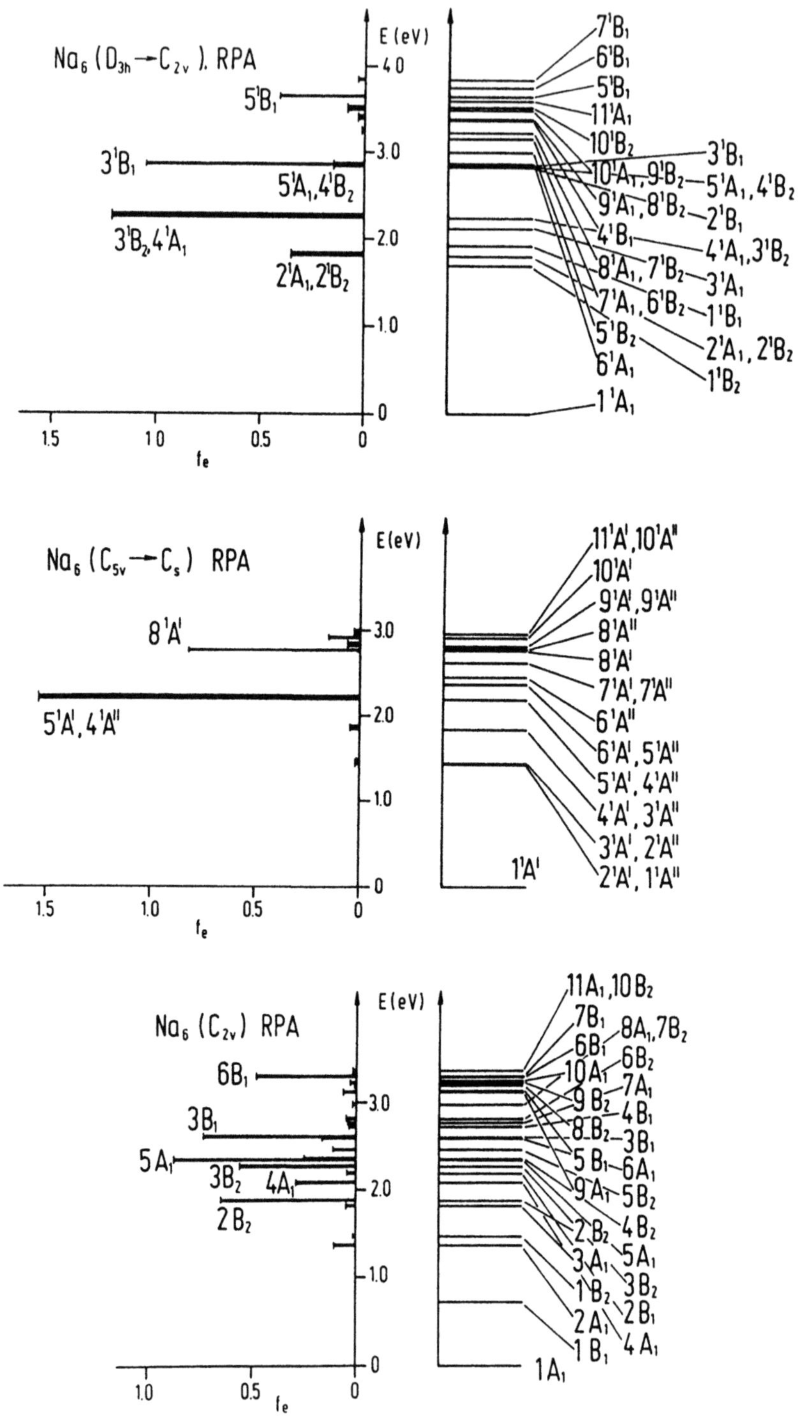

Fig.3: The RPA (based on ECP–CPP–SCF) transition energies (eV) and oscillator strengths f_e for optically allowed states of the three Na_6 structures of Fig. 2.

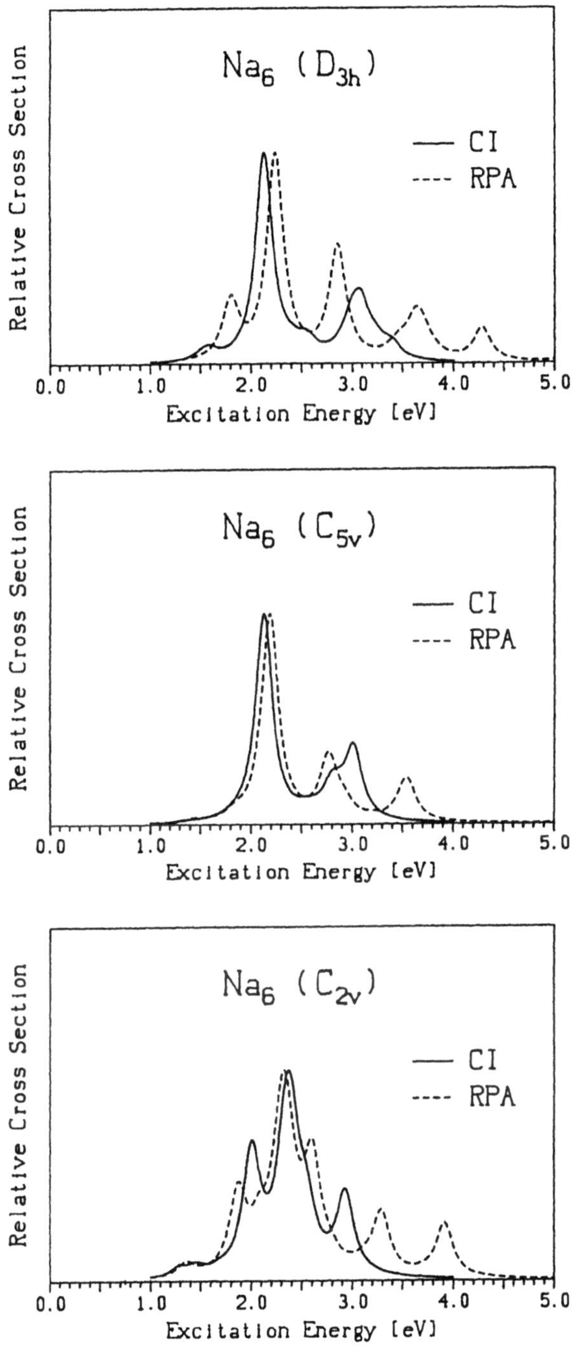

Fig.4: Comparison of the Lorentzian simulations based on the CI (full line) and RPA (dashed lines) results from Figures 2 and 3 for the three Na_6 structures.

position of the recorded weaker band, this result is fortuitous. All other transitions with considerable intensities are strongly blue shifted with respect to the CI results. From a careful analysis it is not possible to conclude that the RPA yields, in general, better results for the energetically lower lying states, although this is the case for the pentagonal pyramid.

Differences between the CI and the RPA results are even more pronounced for the C_{2v} structure. It seems that the splitting of degenerate one electron levels due to lower symmetry leads to the necessity for introducing a more balanced interaction among single and double excitations, since the lack of the appropriate inclusion of double or higher excitations in RPA leads to deviations in values of transition energie and oscillator strengths with respect to the CI values randomly distributed over the whole energy interval of interest. It seems that, if the CI method yields one dominant transition, which is usually connected with highly symmetrical structure, this might be more easily reproduced by the simplified methods.

The analysis of the wavefunctions of excited states in terms of configurations with single and double excitations shows that very few leading configurations characterize the states to which intense transitions occur similarily as in many molecular excited states. A comparison of the dominant features in the wavefunctions of the states to which transitions with small and large oscillator strengths have been calculated indicates the occurence of interference phenomena [1,16,19], since the same configu– rations differing in the amount and the sign of their contributions determine to a large extent the composition of the wavefunctions. This is the case for all Li_{3-8} and Na_{3-8} clusters considered.

It is instructive to point out that the description of the excited states of clusters in terms of interactions of individual excitations is just a way of improving the wrong starting point which is the one–electron picture. However, this discussion can be useful for distinguishing individual single (particle–hole) excitations from many–electron effects which can be qualitatively characterized by interaction of very few leading excitations or by very large number of excitations with comparable weights, respec– tively. In the latter case the notion of collective excitations such as for surface or volume plasmons can be introduced. In this context the question can be raised at which cluster size a molecular picture will disappear. Therefore, the interpretation of Na_{20} spectrum is of particular interest.

Although the structural assignment of Na_{20} absorption spectrum is not yet comple– ted, a comparison of the RPA predictions based on the HF SCF calculations for the two T_d structures with experimental data [8] shown in Figure 5 illustrates three important points: 1) Almost "spherical" compact structure yields one dominant

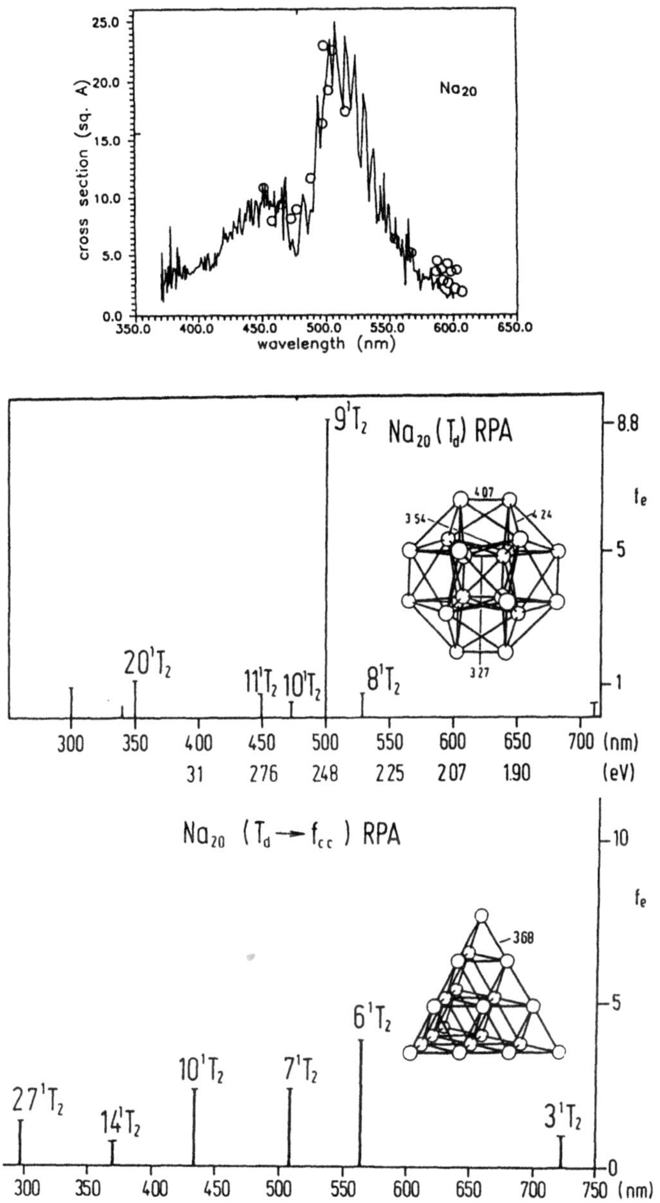

Fig.5: Comparison of the photodepletion spectrum of Na_{20} [8] and Na_8 [7] and the RPA optically allowed transitions (nm) and oscillator strengths f_e a) for the highly symmetrical T_d structures of Na_{20} and b) for the T_d structure of Na_{20} which is a section of fcc lattice [1].

transition which location is in a good agreement with the maximum of the recorded intense band. Several other relatively weak transitions are strongly blue shifted and coincide less well with the measured broad substantially weaker band. The second T_d structure which is the section of the fcc lattice gives rise to entirely different spectroscopic pattern. The locations of calculated intense transitions coincide well with the regions of recorded bands but the values of oscillator strengths do not indicate presence of one dominant band. 2) The analysis of the RPA results for the intense transition obtained for the "spherical" T_d structure contains contributions with comparable weights from the five single excitations illustrating that there might be a long way to go reaching the collective type of of excitations. 3) Our RPA spectra for Na_{20} differ substantially from the RPA predictions obtained using jellium model. It is evident that the two model calculations give rise to different energies of one–electron levels among which single excitations take place. This can be due to the LDA approximation of jellium model as well as to the potential employed. Therefore, the interpretation of the "fragmentation plasmon" due to the presence of two bands in the recorded spectrum based on RPA–jellium calculations [4] might be rather arbitrary. At present ab–initio RPA calculations are in progress for several other structures [24] which might be good candidates for stable isomers.

III. Conclusions

The predictive power of quantum chemical methods, although still limited to small clusters with a small number of valence electrons, has been clearly demonstrated by successful structural assignments and interpretation of the optical response properties. A comparison between accurate theory and experiment makes it possible to gain the information about the geometries and specific electronic properties such as nature of excitations responsible for the spectroscopic pattern.

The appearance of relatively small number of intense transitions and a very large number of weak ones in the depletion spectra of small alkali metal clusters is connected with the symmetry of the geometry given by the position of nuclei and with the role of p functions which participate in the delocalized bonding and in polarization connected with the molecular excitations.

Acknowledgement:

This work has been supported by the Deutsche Forschungsgemeinschaft (Sfb 337, 'Energy transfer in molecular aggregates') and the Consiglio Nationale delle Ricerche (CNR). We thank our coworkers C. Fuchs, J. Pittner and colleagues C. Gatti and S. Polezzo who have substantially contributed to the work included in this contribution. We extend our thanks to M. M. Kappes, M. Broyer, J.–P. Wolf and L. Wöste for providing us with their data prior to publication.

References:

1. V. Bonačić–Koutecký, P. Fantucci and J. Koutecký; Chem. Rev., 91, 1035–1108 (1991).

2. a) W.A. De Heer, K. Selby, V. Kresin, J. Masui, M. Vollmer, A. Chatelain and W.D. Knight; Phys. Rev. Lett., **59**, 1805–1808 (1987).
   b) K. Selby, V. Kresin, M. Vollmer, W.A. de Heer, A. Scheidemann and W.D. Knight; Phys. Rev. B **43**, 4565–4578 (1991).

3. a) W. Ekardt; Phys. Rev. Lett., **52**, 1925–1928 (1984).
   b) W. Ekardt; Phys. Rev. B **31**, 6360–6370 (1985).

4. C. Yannouleas, R.A. Broglia, M. Brack and P.F. Bortignon; Phys. Rev. Lett., **63**, 255–258 (1989); C. Yannouleas, J.M. Pacheco and R.A. Broglia; Phys. Rev. B **41**, 6088–6091 (1990); P.–G. Reinhardt, M. Brack and O. Genzken; Phys. Rev. B **41**, 5568–5582 (1990); C.Yannouleas, R.A. Broglia; (1991) to be published.

5. a) H. Fallgren and T.P. Martin; Chem. Phys. Lett., **168**, 233–238 (1990).
   b) H. Fallgren, K.H. Bowen and T.P. Martin; Z. Phys. D – Atoms, Molecules and Clusters, **19**, 81–84 (1990).

6. a) C. Bréchignac, P. Cahuzac, F. Carlier, and J. Leygnier; Chem. Phys. Lett., **164**, 433–437 (1989).
   b) C. Bréchignac, P. Cahuzac, F. Carlier, M. de Frutos and J. Leygnier; Z. Phys. D – Atoms, Molecules and Clusters, **19**, 1–6 (1991).

7. C. Wang, S. Pollack, D. Cameron and M.M. Kappes; J. Chem. Phys., **93**, 3787–3801 (1990).

8. S. Pollack, S. Wang and M.M. Kappes; J. Chem. Phys. **94**, 2496–2501 (1991).

9. C.R.Ch. Wang, S. Pollack, T. Dahlseid, G.M. Koretzky and M.M. Kappes; J. Chem. Phys., in press.

10. M. Broyer, J. Chevaleyre, P. Dugourd, J.–P. Wolf and L. Wöste; Phys. Rev. A **42**, 6954–6957 (1990).

11. J. Blanc, V. Bonačić–Koutecký, M. Broyer, J. Chevaleyre, P. Dugourd, J. Koutecký, C. Scheuch, J.–P. Wolf and L. Wöste; J. Chem. Phys., in press

12. P. Dugourd, J. Blanc, V. Bonačić–Koutecký, M. Broyer, J. Chevaleyre, J. Koutecký, J. Pittner, J.–P. Wolf and L. Wöste; Phys. Rev. Lett., 67, 2638 (1991)

13. S. Pollack, C.R.Ch. Wang, T. Dahlseid and M.M. Kappes; J. Chem. Phys., in press.

14. V. Bonačić–Koutecký, P. Fantucci and J. Koutecký; Chem. Phys. Lett. **166**, 32–38 (1990).

15. V. Bonacic–Koutecky, M.M. Kappes, P. Fantucci and J. Koutecký; Chem. Phys. Lett. 170, 26–34 (1990).

16. V. Bonačić–Koutecký, P. Fantucci and J. Koutecký; J. Chem. Phys. **93**, 3902–3825 (1990).

17. V.Bonačić–Koutecký, P. Fantucci and J. Koutecký; Chem. Phys. Lett. **146**, 518–523 (1988).

18. V. Bonačić–Koutecký, J. Gaus, M.F. Guest and J. Koutecky; J. Chem. Phys., in press.

19. V. Bonačić–Koutecký, J.Pittner, C. Scheuch, M.F. Guest and J. Koutecký; J. Chem. Phys., in press.

20. C. Bréchignac, Ph. Cahuzac, F. Carlier and M. de Frutos; Chem. Phys. Lett., in press.

21. C. Bréchignac, Ph. Cahuzac, N. Kebaili, J. Leygnier and A. Safari; Phys. Rev. Lett., submitted.

22. a) I. Boustani, W. Pewestorf, P. Fantucci, V. Bonačić–Koutecký and J. Koutecký; Phys. Rev. B **35**, 9437–9450 (1987).
    b) J. Koutecký, I. Boustani and V. Bonačić–Koutecký; Int. J. Quant. Chem. **38**, 149–161 (1990).

23. V. Bonačić–Koutecký, P. Fantucci and J. Koutecký; Phys. Rev. B **37**, 4369–4374 (1988).

24. U. Röthlisberger and W. Andreoni; J. Chem. Phys. 94, 8129–8151 (1991).

25. C. Gatti, S. Polezzo and P. Fantucci; Chem. Phys. Lett., 175, 645–654 (1990).

26. C. Gatti S. Polezzo, P. Fantucci, V. Bonačić–Koutecký and J. Koutecký; to be published.

Photoelectron Spectroscopy of Alkali Metal Cluster Anions

J.G. Eaton, L.H. Kidder, H.W. Sarkas, K.M. McHugh, and K.H. Bowen[1]

[1]Department of Chemistry
The Johns Hopkins University
Baltimore, MD 21218 USA

ABSTRACT

We present the photoelectron (photodetacment) spectra of K^-_{2-19} recorded using 2.540 eV photons. These spectra are highly structured, providing a wealth of information about the electronic structure of alkali metal clusters, including adiabatic electron affinities and electronic splittings for each size. Spectral assignments have been made with guidance from theory. For the smaller cluster sizes, detailed assignments of the spectra were aided by *ab initio* quantum chemical calculations. For the larger clusters, assignments were guided by a simple shell model. A substantial correlation is found between the predicted shell model energy levels and the observed patterns and spacings in the spectra.

1. Introduction

Among metal cluster anions, alkali metal cluster anions offer relative simplicity, and their photoelectron spectra exhibit unusual clarity. Alkali atoms are hydrogen-like, each with a s^1 valence electron and a rare gas electron core configuration. Bulk alkali metals are the simplest of metals, being essentially free electron metals. Because of these simplifying characteristics at the extrema of size, alkali metal clusters in the vast size regime between a single atom and the condensed phase offer unique opportunities for better understanding the electronic properties of metal aggregates generally.

Interest in alkali metal clusters has generated a rich literature pertaining not only to theoretical[1-11] and experimental[12-20] studies of neutral and cationic alkali clusters, but also to a lesser extent to theoretical[21-25] and experimental[26,27] work on alkali cluster anions. Negative ion photoelectron studies of metal cluster anions comprised of s^1 atoms have included work on metal cluster anions of the coinage metals by Lineberger[28,29], Meiwes-Broer[30], and Smalley[31] and on small alkali cluster anions by our group[27]. Here, we extend our previous work on alkali cluster anions and present the phototelectron (photodetachment) spectra of potassium cluster anions, K^-_n, with n = 2 -19.

2. Experimental

Negative ion photoelectron spectroscopy is conducted by crossing a mass-selected beam of negative ions with a fixed-frequency photon beam and energy analyzing the resultant photodetached electrons. The main elements of our negative ion photoelectron spectrometer are (a) an ion beam line along which negative ions are formed, transported, and mass-selected by an

$E \times B$ Wien filter, (b) an argon ion laser which is operated intracavity in the ion/photon interaction region, and (c) a magnetically shielded, high resolution hemispherical electron energy analyzer, located below the plane of the crossed ion and photon beams. A particularly important attribute of this technique is its ability to size-select cluster anions before photodetachment so that photoelectron spectra of individual cluster anion sizes can be recorded.

To generate beams of potassium cluster anions, we use a heated supersonic expansion ion source. In this device, the stagnation chamber is divided into separately heated oven reservoir and nozzle channel sections. The alkali metal is heated in the oven to a temperature corresponding to several hundred torr equilibrium vapor pressure. The vapor is then coexpanded with several hundred torr of argon into high vacuum through the nozzle, which is maintained ~50 degrees hotter than the oven. Finally, a negatively biased hot filament injects low energy electrons directly into the expanding jet in the presence of axial magnetic fields, producing potassium cluster anions.

3. Results

The photoelectron spectra of K^- and the potassium cluster anions, K^-_{2-19} are presented in Figure 1. These spectra were recorded with 2.540 eV photons. The photoelectron spectra of alkali cluster anions are highly structured, an attribute that derives both from the inherent properties of the alkalis and from the available electron energy resolution (~30 meV) of our spectrometer. The individual peaks in these spectra arise due to photodetachment transitions between the ground electronic state of a given cluster anion and the ground and various energetically-accessible excited electronic states of its corresponding neutral cluster. Thus, one learns two main kinds of information from an analysis of these spectra, ie., adiabatic electron affinity values and electronic structure data, both as a function of cluster size. Adiabatic electron affinities for metal clusters are expected to increase with cluster size, reaching the work function of the bulk metal at infinite cluster size. There are two ways of looking at the electronic structure information in these spectra. In the picture alluded to above, the spacings between peaks in a given spectrum reflect the energy splittings between the various electronic states of the neutral cluster, albeit at the geometry of its corresponding cluster anion. In the other view, which is equivalent to the first picture at the Koopmans' theorem level of approximation, the peaks in a given spectrum are ascribed to electrons being removed or plucked from the occupied energy levels of the cluster anion. In this picture, the peak spacings and patterns directly reflect the upper portion of the occupied energy level structure of the cluster anion. In either picture, however, the observed peak structure in these spectra is a manifestation of how the electronic structure of the alkali metal cluster system is evolving as a function of cluster size. Eventually at some larger cluster size, this electronic structure is expected to organize itself into the essentially structureless valence band of bulk potassium metal.

4. Electron Affinities vs. Cluster Size

Adiabatic electron affinities (EA's) were extracted from the spectra and are presented as a function of cluster size in Figure 2. These results exhibit largely an odd-even alternation of electron affinities with cluster size. Using *ab initio* quantum chemistry methods, Bonačić-Koutecký [4] has calculated the adiabatic electron affinities for potassium clusters from sizes n = 2-6. The quantitative agreement of her calculations with our experimental results is rather good. For theoretical guidance at larger potassium cluster sizes, we turned to a qualitative comparison with the predictions of simple shell models. Figure 3 compares our experimentally-determined EA vs. n trend with the qualitative EA vs. n trend predicted by an ellipsoidal shell model. The latter plots the scaled energy of the highest occupied energy level of the cluster containing n + 1 electrons as a function of cluster size, n. Since a large energy for the highest occupied energy level in the anion implies a relatively low electron affinity, we have inverted the energy scale on our experimental EA vs. n plot to facilitate comparison of these two graphs. It is evident that this shell model reproduces much of the EA vs. n trend found experimentally. Note that an ellipsoidal shell model predicts "peaks" in the EA vs. n trend in Fig. 3 at n = 8, 10, 14, and 18, and that we observe these experimentally. Additionally, however, we also see "peaks" at n = 4 and 12 which are not predicted by this shell model. (Note, of course, that "peaks" in the upper panel of Fig. 3 appear as "dips" in Fig. 2.)

5. Electronic Structure and Spectral Assignments

The photoelectron spectra of alkali metal cluster anions are laden with information about electronic structure, yet their interpretation is not obvious by inspection. Thus, our primary goal at this stage has been to assign these spectra in some rudimentary but useful way. In order to do this we need theoretical support. At relatively small alkali cluster anion sizes, the *ab initio* calculations of Bonačić-Koutecký et al. have been an invaluable guide. Two years ago Bonačić-Koutecký [21] modelled our photoelectron spectra of sodium cluster anions over the size range n = 2-5, and very recently she and her group performed similar calculations on potassium cluster anions ranging in size from n = 2-6. This was done by calculating the photodetachment transitions between several possible (energetically close) geometries of a given cluster anion and the ground and energetically-accessible excited electronic states of its corresponding neutral cluster, with the latter being taken in each case at the geometry of the particular isomer of the cluster anion in question. That is, high level quantum mechanical calculations were separately performed on both the cluster anion and its corresponding neutral cluster, followed by the computation of vertical photodetachmment transitions between them. By comparing the photoelectron spectra predicted in this way with a given experimentally observed alkali cluster anion photoelectron spectrum, a wealth of valuable information was obtained about the geometry (or isomeric geometries) of the alkali cluster anion and about the states involved in each of the observed photodetachment transitions. The success of this approach effectively led to an assignment of the spectra in that it identified the transitions which were involved in each observed spectral peak.

As larger cluster sizes are considered beyond the present range of *ab initio* quantum chemistry treatments, however, one must rely on some other form of theoretical guidance. In our earlier work with alkali cluster anions, we had noticed some intriguing correlations between the peak patterns and spacings of our spectra and the predictions of simple shell models. Consider K_7^- for example, an eight valence electron system. Imagine putting two of these electrons into a "s-like" shell and then the remaining six into a "p-like" shell. Further imagine that in such a small system structural effects (or perhaps crystal field effects) lift the degeneracies of the three p sub-levels, leaving two electrons in each of three closely spaced p levels. Next, imagine removing each of these eight electrons from the cluster anion to their common ionization continuum limit, the energy required to do this in each case being its electron binding energy. That predicts a photoelectron spectrum with three closely spaced peaks at low electron binding energy, followed by a gap, followed by another peak at higher electron binding energy. This is what is observed, and furthermore the 1s-1p splitting is quantitatively close to the prediction of the Clemenger-Nilsson[6,8] model for K_8.

With these correlations to encourage us, we set out to see if a simple shell model could work well enough to aid in the assignment of our spectra. To calculate energy levels from which a predicted shell model "stick spectrum" could be generated, we used a program written by Saunders and Knight[20,32]. Essentially, this calculation treats a three dimensional harmonic oscillator having a single perturbation parameter. In addition to its relative simplicity, it has the advantage of easily being able to accommodate triaxial distortions. Only two parameters had to be fixed, one to set the energy scale and one to set the magnitude of the perturbation. These same two parameter values were then used for calculations on all cluster anion sizes. In this model the value of one quanta of energy is equal to the 1s-1p splitting in the closed shell system with eight valence electrons, and we set it in accordance with the value of the tentatively assigned 1s-1p splitting in our spectrum of K_7^-. Also in this model, the value of the perturbation parameter is equal to the 1d-2s splitting in the closed shell system with twenty valence electrons, and we chose it in accordance with the tentatively assigned 1d-2s splitting in the spectrum of K_{19}^-. Basically, this is a triaxial shell model parameterized for potassium cluster anions.

This approach allowed us to generate predictive "stick spectra", with distinctive spectral patterns and quantitative splittings, for first approximation comparisons with the observed photoelectron spectra. Figures 4-7 present these calculated "stick-spectra" (along with their level designations and electron occupations) as insets on the photoelectron spectra of K_n^- over the cluster size range, n = 4-19. The correlation between the shell model's "stick spectrum" and the actual spectrum for n = 4 is rather good. For n = 5, however, the correlation is not as good. The predicted 1p peak actually appears as two peaks in the real spectrum. Conceivably, the two degenerate 1p sub-levels could be split by some interaction not in the model to yield the two observed peaks. If so, the predicted value of the 1s-1p splitting would agree fairly well with that in the photoelectron spectrum. For n = 6, the correlation is somewhat more convincing. Two peaks deriving from 1p levels are predicted, but the lifting of the degeneracy of two of the 1p sub-levels would result in three 1p peaks. This is consistent with the observed spectrum. The case of n = 7 has been previewed above. The observed triplet of peaks is suggestive of the 1p level's three-fold degeneracy having been lifted. The case of n = 8 is rather convincing. The above kinds of arguments about 1p sub-levels and peaks are applicable here, but in addition the peak for a 1d level appears on cue at the predicted energy due to the opening of a new shell. Inspection of the n = 9 case is also fairly convincing if one allows the doubly degenerate 1p level to split into a doublet of peaks in the spectrum. The n = 10 case is consistent with the model but by itself is not convincing. The n = 11 case has the predicted number of peaks in the combined 1p and 1d manifolds. In most cases, the suspected 1s-1p splittings observed in the spectra are in reasonable accord with those predicted by the model. On the other hand, the model often does poorly in predicting the lifting of 1p sub-level degeneracies.

At n = 12, we lose sight of the 1s peak, ie., it shifts out beyond our photon range. Also here, 1p and 1d manifolds start to segregate into separate groupings of peaks, the beginning of an important trend. Within the 1p grouping of peaks there are three discernible peaks as predicted by the model. The same situation holds for the 1d grouping of peaks. By n = 13, the valley between the 1p and 1d groupings has become deeper and the two groupings more obvious, even though the spectral sub-structure on each grouping is more difficult to interpret. At n = 14, the spectral sub-structure on each of the two peak groupings is practically gone, but the 1p peak grouping has become clearly distinct from the 1d grouping, resulting in a spectrum having two broadened peaks. At n = 15, the pace of 1p and 1d segregation quickens. This is apparent in the predictions of the model and in the observed spectrum, and the valley between the two peak groupings deepens. The n = 16 case continues the trend as does the n = 17 case. The model predicts n = 17 (an 18 valence electron system) to be a spherical closed shell. By n = 18, the model predicts the appearance of a singly occupied 2s level. It faintly presents itself as a shoulder on the low electron binding energy side of the 1d shell peak grouping. At n = 19 (a 20 valence electron system), the model again predicts a spherical closed shell system. Now, however, the doubly occupied 2s level pops out dramatically as a new peak in the spectrum. The size range between n = 14 and 19 provides the most convincing set of correlations between the spectra and the shell model. We observe that, all else being equal, the correlation between the shell model and the observed spectra tends to be better (1) near shell closings, ie., near the 8, 18, and 20 valence electron systems, and (2) for larger cluster anions than for smaller ones. Indeed, both trends are what one would expect from a shell model for metal clusters. Viewing the evidence as a whole, we are led to the conclusion that there is something to a first approximation assignment of these photoelectron spectra in terms of a simple shell model. This is the first time that shell models have been successfully put to the test at the level of single particle energy levels. Always before, shell models were tested via consequential properties such as EA vs n trends, IP vs n trends, and mass spectral abundance patterns.

Shell models have been the subject of intense interest in cluster physics in recent years, the most famous shell model being the celebrated jellium model. Shell models generally arise as a result of fermions moving in attractive smooth potentials. Thus, even though the Saunders/Knight shell model that we have used here is not a jellium shell model, it is reasonable to expect that it should reproduce some of the gross features of a jellium calculation. To date, however, the predictions of a true jellium calculation have not been compared to the photoelectron spectra of potassium cluster anions. It would be interesting to do so, and we look forward to the availability of theoretical results that will make this possible.

Acknowledgements

It is a pleasure to acknowledge valuable conversations concerning this work with V. Bonačić-Koutecký , W. Knight, W. Saunders, D. Lindsay, J. Koutecky, W. Ekardt, M. Kappes, W. deHeer, E. Poliakoff, J. Pacheco, C. Yannouleas, B. Judd, L. Madansky, and K.-H. Meiwes-Broer. This work was supported by the (U.S.) National Science Foundation under grant number CHE-9007445.

References

1. Konowalow, D.D., and Rosenkrantz, M.E., in J.L. Gole and W.C. Stwalley (eds.), Metal Bonding and Interactions in High Temperature Systems, American Chemical Society, Washington D.C. 3-17 (1982)

2. Cocchini, F., Upton, T.H., and Andreoni, W., J. Chem. Phys. 88 6068-6077 (1988)

3. Martins, J.L., Car, R., and Buttet, J., J. Chem. Phys. 78 5646-5655 (1983)

4. Bonačić-Koutecký V., Fantucci, P., and Koutecký, J., Chem. Rev. 91 1035-1108 (1991)

5. Rao, B.K., Khanna, S.N., and Jena, P., Phys. Rev. B 36 953-960 (1987)

6. Clemenger, K., Phys. Rev. B 32 1359-1362 (1985)

7. Cohen, M.L., Chou, M.Y., Knight, W.D., and de Heer, W.A., J. Phys. Chem. 91 3141-3149 (1987)

8. de Heer, W.A., Knight, W.D., Chou, M.Y., and Cohen, M.L., Solid State Physics 40 93-181 (1987)

9. Röthlisberger, U., and Andreoni, W., J. Chem. Phys. 94 8129-8151 (1991)

10. Ekardt, W., Phys. Rev. B 29, 1558-1564 (1984)

11. Ekardt, W., and Penzar, Z., Phys. Rev. B 38 4273-4276 (1988)

12. Ross, A.J., Crozet, P., Effantin, C., d'Incan, J., J. Phys. B: At. Mol. Phys. 20 6225-6231 (1987)

13. Heinze, J., Schöhle, U., Engelke, F., and Caldwell, C.D., J. Chem. Phys. 87 45-53 (1987)

14. Gole, J.L., in M. Moskovits (ed.), Metal Clusters, Wiley, New York 131-184 (1986)

15. Kappes, M.M., Kunz, R.W., and Schumacher, E., Chem. Phys. Lett. 91 413-418 (1982)

16. Pollack, S., Wang, C.R.C., and Kappes, M.M., J. Chem. Phys. 94 2496-2501 (1991)

17. Brechignac, C., Cahuzac, Ph., and Roux, J. Ph., J. Chem. Phys. 87 229-238 (1987)

18. Lindsay, D.M., Herschbach, D.R., and Kwiram, A.L., Mol. Phys. 32 1199-1213 (1976)

19. Delacretaz, G., Grant, E.R., Whetten, R.L., Woste, L., and Zwanziger, J.W., Phys. Rev. Lett. 56 2598-2601 (1986)

20. Selby, K., Kresin, V., Masui, J., Vollmer, M., de Heer, W.A., Scheidemann, A., and Knight, W.D., Phys. Rev. B 43 4565-4572 (1991)

21. Bonačić-Koutecký V., Fantucci, P., and Koutecký, J., J. Chem. Phys. 91 3794-3795 (1989)

22. Rao, B.K., and Jena, P., Int. Jour. of Quant. Chem.: Quant. Chem. Symp. 22 287-296 (1988)

23. Lindsay, D.M., Chu, L., Wang, Y., George, T.F., J. Chem. Phys. 87 1685-1689 (1987)

24. Ortiz, J.V., J. Chem. Phys. 89 6353-6356 (1988)

25. Gole, J.L., Childs, R.H., Dixon, D.A., and Eades, R.A., J. Chem. Phys. 72 6368-6375 (1980)

26. Lelyter, M., and Joyes, P., J. Phys (Paris) 35 L85-L88 (1974)

27. McHugh, K.M., Eaton, J.G., Lee, G.H., Sarkas, H.W., Kidder, L.H., Snodgrass, J.T., Manaa, M.R., and Bowen, K.H., J. Chem. Phys. 91 3792-3793 (1989)

28. Ho, J., Ervin, K.M., and Lineberger, W.C., J. Chem. Phys. 93 6987-7002 (1990)

29. Leopold, D.G., Ho, J., and Lineberger, W.C., J. Chem. Phys. 86 1715-1726 (1987)

30. Ganteför, G., Gausa, M., Meiwes-Broer, K.H., and Lutz, H., Farad. Disc. Chem. Soc. 86 1-12 (1988)

31. Pettiette, C.L., Yang, S.H., Craycraft, M.J., Conceicao, J., Laaksonen, R.T., Chesnovsky, O., and Smalley, R.E., J. Chem. Phys. 88 5377-5382 (1988)

32. Saunders, W.A., Ph. D. Thesis, University of California, Berkeley (1986)

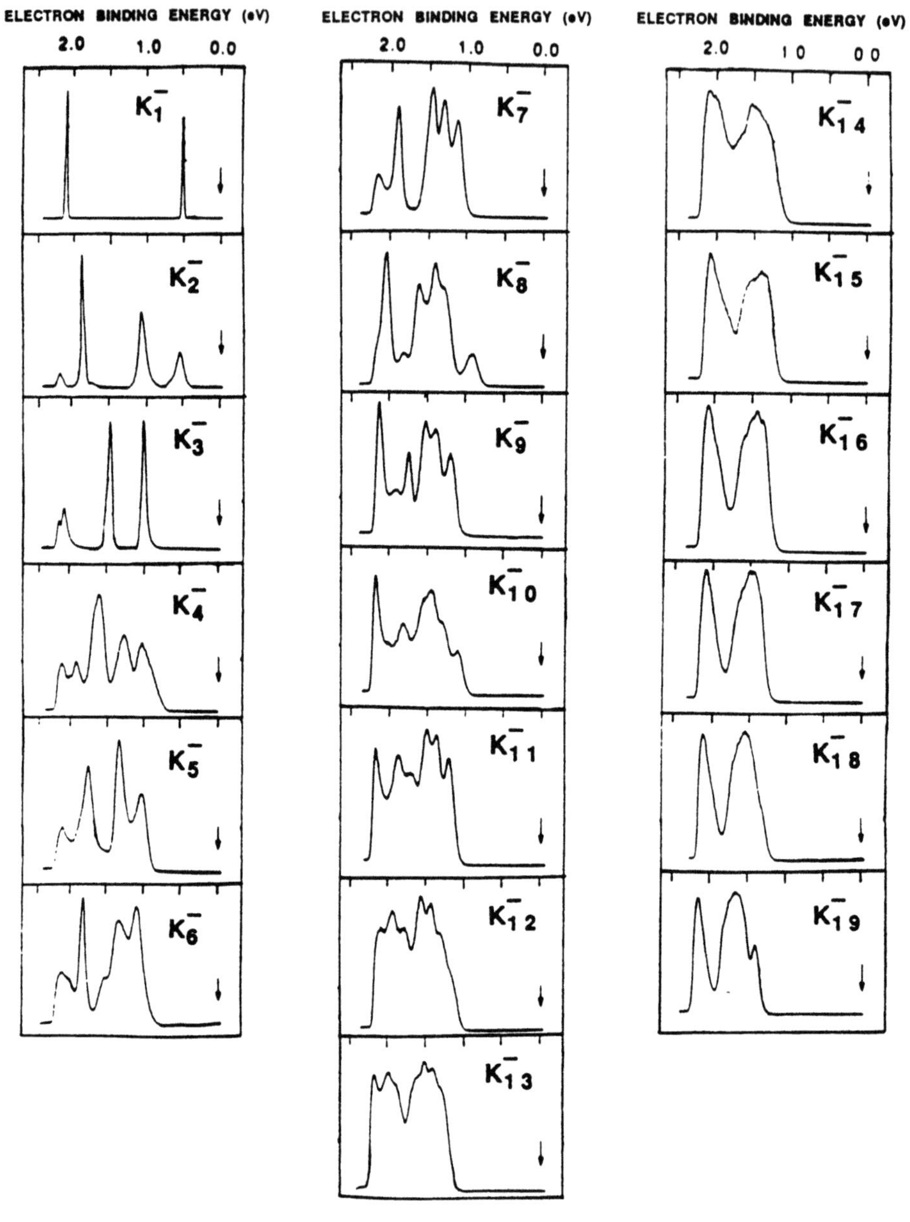

Figure 1. The photoelectron spectra of $K^-_{1\text{-}19}$ recorded with 2.540 eV photons.

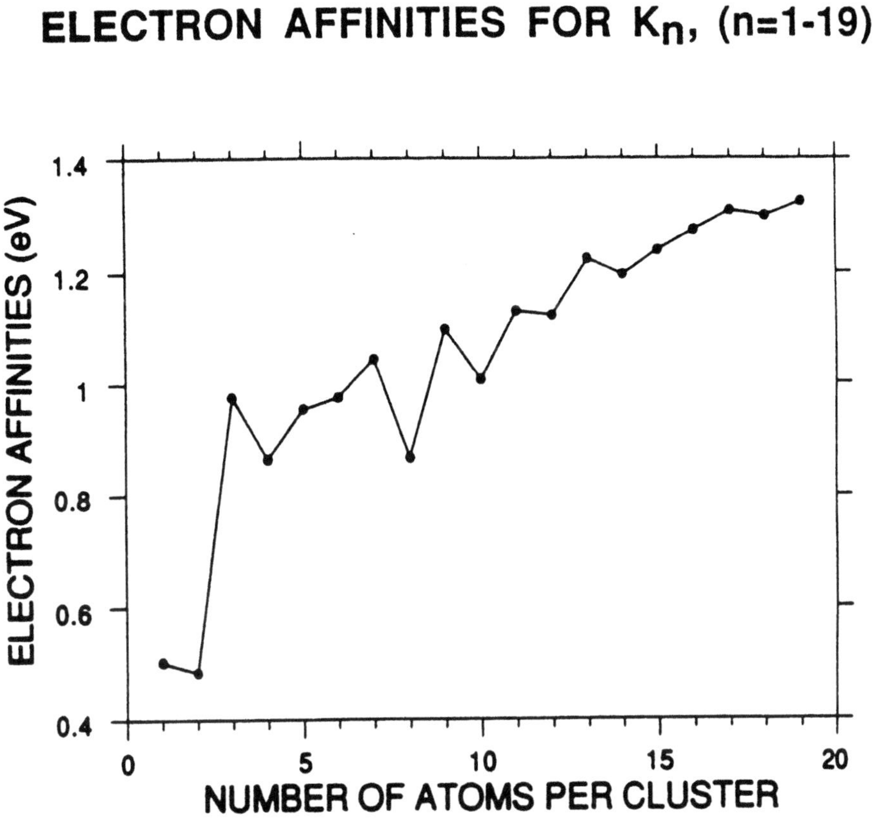

Figure 2. Measured adiabatic electron affinities of $K_{1\text{-}19}$ as a function of cluster size.

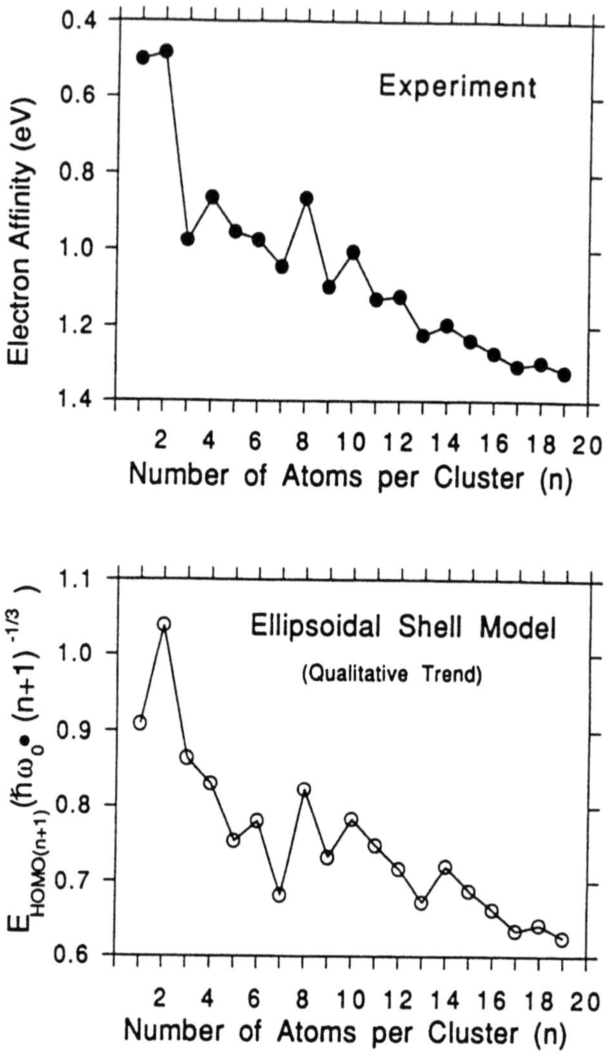

Figure 3. Comparison of the experimentally determined EA vs. n trend for $K_{n=1-19}$ (top) with the qualitative EA vs. n trend predicted by the ellipsoidal shell model (bottom). In the latter, the ellipsoidal shell model energy of the highest occupied level for the n+1 electron system is plotted as a function of cluster size, n. Note that the energy scale of the top plot has been inverted to facilitate comparison.

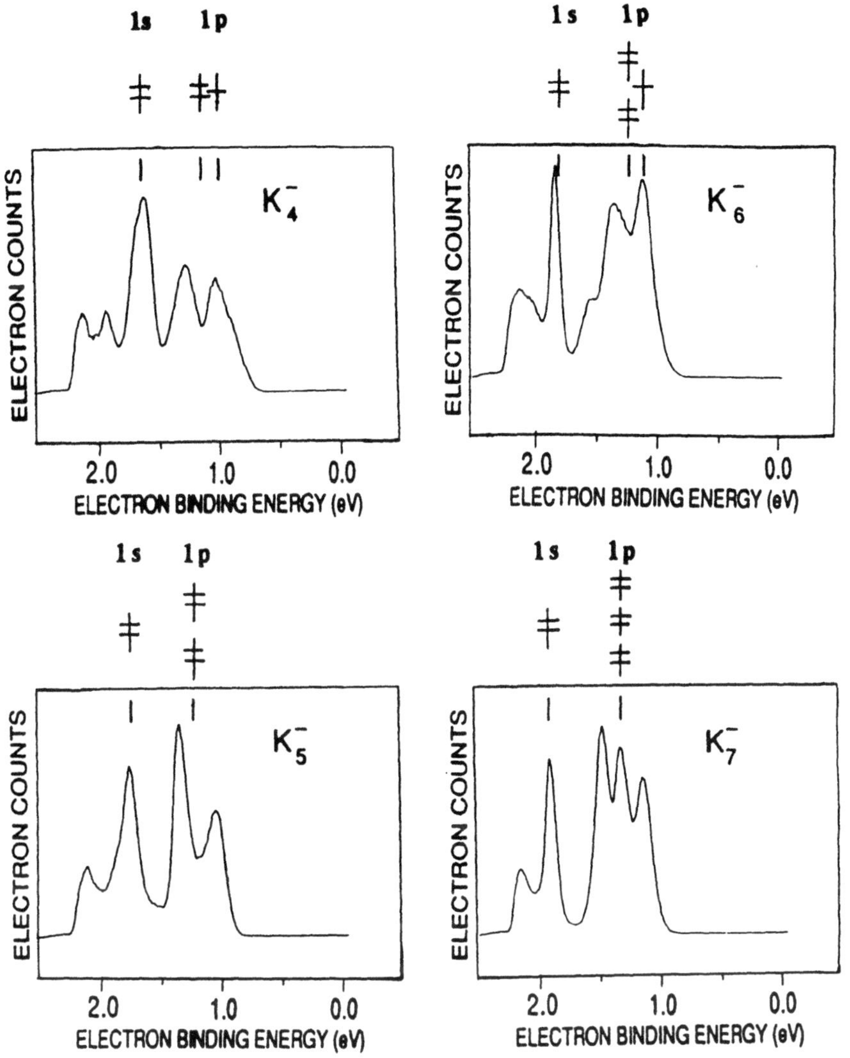

Figure 4. Comparisons of calculated shell model "stick spectra" with the corresponding observed photoelectron spectra of K_n^- (n=4-7). The "stick spectra" are shown as insets on each photoelectron spectrum. Designations and electron occupations for individual levels are shown above each "stick spectrum".

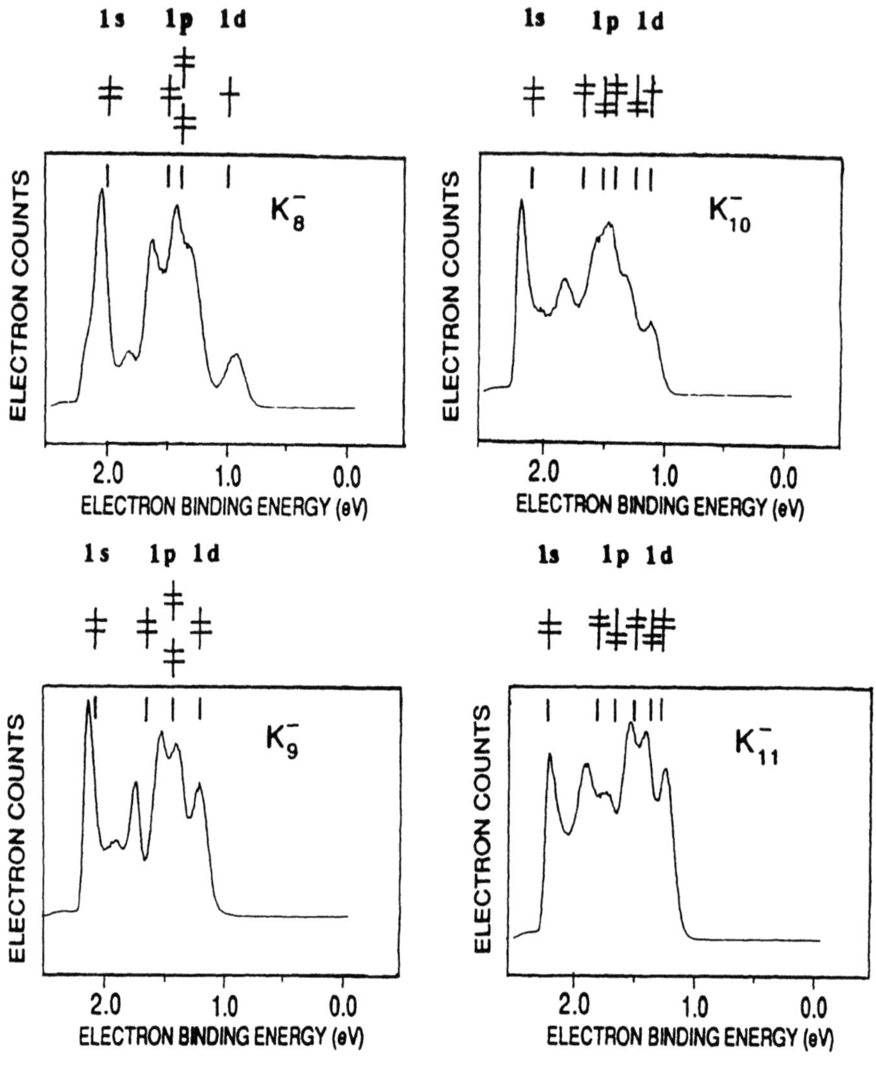

Figure 5. Comparisons of calculated shell model "stick spectra" with the corresponding observed photoelectron spectra of K_n^- (n=8-11). The "stick spectra" are shown as insets on each photoelectron spectrum. Designations and electron occupations for individual levels are shown above each "stick spectrum".

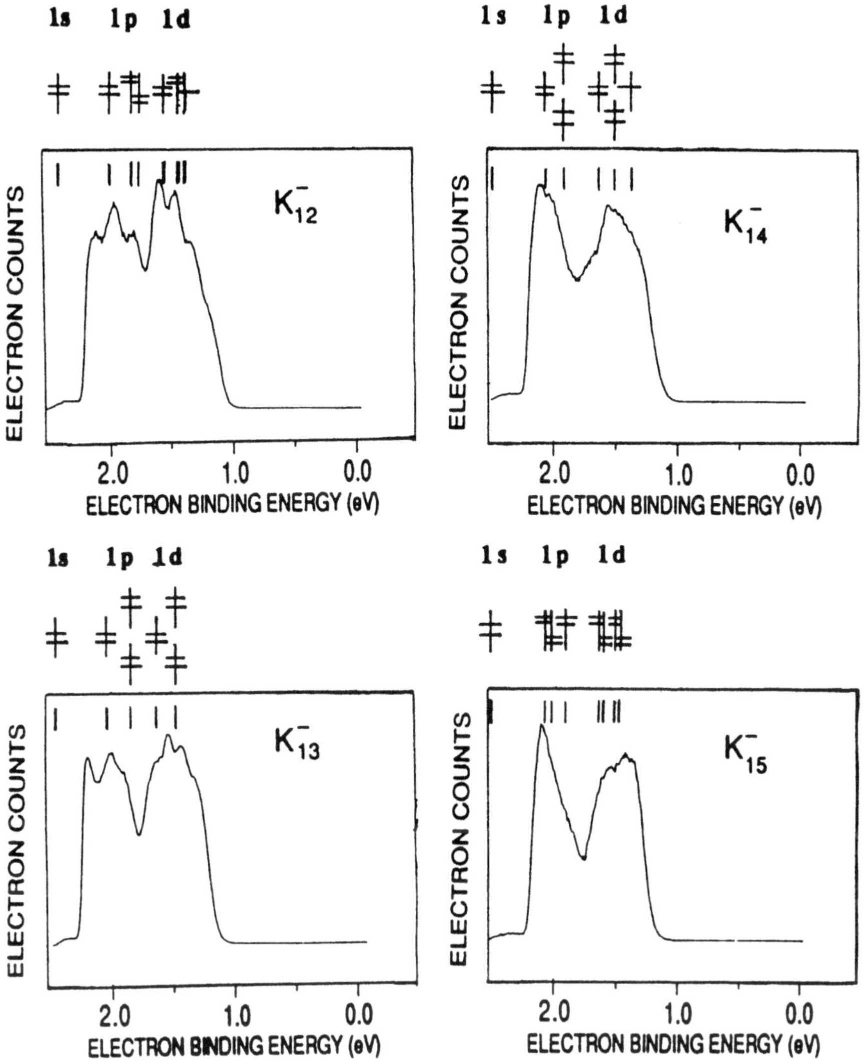

Figure 6. Comparisons of calculated shell model "stick spectra" with the corresponding observed photoelectron spectra of K_n^- (n=12-15). The "stick spectra" are shown as insets on each photoelectron spectrum. Designations and electron occupations for individual levels are shown above each "stick spectrum".

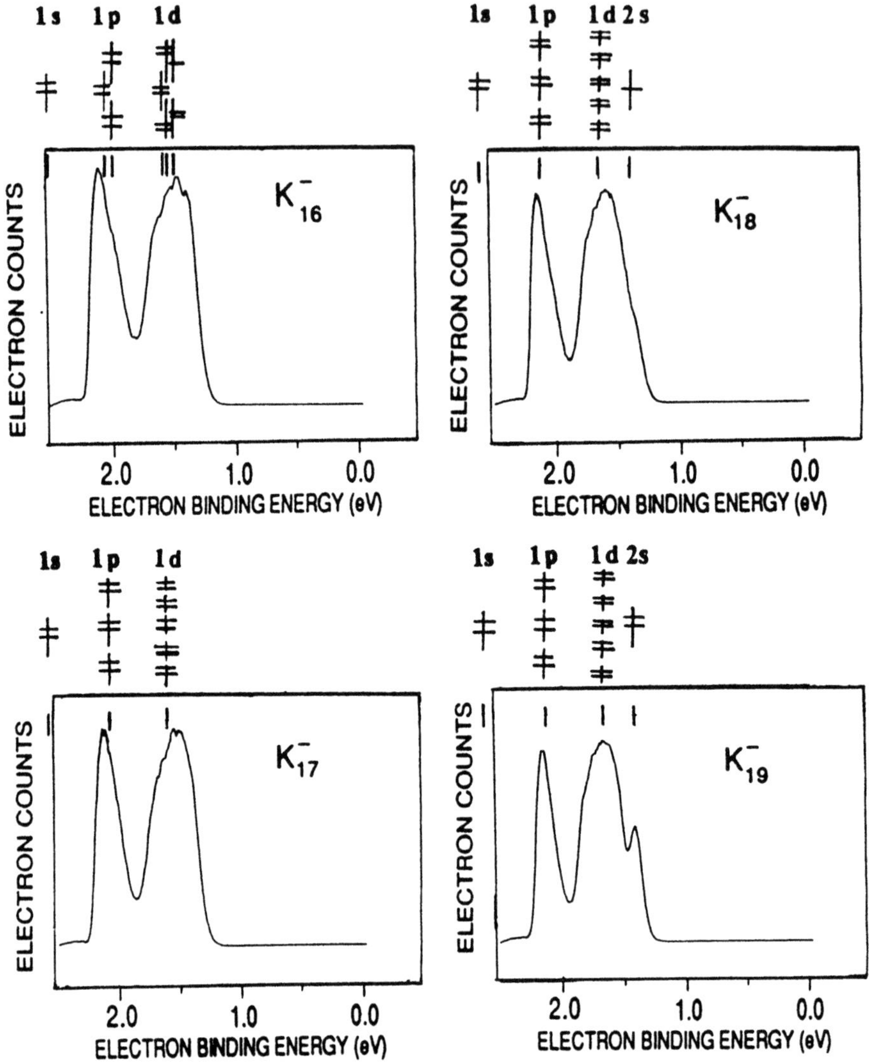

Figure 7. Comparisons of calculated shell model "stick spectra" with the corresponding observed photoelectron spectra of K_n^- (n=16-19). The "stick spectra" are shown as insets on each photoelectron spectrum. Designations and electron occupations for individual levels are shown above each "stick spectrum".

Structure and bonding in aluminium and gallium clusters

R.O. Jones

Institut für Festkörperforschung, Forschungszentrum Jülich
D-5170 Jülich, Federal Republic of Germany

The geometrical structure is one of the most basic properties of any aggregate of atoms. In principle, it is easy to calculate, as one "only" has to find the structure with the lowest total energy. This is seldom practicable, as we must both calculate the energy surfaces and avoid unfavourable minima in them. We outline a method that addresses both problems, and show that it leads to interesting and unexpected results for clusters of aluminium and gallium up to $n = 10$.

1. Introduction

Interest in clusters has grown dramatically in the last decade. With the advances in theory and experiment, we can expect an improved understanding of many of their properties, particularly those reflecting the transition from atomic $\rightarrow$ molecular $\rightarrow$ bulk behaviour. The geometrical arrangement of a cluster, one of its most important properties, is often difficult to determine experimentally. In principle, the most stable forms of a system of electrons and ions can be found by calculated by locating the lowest-lying minima in the energy surface. However, the numerical effort required to determine the energy E from the exact wave function Ψ increases dramatically with increasing electron number, and the number of minima (and possible structures) increases very rapidly (probably exponentally) with increasing number of atoms n.

The approach we use addresses both problems. First, the density functional (DF) formalism provides a numerically efficient scheme for calculating the energy E for a system of ions and electrons. Second, "simulated annealing" – based on finite temperature molecular dynamics (MD) – allows us to avoid unfavourable energy minima. We give results for small aluminium and gallium clusters, and discuss trends in binding and ionization energies.

2. Density functional calculations with molecular dynamics

The two basic theorems of the DF formalism are:

(1) Ground state (GS) properties of a system of electrons and ions in an external field V_{ext} can be determined from the electron density $n(\mathbf{r})$ *alone*. The total energy E is such a functional of the density, $E = E[n]$.

(2) $E[n]$ satisfies the variational principle $E[n] \geq E_{GS}$. The equality holds for the ground state density, n_{GS}.

Two minimization problems must be solved to determine the most stable structures: The total energy E must be minimized for each geometry by varying the density, and we must find the geometry with the lowest E. If there are many local minima, finite-temperature "simulated annealing", whether based on a Monte Carlo sampling[1] or on molecular dynamics (MD),[2] is an appropriate way to avoid high-lying local minima in the energy surface. The combined MD/DF scheme,[2] which views E as a function of interdependent sets of degrees of freedom related to the electronic and ionic motions, is free of adjustable parameters and makes no assumptions about ground state geometries. We use large unit cells with periodic boundary conditions, and the scattering properties of the atoms are described by pseudopotentials. Further details are given in the original papers.[2,3]

3. Clusters of aluminium and gallium

Aluminium has been a favourite element for cluster studies. The electronic structure of the bulk material is characterized by small departures from free-electron behaviour, and several theoretical studies of Al_n clusters have adopted the "spherical jellium" model,[4] where both the electronic charge and positive background distributions are uniform within a sphere of appropriate size. Particular attention has been paid to the existence of prominent

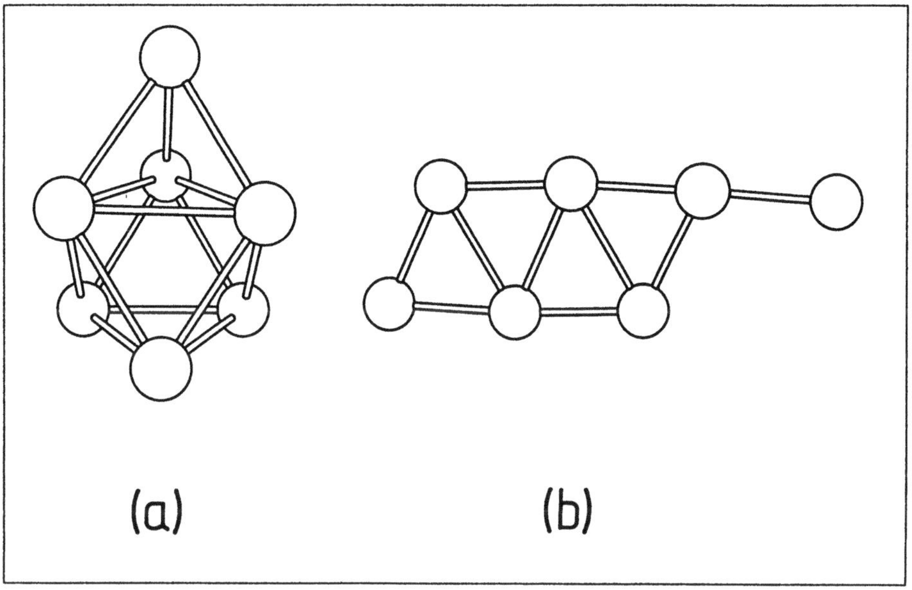

Fig. 1: Structures of Al_7.

or unusually stable clusters, e.g., Al_7^+, with "magic numbers" of atoms or electrons. There are, however, many experimental data that cannot be explained by this model, and detailed calculations of geometries are essential to explain measured properties of clusters with less than ~40 atoms. We describe here calculations of Al_n and Ga_n clusters up to $n = 10$.[3]

There is excellent agreement with experimental values of equilibrium separations r_e and vibration frequencies ω_e for both the $^3\Pi_u$ and $^3\Sigma_g^-$ states of Al_2.[5] The near degeneracy between these states reflects the ease of transfer between σ- and π-electrons. The most stable form of Al_3 is an equilateral triangle ($^2A_1$, $r_e = 4.65$ a.u.),[6] and the calculations lead in Al_4 to a planar rhombus structure [$r_e = 4.75$ a.u., bond angle $\alpha = 56.5°$]. In larger clusters we find a rich variety of structures and spin multiplicities, and it is not surprising that previous results have been limited and often contradictory.[3] We observe transitions from planar to non-planar structures at Al_5 – where a planar C_{2v} structure is almost degenerate with a C_s structure with similar bond lengths – and to states with minimum spin degeneracy at Al_6, where the lowest lying singlet (D_{3d}) and triplet (D_{2d}) states are nearly degenerate.

Two stable isomers of Al_7 [Fig. 1] – a compact structure reminiscent of P_4S_3, and a planar form ~ 0.8 eV higher in energy – are each the closest

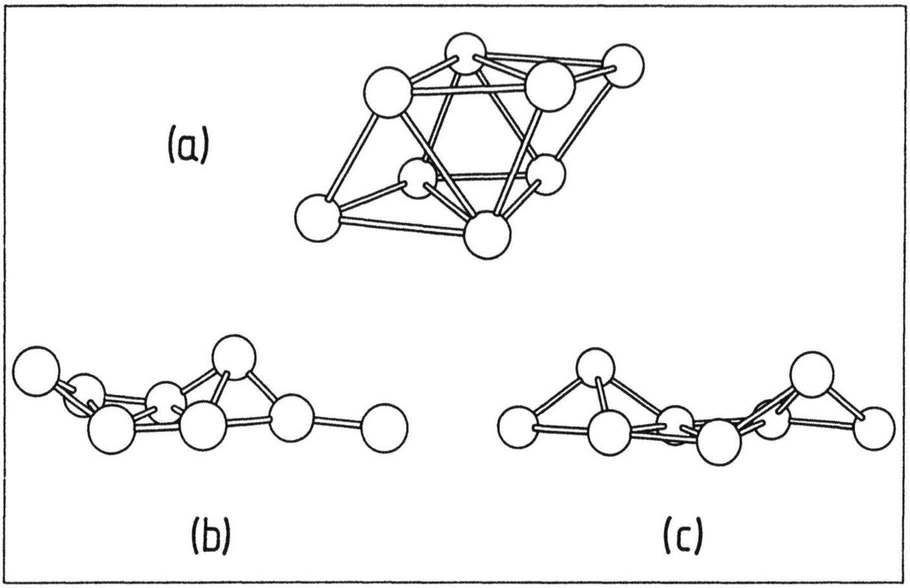

Fig. 2: Three structures of Al_8.

minimum for a large region of configuration space. In Al_8 to Al_{10}, there are many more stable isomers, and examples are shown in Fig. 2 for Al_8. In addition to a "bulklike" fragment [Fig. 2(a)], there are numerous buckled planar structures, such as shown in Figs. 2(b) and 2(c). *All* of the structures conform to the pattern found in Al_5: almost equilateral triangles connected so that they are either coplanar [dihedral angle $\gamma \sim 0°$] or with $\gamma \sim 50°$. Interconversion between these structures can be observed in simulations at 300-500K, and they also occur with a single extremal atom, as in Fig. 2(c). The energies of the isomers 2(b) and 2(c) are almost identical, and they are the more stable than the isomer shown in Fig. 2(a).

In Al_9 and Al_{10}, "bulklike" structures are more stable than the buckled planes. The most stable isomer in Al_9 (C_{3v}) has the C_{2h}-structure of Al_8 with an additional "dangling" atom, and capped structures of Al_9 are the most stable in Al_{10}. In Al_{10}, the "open" arrays of triangles in the buckled planes distort towards "closed" bulklike structures. We observed interconversion between different capped structures, and found a stable structure comprising Al_3 and Al_7 units. The large number of local minima in the energy surface for Al_{10} indicates how difficult a detailed study of larger clusters would be.

We have calculated ionization energies I_p for ionic structures that are: (a) the same as in the most stable isomer found for Al_n, and (b) fully relaxed.[3] The experimental trends[7,8] are reproduced well: I_p increases initially to a maximum at Al_6,[8] and has a sharp minimum at Al_7,[7] and the dissociation energies of the ions decrease in the order $Al_7^+ > Al_9^+ > Al_{10}^+ > Al_6^+ > Al_5^+ > Al_4^+, Al_3^+ > Al_8^+$. The Al_7^+ ion is one of the most prominent in beam experiments, and the energy change on relaxation is much larger (~0.5 eV) than in the other ions (< 0.2 eV). The final structure (C_{3v} symmetry) is very compact, with bonds from an atom in the central layer to atoms in base and apex of equal length (4.78 a.u.). In fact, each atom is almost equidistant from the centre of mass of the cluster.

The calculated dissociation energy of Al_7 is larger than those of Al_6 and Al_8, and there is a separation of 1.15 eV between the highest (singly-) occupied orbital and the next highest (doubly degenerate) orbital. The large gap between the energy eigenvalues of valence electrons 20 and 21 is consistent with the low ionization energy in Al_7. The gap between the corresponding eigenvalues in the planar structure (~0.8 eV) is smaller, but larger than typical energy differences in other *Al*-clusters. These large gaps are in accord with the pronounced stabilities of both Al_7^+ isomers.

Apart from minor differences in the stabilities of the isomers, the results for gallium clusters show only one important difference from those in the corresponding *Al* clusters; the bonds are ~ 5% *shorter*. The transferability of structures within a group of the periodic table has been evident in earlier work.[9] In group VI, the "atomic radii" so determined are ~ 15% larger in *Se* than in *S*, and 10% larger in *As* than in *P*. There are corresponding changes in the radial valence functions for the atoms, which we show for *P* and *As* in Fig. 3(a). This trend continues as we proceed to groups IV (*Si*, *Ge*) and III (*Al*, *Ga*). It is unusual that clusters of a heavier element (*Ga*) are more compact than those of an element of the same group with lower atomic number (*Al*), but this simply reflects the nature of the valence orbitals [Fig. 3(b)].

The variety of structures found in *Al* and *Ga* clusters is consistent with the "metallic" nature of the elements: The valence *sp*-shell is less than half-

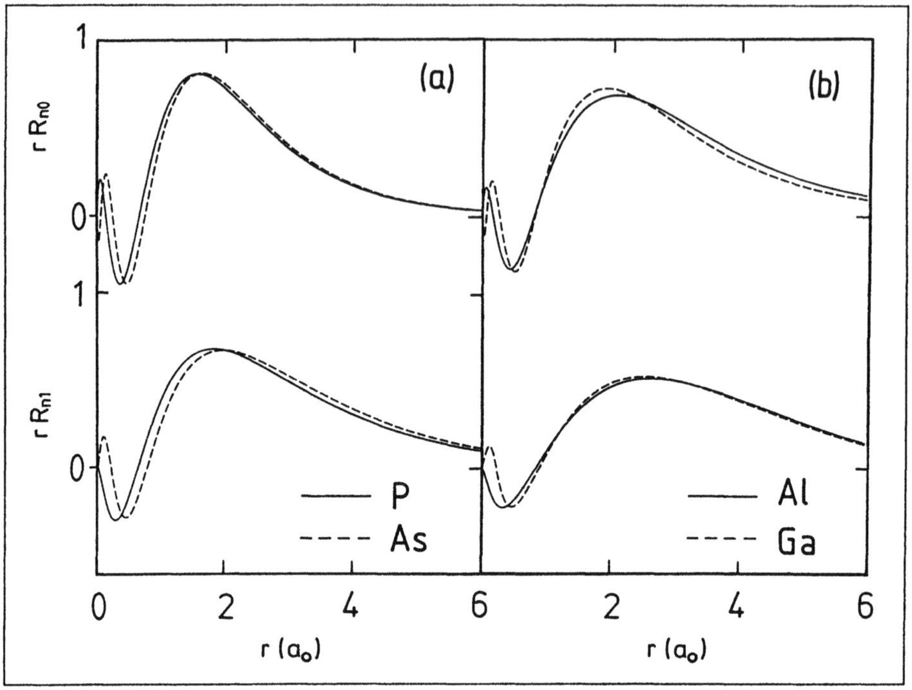

Fig. 3: Radial functions for valence (s and p) electrons in (a) Al and Ga, (b) P, As.

filled, and there are usually unoccupied bonding orbitals near the highest occupied orbital. Furthermore, it is easy to transfer electrons between π-orbitals (which dominate in the bonding in planar structures) and σ-orbitals. The remarkably uniform patterns of bond and dihedral angles are also found in bulk aluminium and in α–gallium,[10] and MD simulations of liquid *Al* show peaks in the bond angle distributions at 60° and near 110°.[11]

The prediction of new structures in clusters of *Al* and *Ga* underlines the advantages of finite-temperature simulated annealing. The departures from regular trends, such as the low dissociation energy in Al_8 and the low I_p in Al_7, are also interesting. The unusually high stability of Al_7^+, with the "magic" number of 20 electrons, suggests some contact to the results of jellium calculations, although the buckled planar structures may not be accessible to calculations that assume a spherical cluster form. Finally, the large relaxation energy found in Al_7^+ indicates that precise measurements of the variation in ionization energies should provide information about the relaxation of ionic structures.

References

1. S. Kirkpatrick, C.D. Gelatt and M.P. Vecchi, Science **220**, 671 (1983).
2. R. Car and M. Parrinello, Phys. Rev. Lett. **55**, 2471 (1985).
3. R.O. Jones, Phys. Rev. Lett. **67**, 224 (1991).
4. See W. A. de Heer, W. D. Knight, M. Y. Chou, and M. L. Cohen, Solid State Phys. **40**, 94 (1987) and references therein.
5. $^3\Pi_u$: (r_e=5.104 a.u., ω_e=290 cm^{-1}), expt: (5.10 a.u., 284.2 cm^{-1}); $^3\Sigma_g^-$: (4.675 a.u., 340 cm^{-1}), expt: (4.660 a.u., 350.01 cm^{-1}). See Ref. 4 for details.
6. The $^2B_1$ state (4.83 a.u., 60°) is 0.32 eV above the ground state, with the $^4A_2$ and $^4B_1$ states $\sim$ 0.1 eV higher.
7. D. M. Cox, D. J. Trevor, R. L. Whetten, E. A. Rohlfing and A. Kaldor, J. Chem. Phys. **84**, 4651 (1986) [$n = 2 - 25$].
8. M. F. Jarrold, J. E. Bower and J. S. Kraus, J. Chem. Phys. **86**, 3876 (1987) [$n = 3 - 26$]; L. Hanley, S. A. Ruatta and S. L. Anderson, J. Chem. Phys. **87**, 260 (1987) [$n = 2 - 7$].
9. See, for example, R.O. Jones, Angew. Chem. **103**, 647 (1991); Angew. Chem. Int. Ed. Engl. **30**, 630 (1991).
10. J. Donohue: *The Structures of the Elements*, (Wiley, New York, 1974), Chap. 5. The f.c.c. structure comprises equilateral triangles with dihedral angles 0°, 54.7°, or 109.5°. The *α-Ga* structure has dihedral angles of 0°, 40°, and 76°.
11. V. A. Polukhin and M. M. Dzugotov, Phys. Met. Metall. **51**, 50 (1981); J. Hafner, J. Non-Crystalline Solids **117/118**, 18 (1990).

Electronic Shell Structures in Aluminum and Noble Metal Clusters

S.Ohnishi

Fundamental Research Laboratories, NEC Corp.
Miyukigaoka 34, Tsukuba, Ibaraki 305, Japan

Abstract. Electronic shell structures of $Au_{12,13}$, $Ag_{12,13}$, $Al_{12,13}$ and $Al_{12}Si$ of icosahedral symmetry are discussed based on self-consistent local density functional calculations using the norm-conserving pseudopotential in the linear combination of atomic orbitals method. The formation of cluster dimers is studied by applying the Harris energy functional scheme.

1.Introduction

Although experimentally observed electronic shell structures in free electron-like metal clusters are well understood by the giant atom model based on the self-consistent jellium background calculations[1,2], those in noble metal clusters[3] characterized by the s-d hybridization are not well studied on the basis of the first principles calculation[4]. It is of interest to elucidate the origin of shell structures for electrons of 5s, 6s, and 3p orbitals in Ag, Au and Al clusters. The discussion here will focus on the electronic shell structures of $Ag_{12,13}$ and $Au_{12,13}$ of icosahedral symmetry based on the self-consistent local density functional scheme by the linear combination of atomic orbitals(LCAO) method using the norm-conserving pseudopotentials [5]. Stability of electronic shell structures in noble metal clusters and $Al_{12,13}$ and $Al_{12}Si$ clusters are also discussed. Using self-consistently determined charge densities of single cluster, the formation of cluster dimers is studied by applying the Harris functional method within an approximation of a weak interaction between clusters[6].

2.Method

The nodeless feature of atomic radial wave functions given by the norm-conserving pseudopotential in the LCAO scheme made it feasible to perform numerical integrations for Hamiltonian matrix elements for even heavy-atom systems. In applying the Harris functional method for bonding of clusters, the potential energy of the weakly coupled system is approximately given by superimposing self-consistently determined charge densities of the separated systems. The kinetic energy is calculated from eigenvalues of the coupled system.

Figure 1 shows calculated results of binding energies for uniform scaling of $Au_{12,13}$, $Ag_{12,13}$, $Al_{12,13}$, and $Al_{12}Si$ vs bond distance by the Kohn-Sham and the Harris functional scheme. In the Harris functional case, the binding energy is given by superimposing atomic charge densities.

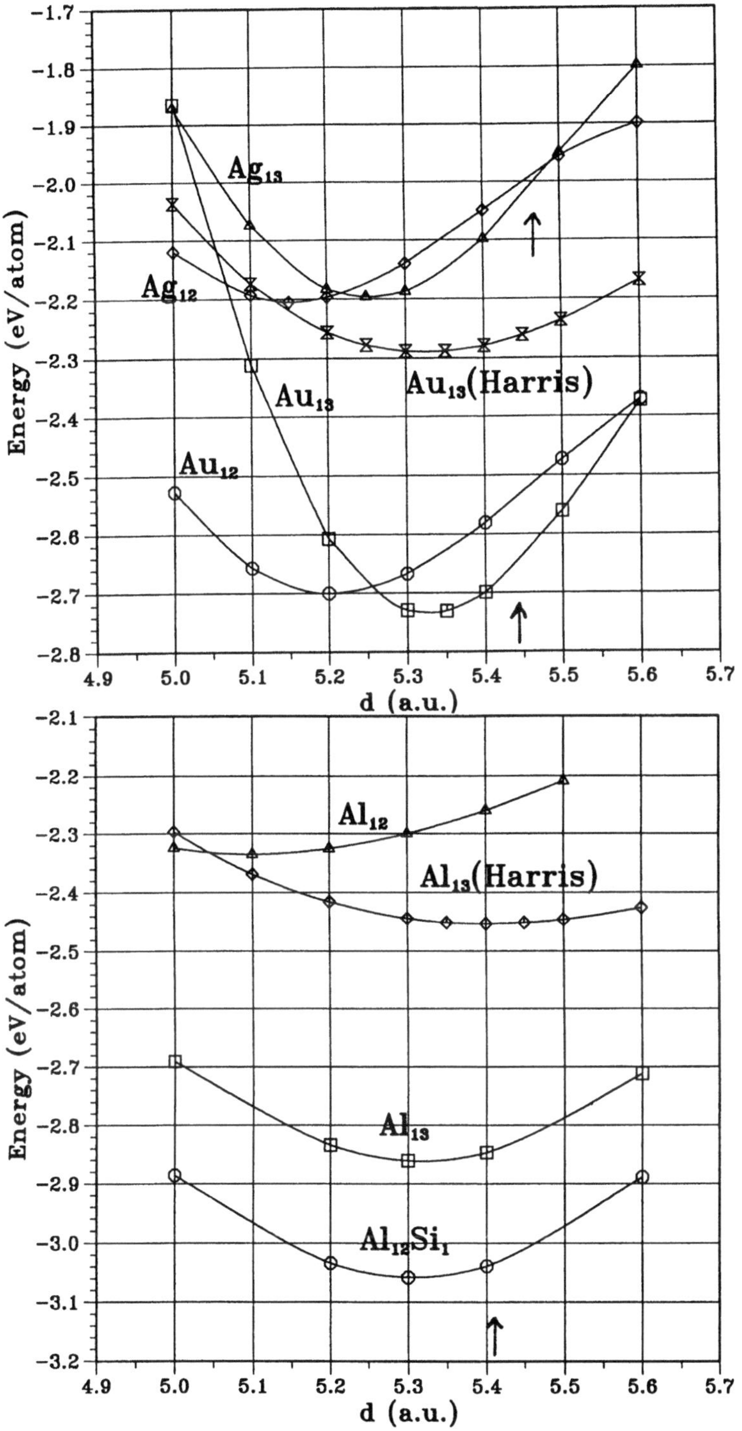

Figure 1. Binding energy curves of clusters of icosahedral symmetry calculated by the Kohn-Sham and the Harris functional scheme.

In Figures 2-1 and 2-2, the density of states(DOS) of Au_{13} and Al_{13} by the Harris functional methods are compared to those by the Kohn-Sham scheme. Since equilibrium distances and electronic structures are almost same in both cases, the Harris functional method is found to be quite reliable in predicting the nature of bonding in cluster molecules.

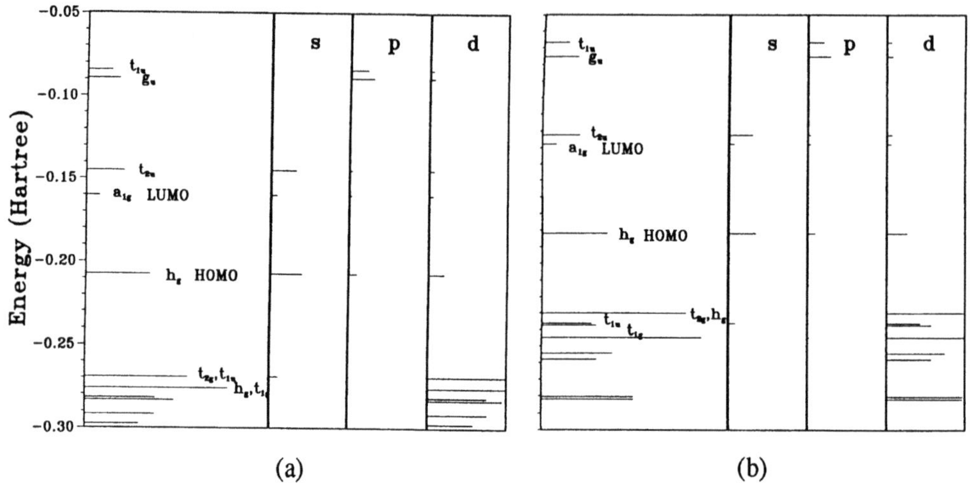

Figure 2-1. DOS's of Au_{13} by the Kohn-Sham(a) and Harris(b) functional method.

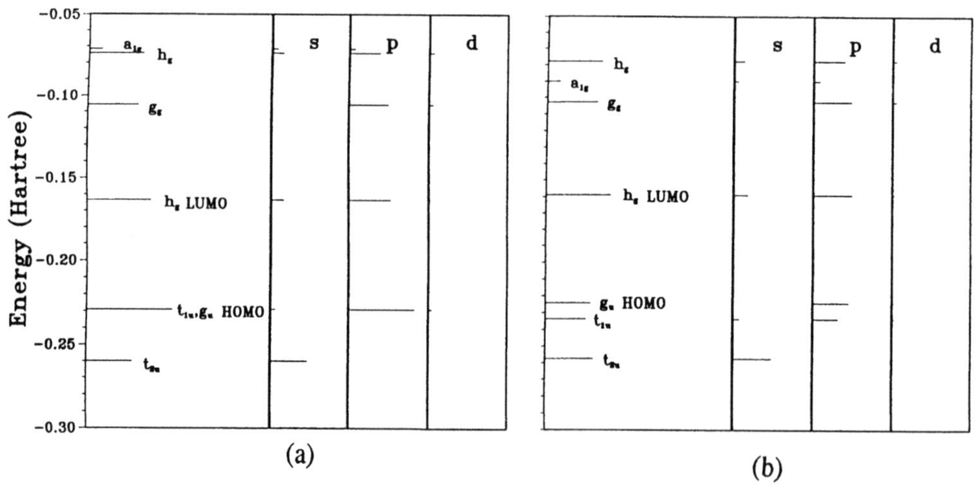

Figure 2-2. DOS's of Al_{13} by the Kohn-Sham(a) and Harris(b) functional method.

The binding energy for cluster dimers are calculated by superimposing self-consistently determined charge densities of clusters. Figure 3 shows binding energy curves of Au_{13} and Al_{13} cluster dimers. Within assumptions of no charge transfer between clusters and no geometrical changes in each cluster, both Au_{13} and Al_{13} clusters are found to form diatomic molecules. HOMO's of these dimers comprise HOMO's of each cluster.

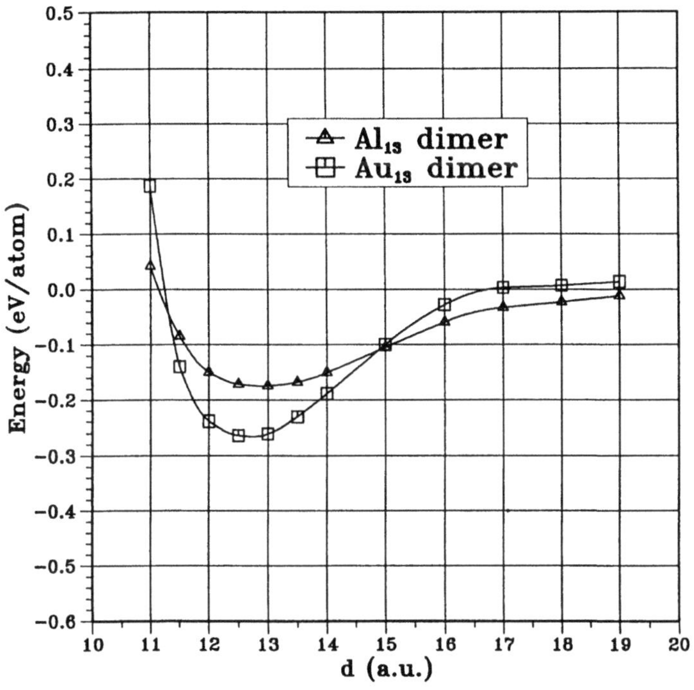

Figure 3. Binding energy of cluster dimers.

3 Shell structures.

The icosahedral symmetry of the cluster induces the splitting of energy levels of the shell structure being higher than the f state: $a_g(1s)$, $t_{1u}(1p)$, $h_g(1d)$, $a_g(2s)$, $t_{2u}+g_u(1f)$, $t_{1u}(2p)$, $h_g+g_g(1g)$...In the monovalent alkali metal clusters, however, the non-spherical effect is found to be not strong enough to change the energy level sequence[5]. In 13 atom clusters of noble metals, relevant energy levels are associated with 1s, 1p, and 1d states of the shell structure. Figures 4 and 5 show energy levels and DOS's for $Au_{12,13}$ and $Ag_{12,13}$ clusters. The DOS is evaluated by the augmentation of molecular orbitals at the center of the cluster. In the cluster of 13 atoms, the highest occupied and the lowest unoccupied molecular orbitals (HOMO and LUMO, respectively) comprise 5s(Ag) and 6s(Au) atomic orbitals of the peripheral and the central atoms, respectively. In 12 atom clusters, energy levels are not different from the case of the 13 atom cluster except the LUMO. Taking out the central atom from Au_{13} and Ag_{13} induces changes in the occupation number of the HOMO(one electron removal) and the character of the LUMO. The shell structure in these noble metal clusters is evident because the HOMO being the h_g state comprises mainly one-center d orbitals. Three levels below HOMO, t_{2g}, t_{1g}, and h_g comprise mainly 5d orbitals. The total amount of the spin-orbit splitting which is originated from each 5d atomic state is about 0.7eV in them, which does not overshoot the HOMO lying 1.7 eV higher.

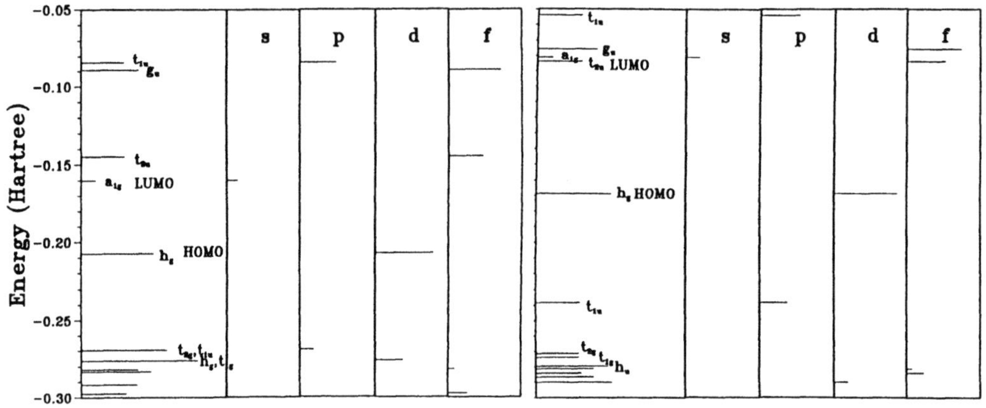

(a) (b)

Figure 4. DOS's of Au_{13}(a) and Au_{12}(b) clusters.

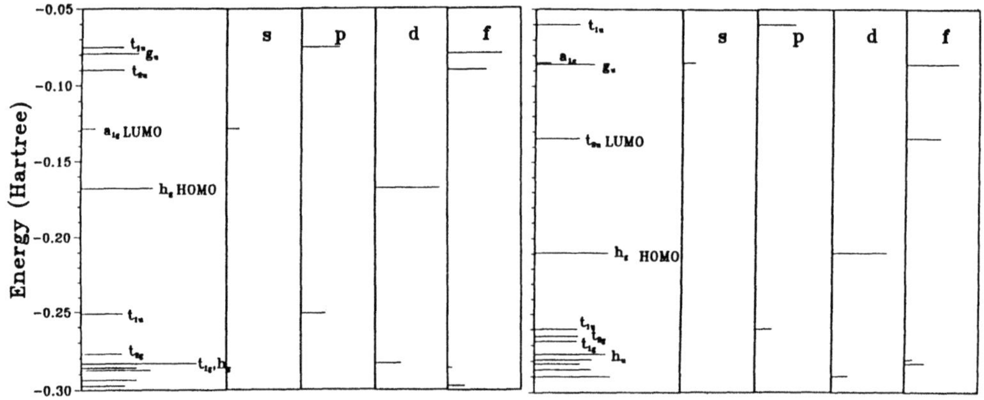

(a) (b)

Figure 5. DOS's of Ag_{13}(a) and Ag_{12}(b) clusters.

In the Al_{13} cluster, relevant shell structure energy levels are 2p and 1f states as shown in Figure 6. The t_{1u}(HOMO) and g_u states comprise 2p orbitals of the peripheral Al atoms. There are 5 electrons in the t_{1u} state. Putting one electron more into the HOMO may stabilize the electronic energy of the cluster. In order to provide one more electron into the HOMO, the central Al was replaced with the Si atom. As shown in Figure 1, the binding energy of the $Al_{12}Si$ is much lower than that of Al_{13}.

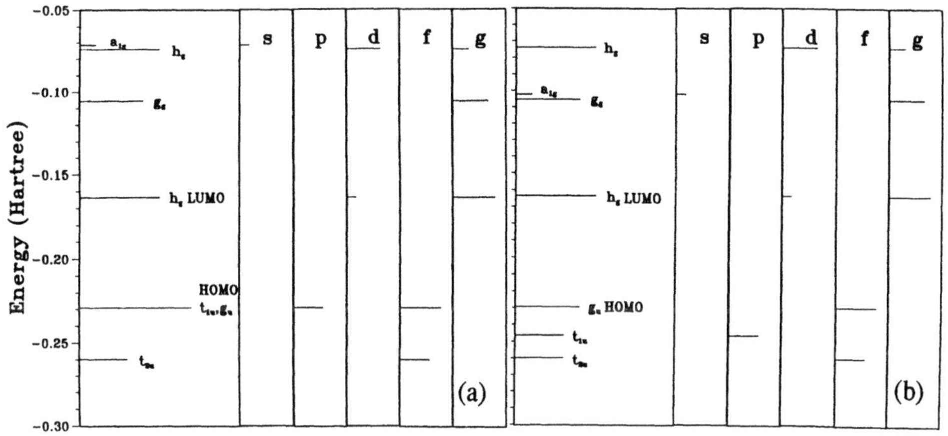

Figure 6. DOS's of Al_{13}(a) and $Al_{12}Si$(b) clusters.

4.Summary.

Electronic shell structures formed by peripheral atoms in Au_{12}, Ag_{12}, and Al_{12} clusters of icosahedral symmetry are found to be stable enough to accept one atom into the cluster center. Doping Si into Al_{12} stabilizes the cluster considerably. Possibility of forming a cluster molecule is proposed within assumptions of fixed charge and geometry.

Acknowledgements.

The author would like to thank Dr.Y.Ishii(Himeji) for fruitful discussions and Mr.T.Ikeda and Mrs.N.Watari in NEC Scientific Information System Development LTD for developing the program and calculations.

References.

[1] W.D.Knight, K.Clemenger, W.A.de.Heer, W.A.Saunders, M.Y.Chou, and M.L.Cohen, Phys.Rev.Lett.**52** 2141(1984).;Y.Ishii, S.Ohnishi, and S.Sugano, Phys.Rev.B **33** 5271(1986).;M.L.Cohen, in Microclusters, Springer Series in Material Science 4, 1987 ed.by S.Sugano, Y.Nishina, and S.Ohnishi,p2.
[2]T.H.Upton,Phys.Rev.Lett.**56** 2168(1986).;H.-P.Cheng,R.S.Berry, and R.L.Whetten, Phys.Rev.B,**43** 10647(1991).
[3] I.Katakuse, T.Ichihara, Y.Fujita, T.Matso, T.Sakurai, and H.Matsuda, Int.J.Mass Spectrom.Ion Processes **67**, 229(1985).;I.Katakuse,in Microclusters,same as [1],p10.
[4]A.F.Ramos, R.Arratia-Perez, and G.L.Malli, Phys.Rev.B,**35** 3790(1987).
[5]G.B.Bachelet,D.R.Hamann,M.Schluter, Phys.Rev.B,**26** 4199(1982).
[6]J.Harris,Phys.Rev.B,**31** 1770(1985).

Correlated Electron Pairs in Metal Clusters

F. Iachello[1], E. Lipparini[2] and A. Ventura[3]

[1] *Center for Theoretical Physics, Yale University, New Haven, Connecticut 06511.*

[2] *Dipartimento di Fisica, Universitá di Trento, 38050 Povo, Italy and INFN Trento.*

[3] *Comitato Nazionale per l'Energia Nucleare e le Energie Alternative,I-40138 Bologna, Italy*

Abstract. *We show that the experimental data on ionization energies and photoabsorption cross sections of alkali metal clusters are consistent with a model of clusters in terms of a system of interacting electron pairs with L=0 and L=2.*

In the last few years several authors have discussed electronic properties of alkali metal clusters in terms of an independent particle motion in which the active electrons move in some average field $V(r)$. As it has been shown by many calculations in the framework of the self-consistent jellium-background model[1], this mean field is dominated by the attractive short range part of the electron-electron interaction responsible for the exchange-correlation term ϵ_{xc} in the jellium energy functional . In fact, there is a strong cancellation between the external, attractive electric field due to the jellium (V^j_{ext}), and the repulsive mean field originating from the Coulomb direct electron-electron interaction ($\int \frac{\rho_e(\mathbf{r}')}{|\mathbf{r}-\mathbf{r}'|}d\mathbf{r}'$). This cancellation yields to a negligible electrostatic contribution to the total energy of the cluster.

The situation is similar to the nuclear shell model where the mean field is originated by the nucleon-nucleon interaction and different from the atomic case where the mean field is dominated by the external Coulomb field of the nucleus. In fact, in the last case the Coulomb direct term compensates only about one-half of the contribution of the external field to the total binding energy. This fact is reflected in the dependence of the binding energy on the number N of costituents: in atoms the binding energy goes like $N^{\frac{7}{3}}$ whereas in nuclei and clusters like N.

In nuclei one derives the mean field from a two-body effective density dependent interaction which is short range and attractive. The binding energy goes like N because each nucleon interacts only with the neighbouring. The effective interaction gives rise, beyond the mean field, to a residual interaction V_{res} which is responsible for correlations beyond Pauli correlations. Due to the strong screening effect, the situation in metal clusters seems to be similar to the nuclear case. In the following we will hence assume a short range attractive effective residual interaction of the form

$$V_{res} = \frac{-V_0}{\rho_0}\delta(\mathbf{r}_1 - \mathbf{r}_2), \tag{1}$$

active between electron pairs. Exploiting further the analogy with the nuclear case we will assume that the residual interaction is active only among electrons belonging to open shells, whereas closed shells corresponding to magic numbers remain inert.

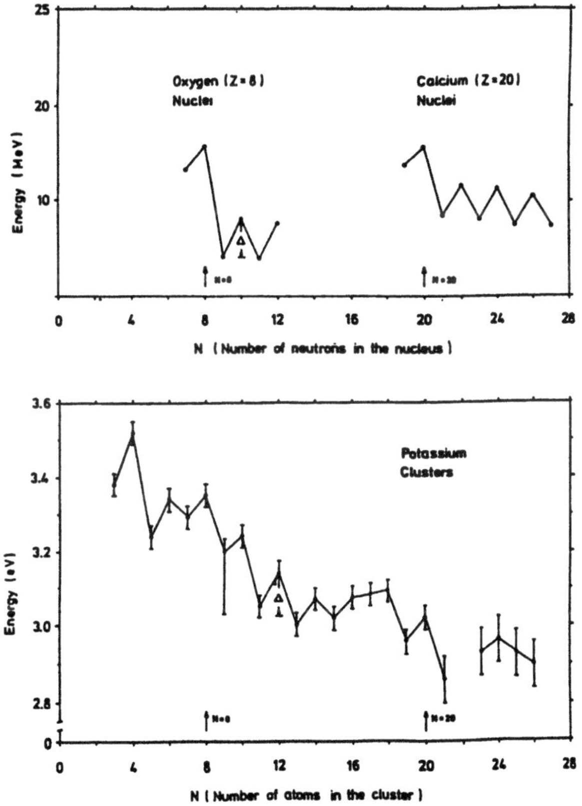

Figure 1

Some experimental evidence of short-range correlations in metal clusters induced by a residual interaction can be found in: i) dissociation experiments[2] where one has observed a competition between the evaporation of monomers and dimers in clusters with an even number of electrons, and emission of only monomers in clusters with an odd number of electrons; ii) the strong resemblance between the odd-even effect seen in the neutron separation energies of nuclei[3] which is explained by pairing effects, and the odd-even alternation seen in the ionization energies of metal clusters[4] (see fig.1). Further evidence for correlated electron pairs could come for example from double charge exchange experiments where one observes the transfer of two particles [5].

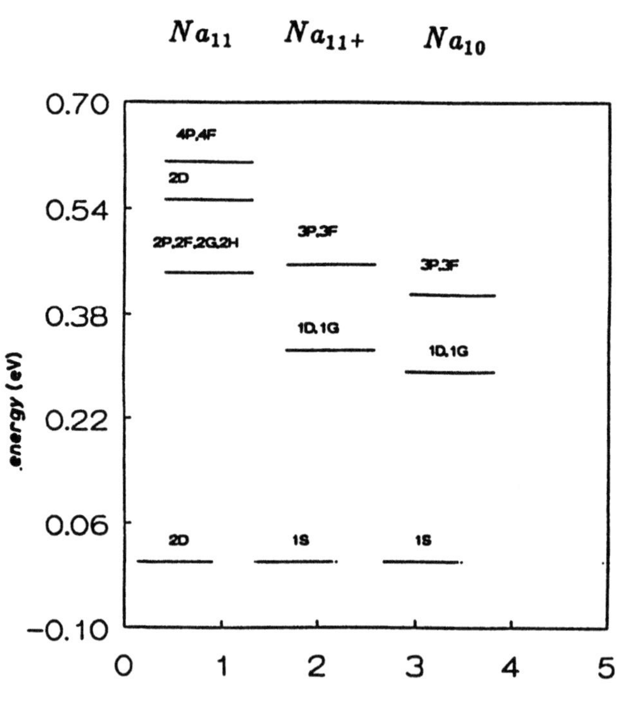

Figure 2

In fig.2 we report a preliminary configuration mixing calculation with the delta-force (1) with V_0=1.5 eV in Na_{10},$Na_{11}+$ and Na_{11}. We have generated all the possible $|\ell^n, s, L, S>$ particle states , starting from n particles in the 1d shell (ℓ=2) outside the inert closed-shell core of 8 electrons in the 1s, 1p shells (s is the seniority number and L,S the total angular momentum and spin respectively). We have then diagonalized the residual interaction on these states. The single particle states used in the calculation are those generated by the spherical jellium model. The ionization energies for N=9,10,11 are 3.547, 4.149 and 4.049 eV respectively, yielding an odd-even staggering $\simeq$ 0.1 eV in agreement with experimental data[6]. The mean field result for the ionization energies is 3.547, 3.738 and 3.889 eV respectively , which do not present any odd-even effect. In the case of Na_{10}, which has two electrons in the 1d level outside the inert core, the residual interaction breaks the degeneracy of the 3P,3F,1D,1G,1S states. As one can see from fig.2 the L=0,S=0 state 1S is much more bound than the others. Short range attractive forces favour the formation of correlated electron pairs in L=0,S=0 states. The most simple residual interaction inducing these pairing correlations is the pure pairing force

$$H_P = -G \sum_{k,k'>0} a_k^\dagger a_{\overline{k}}^\dagger a_{k'} a_{\overline{k}'} \tag{2}$$

where G is the strength of the interaction and $a^\dagger$, a are electron creation and annihilation operators. The coniugate state $\overline{k}$ is the time reversed state (in a spherical basis if$|k>=|n\ell jm>$, $|\overline{k}>=|n\ell j-m>, m>0$). The pairing interacion (2) acts only in L=0, S=0 two-particle states and its strength G can be related to the strength V_0 of the delta-interaction (1) by imposing that they give the same value for for the splitting of the L=0 and S=0 state. One then gets

$$G =\simeq \frac{V_0}{N}, \tag{3}$$

where N is the total number of electrons in the cluster. The advantage of using interaction (2) lies in the possibility of an exact diagonalization of the total Hamiltonian $H = H_0 + H_P$ where H_0 is the one-body Hamiltonian. For n electrons in a single $2(2\ell+1)$-fold degenerate ℓ-shell interacting through the pairing force (2) the ground state, in the case of even n, is given by the pair coupled wave function

$$(S^\dagger)^{\frac{n}{2}}|- > \tag{4}$$

and with n odd by

$$(S^\dagger)^{\frac{n}{2}} a^\dagger_{l,m_l;\frac{1}{2},m_s}|- > . \tag{5}$$

where $|->$ is the inert core wave function and $S^\dagger$ is the pair operator which creates the two-particle L=0,S=0 state . We have then that the ground state is a condensate of $(\frac{n}{2})$ S-pairs (n=even) or a condensate plus an unpaired particle in the orbital l, m_l, s, m_s (n=odd). The eigen-energies are given by

$$E(n,s) = -\frac{G}{4}\{(n-s)(2\Omega - s - n + 2)\} \tag{6}$$

where $\Omega = 2\ell + 1$ and s is the seniority quantum number which gives the number of unpaired particles and takes the values $s = 0, 2, 4, ...n$ for n even and $s = 1, 3, ...n$ for n odd. The ground state has s=0 or 1 for n even or odd. Application of eqs. (6) to clusters with N active electrons distributed in a closed core plus a ℓ-shell with n electrons , yields to the following contribution of the pairing interaction to the ionization energies S(N):

$$\begin{aligned} S(N) &= E(N-1) - E(N) = S_0 - AN \quad , N = even \quad , \\ S(N) &= E(N-1) - E(N) = S_0 - AN - \Delta \quad , N = odd \quad , \end{aligned} \tag{7}$$

whereΔ,A and S_0 are quantities that depend on the strength of the pairing interaction, G, and from the degeneracy Ω of the last unfilled shell($S_0 = G(\Omega+1)$, $A = \frac{G}{2}$ and $\Delta = G(\Omega + \frac{1}{2})$. Note, incidentally, that in the limit N$\rightarrow \infty$ one has $\Delta \rightarrow 0$, $A \rightarrow 0$ and $S_0 \rightarrow 0$ (see eq.(3)). In particular the odd-even staggering Δ decreases as N increases according to $\Delta = const/N^\alpha$, with $1 > \alpha > 0$, (in nuclei $\alpha \sim \frac{1}{2}$).

Figure 3

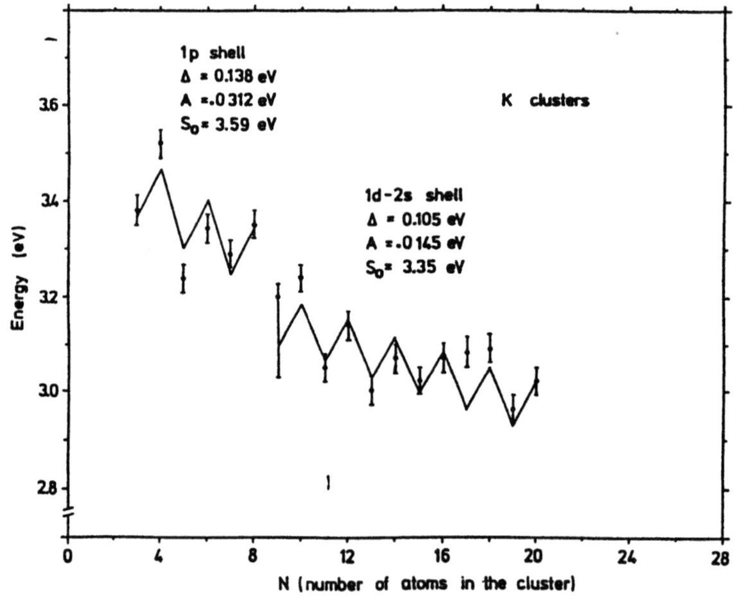

One can see in fig.3 that the ionization energies of potassium clusters[4] do indeed approximately satisfy Eqs.(7) (mean field effects of the type $W + \frac{1}{2}\frac{e^2}{r_s}N^{-\frac{1}{3}}$ are incorporated in S_0). The experimental data of ref. 4, do show that the magnitude of the odd-even staggering decreases as N increases, but in view of their accuracy ($\simeq \pm 0.06$ eV) it is still not possible to determine at the present time the value of α. A value α=1 would give a staggering $\Delta \simeq 0.01$ eV for N=100.

Ionization energies data have been explained by other authors with deformation effects. Mean field calculation show in fact that open shell clusters are strongly deformed. In fig. 4 it is shown the comparison of the spheroidal jellium model calculation[7] and experiments [4,6] for the Na cluster ionization potentials (in eV) in function of N.

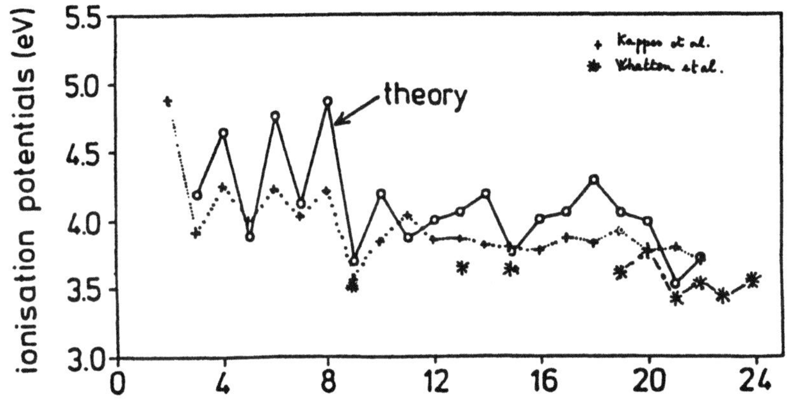

Figure 4

From the figure one can conclude that there is evidence for deformation effects but probably also for pairing correlations as discussed previously.

Effects due to the deformation can be taken into account in the previous model by adding to the pairing Hamiltonian (2) a quadrupole interaction

$$H_Q = -k \sum_{\mu} Q^{\dagger}_{\mu} Q_{\mu}$$

$$Q_{\mu} = \sum_{k,k'} < k|r^2 Y_{2,\mu}|k' > a^{\dagger}_{k} a_{k'}. \tag{8}$$

This quadrupole force originates from a multipole expansion of the delta-interaction of eq.(1). Note that the pairing plus quadrupole Hamiltonian, $H = H_0 + H_P + H_Q$ contains, as a special case, the Nilsson-Clemenger single-particle model used in Ref.8, but it is more general than it, since it includes the effects of pairing. This can be readily seen by using this Hamiltonian in the Hartree-Fock-Bogoliubov theory.

The effect of the quadrupole interaction (8) is to split 1D states in addition to the 1S states splitted by the pure pairing force. It then introduces correlated pairs with orbital angular momentum L=2 (and spin S=0), called D-pairs, in the electronic states. For n even , these states can be written

$$(S^{\dagger})^{n_s}(D^{\dagger})^{n_d}|->, \tag{9}$$

and for n odd

$$(S^{\dagger})^{n_s}(D^{\dagger})^{n_d}a^{\dagger}_{l,m_l;\frac{1}{2},m_s}|-> \quad . \tag{10}$$

where $D^{\dagger}$ is the pair operator which creates the two-particle L=2, S=0 state. The states (9) and (10) are condensates of n_s S-pairs and n_d D pairs, with $n_s + n_d = \frac{n}{2}$ (plus an unpaired electron in odd clusters). These condensates lead to clusters which are no longer spherical but, in general, spheroidal. For spheroidal condensates ,there is an additional contribution to the binding energies which changes Eq.(7). This contribution is proportional to the coupling constant k and must be calculated numerically. An approximate expression can be written in terms of the distorsion parameter, δ , of Ref. 7. The contribution to the binding energy is $E_{DEF}(N) = -c\delta^2(N)$ from which one can obtain

$$S_{DEF}(N) = E_{DEF}(N-1) - E_{DEF}(N) = c[\delta^2(N) - \delta^2(N-1)] \quad . \tag{11}$$

The subscript DEF denotes the contribution due to deformation, the scale of which is determined by the constant c.

Properties of metal clusters can be directly evaluated by diagonalizing $H = H_0 + H_P + H_Q$ in the fermion space for small number of electrons. When N becomes very large, a direct evaluation is not possible in view of the large size of the matrices involved. For these cases, it is convenient, as in the nuclear case, to bosonize the pairs. Replacing the S-D pairs by s-d bosons (we use lower case letters for boson operators and capital letters for pair operators) one can write the states of even clusters as

$$(s^{\dagger})^{n_s}(d^{\dagger})^{n_d}|->, \tag{12}$$

with $n_s + n_d = n_B = \frac{n}{2}$ = total number of pairs. In the case of n electrons in a single degenerate ℓ-shell, the electron Hamiltonian $H = H_0 + H_P + H_Q$ can be rewritten in terms of boson operators as[9]

$$\begin{aligned}
H_B &= E_0 + \epsilon_d \hat{n}_d + k_B \hat{Q}\cdot\hat{Q} + k'_B \hat{L}\cdot\hat{L} \quad , \\
\hat{n}_d &= (d^{\dagger}\cdot\tilde{d}) \quad , \\
\hat{Q} &= [d^{\dagger}\times\tilde{s} + s^{\dagger}\times\tilde{d}]^{(2)} + \chi[d^{\dagger}\times\tilde{d}]^{(2)} \quad , \\
\hat{L} &= [d^{\dagger}\times\tilde{d}]^{(1)} \quad ,
\end{aligned} \tag{13}$$

where $\tilde{d}_\mu = (-)^\mu d_{-\mu}$. The coefficients ϵ_d, k_B, k'_B and χ are obtained from a knowledge of the electron single particle energies ϵ_l and the strengths of the pairing, G, and quadrupole interaction, k. For clusters with an odd number of electrons, states are of the type

$$(s^\dagger)^{n_s}(d^\dagger)^{n_d} a^\dagger_{l,m_l;\frac{1}{2},m_s}|- > . \tag{14}$$

with Hamiltonian $H = H_B + H_F + V_{BF}$ where H_B is the boson Hamiltonian, Eq. (13), and H_F and V_{BF} denote the Hamiltonian of the unpaired particle and its interaction with the s , d bosons. One can also show that the boson Hamiltonian (13) corresponds to the quantization of vibrations and rotations of a classical shape[9]. If the shape is ellipsoidal with radius

$$R = R_0[1 + \sum_\mu \alpha_{2\mu} Y_{2\mu}(\theta, \varphi)] \tag{15}$$

it can be quantized by means of quadrupole bosons (d-bosons), having angular momentum and parity $L^\pi = 2^+$. In addition to the five independent quadrupole degrees of freedom, characterized by the creation operators $d^\dagger_\mu$ ($\mu = -2, -1, 0, +1, +2$), a monopole degree of freedom (s boson) with $L^\pi = 0^+$ is also introduced to take into account the finite size of the system and its volume conservation.

The boson Hamiltonian (13) is particularly well suited for a description of metal clusters, since one can account for several types of shapes[9]: (i) spherical shapes when $\epsilon_d \gg k_B$; (ii) spheroidal shapes when $\epsilon_d \ll k_B$ and $\chi = \pm\sqrt{7}/2$ (plus sign oblate, minus sign prolate) and (iii) deformed shapes with no axial symmetry (γ-unstable shapes) when $\epsilon_d \ll k_B$ and $\chi = 0$. To these different shapes c orrespond different low-energy spectra[9]: (i) vibrational , (ii) rotational, (iii) γ-soft. Besides the spettroscopy of the cluster which remains the final test of our and other models, the shell model and interacting boson models can be also used to make predictions on collective states like for example the plasmon mode or other states not yet observed. An analysis of the fragmentation of the plasmon mode due to the coupling with surface oscillations described by the boson Hamiltonian has been made in ref.10.

Just to give an idea of the quality of the agreement between theory and experiment, in fig.5 we report the calculated[10] and experimental[11] photoabsorption cross sections in Na_{10} and Na_{12}.

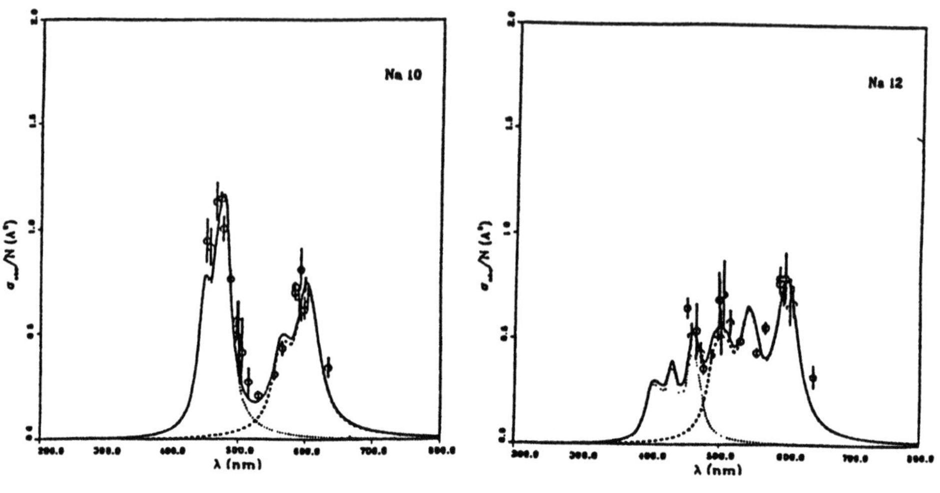

Figure 5

REFERENCES

1. W. Ekardt, Phys. Rev. **B29**, 1558 (1984).
2. C. Brechignac et al, J. Chem. Phys. **90**,1492 (1989); **93**,7449 (1990).
3. J.H.E. Mattauch, W. Thiele, and A.M. Wapstra, Nucl. Phys. **67** ,1 (1965).
4. W.A. Saunders, K. Clemenger, W.A. de Heer, and W.D. Knight, Phys. Rev. **B32**, 1366 (1985);W.D. Knight, W.A. de Heer, and W. Saunders, Z.Phys. **D3**, 109 (1986). M.M. Kappes, M. Schar, U. Rothlisberger, C. Yeretzian and E. Schumacher, Chem. Phys. Lett. **143**, 251 (1988).
5. E. Lipparini and A. Vitturi, to be published.
6. M.L. Homer, J.L. Persson, E.C. Honea and R.L. Whetten, submitted to Z. Phys. D.
7. Z. Penzar and W. Ekardt, Z. Phys. **D17**, 69 (1990).
8. K. Clemenger, Phys. Rev. **B32**, 1359 (1985).
9. For a review, see,F. Iachello and A. Arima, "The Interacting Boson Model", Cambridge University Press, Cambridge (1987).
10. F. Iachello, E. Lipparini and A. Ventura, Phys. Rev.**B** in print.
11. K. Selby et al., Phys. Rev. **B** **43**,4565 (1991).

Electronic and Atomic Structure of Simple-Metal Clusters: Beyond the Spherical Jellium Model

G. Borstel, U. Lammers
Fachbereich Physik
Universität Osnabrück, W-4500 Osnabrück, Germany

A. Mañanes
Departamento de Física Moderna
Universidad de Cantabria, Santander, Spain

J.A. Alonso
Universidad de Valladolid
Valladolid, Spain

Abstract

The ground-state atomic and electronic structure of simple-metal clusters is studied by minimizing the total cluster energy using the density functional formalism. The geometrical structure of the cluster is taken into account in an approximate way by replacing the total (three-dimensional) external potential of the ions by its spherical average around the cluster center when solving the Kohn-Sham equations of density functional theory. When combined with the technique of simulated annealing to find the global minimum in the energy hypersurface, the procedure represents an approximate but very effective approach beyond the spherical jellium model for studying atomic and electronic shell effects, form-isomers and cluster stability. Results are shown for Na- and Cs-clusters.

1 Introduction

Ab-initio quantum mechanical calculations of the electronic structure of clusters face two main problems: (i) Usually one is interested in the evolution of cluster properties with the number N of atoms in the cluster. Even for fixed cluster geometries, however, the computing time increases rapidly with N, such that finite computer resources will quickly become exhausted. (ii) The geometry of clusters at $T = 0$ is usually not known and in principle must be calculated by locating the global minimum for the total energy as a function of the cluster geometry. However, the number of low-lying local minima existing in the energy hypersurface is likely to increase approximately exponentially with increasing N [1], and so far no algorithm is known that grows with time as some power of N and which allows for finding the global minimum in such a case [2]. In particular this second difficulty makes the determination of the true ground state energy and structure of a cluster with N atoms in a strict sense intractable.

For fixed cluster geometries the asymptotic dependence of the computing time t for traditional computational schemes has the form $t \approx cN^{\alpha}$, with $\alpha = 5$ for the configuration interaction calculations, $\alpha = 4$ for the Hartree-Fock method and $\alpha = 3$ for local density functional calculations. A recent new method due to *Car* and *Parrinello* [3] combines molecular dynamics methods with local density functional calculations and has a computing time dependence $t \approx cN^2 \ln N$ per time step. But even in this method for clusters larger than $N \simeq 20$ one usually has to make drastic simplifying assumptions to bring the problem down to a tractable size.

Such simplifying assumptions in particular refer to the shape of the ionic potential. In this respect the spherical jellium model, i.e. a uniform positive background, is quite popular, because of its simplicity and the fairness of some of its predictions [4]. Its computing time dependence is $t \approx cN$, i.e. $\alpha = 1$, which allows for extending the calculations to the range of quite large clusters $N \simeq 3000$. Only recently it has been recognized, that a fixed ionic structure can be included in such a spherical model without enlarging the value of α: *Spina* and *Brack* [5], proposed a simple semi-classical model for ionic structure effects in large metal clusters, where they assumed that the ions are distributed on concentric geometrical shells. *Iñiguez* et al. [6] included the effects of the ionic structure in a density functional approach by treating the Coulomb energy between the point-like ions exactly, while the ionic potential acting on the electrons was replaced by its spherical average. Up to now this latter model represents the simplest method for the ab-initio calculation of electronic properties of medium-sized cluster ($N \lesssim 250$), which still preserves the individual character of the ions constituting the cluster. The purpose of the present work is to give a short summary of this method and to compare the results with other calculations.

2 Theoretical method

The geometric and electronic structure of the cluster is calculated by means of a *spherical average pseudopotential* (SAPS) method. It is based on density functional theory (DFT) and can briefly be described as follows:

For a given set of positions $\{\mathbf{R}_i\}$ of the atoms constituting the cluster the total ionic potential

$$V_I(\mathbf{r}) = \sum_i v_i(\mathbf{r} - \mathbf{R}_i) \tag{1}$$

is constructed from the N individual ionic potentials v_i at the sites $\mathbf{R}_i$. For the N atoms v_i is approximated by *Ashcroft's* empty-core model pseudopotential

[7] or some other suitable potential, like that by *Manninen* [8]. The Ashcroft potential is zero inside and purely coulombic outside the so-called empty-core radius r_c:

$$v_i(\mathbf{r}-\mathbf{R}_i) = \begin{cases} 0, & |\mathbf{r}-\mathbf{R}_i| \le r_c, \\ -Z/|\mathbf{r}-\mathbf{R}_i|, & |\mathbf{r}-\mathbf{R}_i| > r_c. \end{cases} \tag{2}$$

The total ionic potential (1) is now simplified by taking its spherical average around the center of mass of the cluster. For both the Ashcroft and the Manninen potential this average can be done analytically. For the Ashcroft potential we have e.g.

$$V_I^{SA}(\mathbf{r}) = \sum_i v_i^{SA}(\mathbf{r}), \tag{3}$$

with

$$v_i^{SA}(\mathbf{r}) = \begin{cases} 0, & |\mathbf{R}_i| - r_c \le 0 \le r \le r_c - |\mathbf{R}_i| \\ -Z/|\mathbf{R}_i|, & r \le |\mathbf{R}_i| - r_c \\ -Z \cdot \frac{r+|\mathbf{R}_i|-r_c}{2r|\mathbf{R}_i|}, & \Big||\mathbf{R}_i| - r_c\Big| < r < |\mathbf{R}_i| + r_c \\ -Z/r, & r \ge |\mathbf{R}_i| + r_c. \end{cases} \tag{4}$$

The electronic structure of the cluster with $N_e = NZ$ valence electrons is now calculated by solving self-consistently the *Kohn-Sham* equations of the DFT:

$$\left\{-\frac{1}{2}\nabla^2 + v_{eff}(\mathbf{r})\right\}\phi_j(\mathbf{r}) = \epsilon_j\phi_j(\mathbf{r}), \qquad j = 1,\ldots,N_e. \tag{5}$$

The $\phi_j(\mathbf{r})$ are single-particle orbitals from which the electron density of the system is obtained

$$n(\mathbf{r}) = \sum_{occ} |\phi_j(\mathbf{r})|^2, \tag{6}$$

and the effective potential

$$v_{eff}(\mathbf{r}) = V_I^{SA}(\mathbf{r}) + \int d^3r' \, \frac{n(\mathbf{r}')}{|\mathbf{r}-\mathbf{r}'|} + V_{xc}(\mathbf{r}) \tag{7}$$

is the sum of the total ionic potential of the cluster, the usual Hartree and the electronic exchange-correlation potential V_{xc}. The selfconsistent solution of (5)–(7) yields the electronic density for the actual cluster geometry. With this information the total energy $E_{tot}[n(\mathbf{r}),\{\mathbf{R}_i\}]$ of the cluster can be calculated:

$$E_{tot}[n,\{\mathbf{R}_i\}] = \sum_{occ}\langle\phi_j| - \frac{\nabla^2}{2}|\phi_j\rangle +$$

$$+\frac{1}{2}\iint d^3r\, d^3r' \, \frac{n(\mathbf{r})n(\mathbf{r}')}{|\mathbf{r}-\mathbf{r}'|} + \int d^3r \; V_{xc}(\mathbf{r})n(\mathbf{r}) +$$

$$+\int d^3r \; V_I^{SA}(\mathbf{r})n(\mathbf{r}) + \frac{1}{2}\sum_{i\neq i'} \frac{Z_iZ_{i'}}{|\mathbf{R}_i - \mathbf{R}_{i'}|}. \tag{8}$$

The task at hand is now to minimize in a second cycle of calculations E_{tot} as a function of the atomic positions $\{\mathbf{R}_i\}$. As discussed above, a strict algorithm to find the global minimum of E_{tot} does not exist and one thus has to find almost optimal strategies to cope with this problem. A first and very popular strategy is the so-called ***steepest-descent method***, i.e. the successive relaxation of the atoms in the direction of classical forces acting on them. The steepest-descent method, however, is unable to locate a global minimum, which happens to be separated from the initial geometry $\{\mathbf{R}_i^0\}$ by a barrier in the energy hypersurface $E_{tot}[\{\mathbf{R}_i\}]$, and it is thus very likely with this method to become trapped in a local minimum. This problem may be circumvented to a certain extend in a third cycle of calculations of $E_{tot}[n, \{\mathbf{R}_i\}]$ for randomly choosen initial geometries $\{\mathbf{R}_i^{0(k)}\}$. A second and more systematic approach for avoiding local minima is the strategy of ***simulated annealing*** [9]. In this method there exists always a non-zero probability to overcome a barrier between two minima in the energy hypersurface and this strategy is therefore more suited for a systematic search for the global minimum. It goes without saying that simulated annealing requires much more computer resources than the simple-minded steepest-descent method.

3 Results

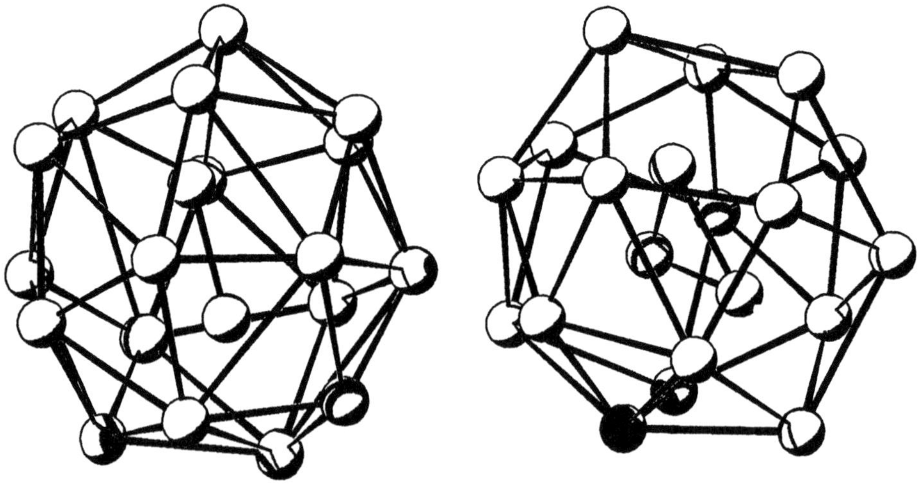

Figure 1: Geometrical structures of two Na_{20} isomers calculated with the SAPS method, using the simulated annealing strategy

In Fig. 1 we show the SAPS ground state geometries of two Na_{20} isomers calculated with the simulated annealing strategy [10]. The difference in energy among these two structures is rather small ($\Delta E \approx 0.06\,\mathrm{eV}$). By

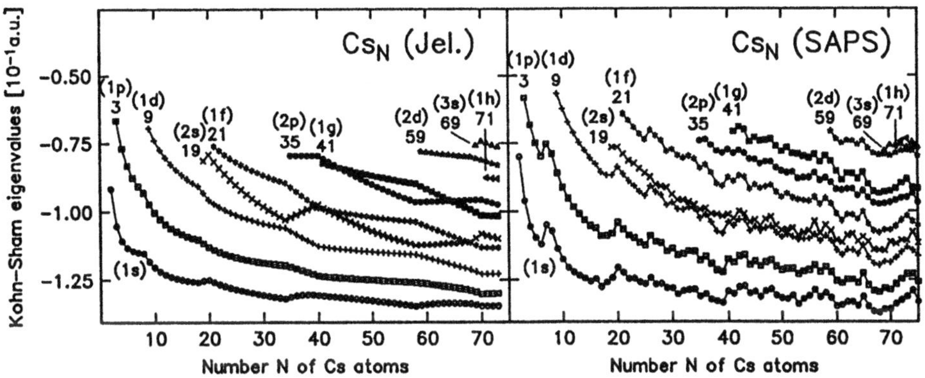

Figure 2: Kohn-Sham eigenvalues for pure Cs_N clusters versus N for the spherical jellium and the SAPS model

comparing this result with the much more elaborated ab-initio molecular dynamics simulations for Na_{20} of *Ballone* et al. [11], one recognizes the tendency of the SAPS method to minimize the eccentricity of the cluster shapes. This is clearly due to the spherical average in Eq. (3), which is equivalent to imposing a spherical symmetry on the electron density. The search of the equilibrium configuration by simulated annealing in the SAPS calculations shows that the energy surface exhibits several local minima, which are very close in energy, and is rather flat around them. The total energy and other average electronic properties are therefore relatively insensitive to the atomic positions, a result which is in accordance with the ab-initio molecular dynamics calculations of Ref. [11] and gives some confidence in the SAPS results as long as mainly electronic properties are concerned.

Fig. 2 shows the Kohn-Sham eigenvalues for Cs_N clusters versus N for both the spherical jellium and the SAPS model [12]. It is seen that the spherical jellium model for large values of the Wigner-Seitz radius (r_s = 5.63 a.u. for Cs) does not result in stable configurations obeying the Aufbau principle for N near 40 and 70. This shortcoming is corrected in the SAPS model: The sequence of one-electron levels $1s, 1p, 1d, 2s, 1f, 2p, 1g, 2d, 3s, 1h$ and corresponding magic numbers 2, 8, 18, 20, 34, 40, 58, 68, 70, which have been verified and extended to even larger cluster sizes N recently for Na [13], is also the energetically correct sequence for Cs clusters. Up to $N = 20$ this shell structure has been found experimentally by measurements of ionization energies for Cs_N [14] and relative mass abundances for Cs^+_{N+1} [15].

The present scheme has been applied to pure sp-bonded metal clusters like Na, Mg, Al, Pb [6] and Cs [12, 16], to describe enrichment and segregation in alkali-metal heteroclusters like Na_NCs_M [17, 18] or $Na_{N-x}Li_x$ [19] and to investigate dissociative channels of Na^+_N clusters [20]. The present method can also handle impurities in clusters, like H in Al [21] and O in Cs [12, 16]. As an example we show in Fig. 3 the calculated atomic structure of $Cs_{14}O$

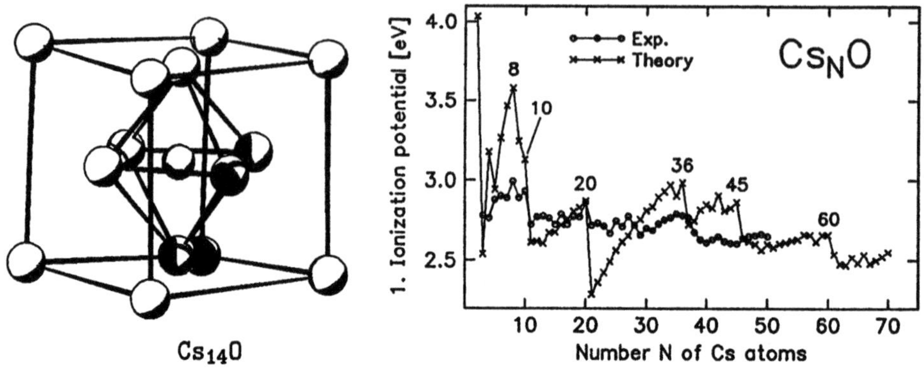

Figure 3: Calculated geometrical structure of $Cs_{14}O$ (left) and first ionization potential IP1 for Cs_NO clusters versus N (right)

[12]. The inner Cs_6O complex with a Cs-O bond length of approximately 5.5 a.u. is highly stable in most of the Cs_NO clusters due to the strong ionic bonding between Cs and O. The six-fold coordination and the bond length in this core agree with the corresponding properties of solid Cs_2O, which is the only known oxidic solid with Anti-$CdCl_2$-structure. Calculated results for the vertical first ionization potential IP1 of Cs_NO clusters [16] are also shown in Fig. 3. Electronic shell-closing effects at $N = 10, 20, 36, 60$ may easily be identified. In addition the calculations predict two geometrical effects in IP1 at $N = 8$ and $N = 45$, which are related to the opening of a new atomic shell of Cs atoms in the cluster. The experimental IP1 data of *Bergmann* et al. [22] for Cs_NO clusters produced by vapor quenching agree quite well with this calculation. In particular the predicted drops in IP1 at $N = 8, 10, 20, 36$ show up clearly in the experimental data.

The question whether there is a contraction of the interatomic distances in metal clusters, when compared to the bulk material, is still under debate, both experimentally [23, 24, 25] and theoretically [26, 27]. At the theoretical side it must be kept in mind that state-of-the-art ab-initio DFT calculations in the framework of the local density approximation give lattice constants, which for bulk alkali metals are around 2–5% smaller than the experimental data [28]. SAPS calculations for Cs_N ($N \leq 78$) [29] and Mg_N ($N \leq 250$) [30] clusters result in nearest-neighbor distances for the outermost atomic shell of the cluster, which are typically 9–11% smaller than the experimental bulk values. However, since this larger contraction could also be due to the additional effect of an insufficiently repulsive core-core interaction in the pseudopotential formalism, we hesitate to interpret this finding as evidence for a global contraction of interatomic distances in metal clusters. On the other hand, these calculations clearly show that the distribution of interatomic distances in larger metal clusters is not uniform, in the sense that atoms in the inner part of the cluster have a smaller nearest-neighbor distance than those in the outer part. This result is in agreement with the explanation of the mass abundance measurements on Cs_NO ($N \leq 600$) [31] and Na_N ($N \leq 1500$) [32, 33] by means of an inhomogeneous spherical jellium model: A positive density that is higher than the bulk average value is needed at the center of the cluster in order to produce the bunching of the electronic levels necessary to give the experimentally observed sequence of magic numbers for $N \geq 100$.

Acknowledgements

We thank C. Meyer for stimulating discussions. This work has been supported by the Deutsche Forschungsgemeinschaft, by Dirección General de Investigación Científica y Técnica of Spain (Grant PB 89-D352) and by NATO.

References

[1] M.R. Hoare and J.A. McInnes, Adv. Phys. **32**, 791 (1983).

[2] L.T. Wille and J. Vennik, J. Phys. A **18**, L419 (1985).

[3] R. Car and M. Parrinello, Phys. Rev. Lett. **55**, 2471 (1985).

[4] W.A. De Heer, W.D. Knight, M.Y. Chou, and M.L. Cohen, Solid State Phys. **40**, 93 (1987).

[5] M.E. Spina and M. Brack, Z. Phys. D **17**, 225 (1990).

[6] M.P. Iñiguez, M.J. López, J.A. Alonso, and J.M. Soler, Z. Phys. D **11**, 163 (1989).

[7] N.W. Ashcroft, Phys. Lett. **23**, 48 (1966).

[8] M. Manninen, Phys. Rev. B **34**, 6886 (1986).

[9] S. Kirkpatrick, C.D. Gelatt, Jr., and M.P. Vecci, Science **220**, 671 (1983).

[10] C. Meyer, Diploma thesis, Universität Osnabrück, 1991

[11] P. Ballone, W. Andreoni, R. Car, and M. Parrinello, Europhys. Lett. **8**, 73 (1989).

[12] U. Lammers, G. Borstel, A. Mañanes, and J.A. Alonso, Z. Phys. D **17**, 203 (1990).

[13] S. Bjørnholm, J. Borggren, O. Echt, K. Hansen, J. Pedersen, and H.D. Rasmussen, Phys. Rev. Lett. **65**, 1627 (1990).

[14] H.G. Limberger and T.P. Martin, Z. Phys. D **12**, 439 (1989).

[15] N.D. Bhaskar, C.M. Klimcak, and R.A. Cook, Phys. Rev. B **42**, 9147 (1990).

[16] U. Lammers, A. Mañanes, G. Borstel, and J.A. Alonso, Solid State Commun. **71**, 591 (1989).

[17] M.J. López, A. Mañanes, J.A. Alonso, and M.P. Iñiguez, Z. Phys. D **12**, 237 (1989).

[18] A. Mañanes, M.P. Iñiguez, M.J. López, and J.A. Alonso, Phys. Rev. B **42**, 5000 (1990).

[19] M.J. López, M.P. Iñiguez, and J.A. Alonso, Phys. Rev. B **41**, 5636 (1990).

[20] M.P. Iñiguez, J.A. Alonso, A. Rubio, M.J. López, and L.C. Balbás, Phys. Rev. B **41**, 5595 (1990).

[21] J. Robles, M.P. Iñiguez, J.A. Alonso, and A. Mañanes, Z. Phys. D **13**, 269 (1989).

[22] T. Bergmann, H.G. Limberger, and T.P. Martin, Phys. Rev. Lett. **60**, 1767 (1988), and private communication

[23] G. Apai, J.F. Hamilton, J. Stöhr, and A. Thompson, Phys. Rev. Lett. **43**, 165 (1979).

[24] P.A. Montano, J. Zhao, M. Ramanathan, G.K. Shenoy, and W. Schulze, Z. Phys. D **12**, 103 (1989).

[25] P.A. Montano, J. Zhao, M. Ramanathan, G.K. Shenoy, W. Schulze, and J. Urban, Chem. Phys. Lett. **164**, 126 (1989).

[26] L.B. Hansen, P. Stoltze, J.K. Nørskov, B.S. Clausen, and W. Niemann, Phys. Rev. Lett. **64**, 3155 (1990).

[27] M.Ya. Gamarnik, Phys. Stat. Sol.(b) **160**, K1 (1990).

[28] M. Sigalas, N.C. Bacalis, D.A. Papaconstantopoulos, M.J. Mehl, and A.C. Switendick, Phys. Rev. B **42**, 11637 (1990).

[29] A. Mañanes, J.A. Alonso, U. Lammers, and G. Borstel, Phys. Rev. B **44**, 7273 (1991).

[30] M.D. Glossman, M.P. Iñiguez, and J.A. Alonso, Z. Phys. D (in print).

[31] H. Göhlich, T. Lange, T. Bergmann, and T.P. Martin, Phys. Rev. Lett. **65**, 748 (1990).

[32] T. Lange, H. Göhlich, T. Bergmann, and T.P. Martin, Z. Phys. D **19**, 113 (1991).

[33] T.P. Martin, T. Bergmann, H. Göhlich, and T. Lange, Chem. Phys. Lett. **172**, 209 (1990).

Shell-Model and Projected Mean-Field Approach to Electronic Excitations of Atomic Clusters

M. Koskinen[1], *E. Hammarén*[1], *M. Manninen*[1], *P. O. Lipas*[1] *and K. W. Schmid*[2]

[1]Department of Physics, University of Jyväskylä, SF-40351 Jyväskylä, Finland

[2]Institut für Theoretische Physik, Universität Tübingen, W-7400 Tübingen, Germany

We apply microscopic nuclear-structure methods to calculate electronic excitations of atomic clusters. The single-particle part of the Hamiltonian is derived for a spherical ionic background ("jellium"). The two-particle interaction is the Coulomb interaction between the electrons. As a consequence the Hamiltonian is parameter free. The standard shell model (JT coupling scheme, code MSH7000 [1]) is used in 1s 1p 2s 1d space for clusters with $N = 1$ to 10 valence electrons. We have calculated for several total angular momenta and both parities the level structures, ionisation potentials and electric-dipole (E1) transition strengths. Larger clusters with $N \gg 10$ are tedious within the shell model because of the large model spaces required. We treat them with a family of particle-number and angular-momentum projected mean-field models (MONSTER, VAMPIR [2]) using the same Hamiltonian. With MONSTER we study states which have essentially a structure of one or two (quasi)electrons.

The independent electron model explains surprisingly successfully certain observed features of metallic clusters such as the (spherical) magic numbers and the (deformed) subshell structure of small and medium-size clusters [3] as well as the recent triumph [4, 5] of super-shell behaviour. To explain in more detail e.g. the ionisation energies already requires a more elaborate picture. These phenomena are related to the electronic ground states.

The first indications of many-electron excited states came from photoabsorption spectra [6]. Especially for small clusters a number of theoretical approaches ranging from microscopic quantum-chemistry and density-functional methods [7, 8] to more phenomenological RPA-type and sum-rule techniques [9, 10] are applied to the *a priori* complex electronic excitations. Inspired by the former methods and, on the other hand, by the success of microscopic many-body approaches in explaining analogous complex multiparticle excitations of atomic nuclei, we have implemented two fully microscopic nuclear-structure models for calculations of spectroscopic properties of metallic clusters.

The many-body Hamiltonian for the valence electrons is

$$H = \sum_i h_i + \sum_{i<j} v_{ij}\,, \tag{1}$$

where we assume that the single-particle potential $V(r)$ is caused by a uniformly charged sphere of the jellium background and that the two-particle interaction is the Coulomb repulsion between the electrons. Thus we have

$$h_i = -\frac{\hbar^2}{2m}\nabla_i^2 + V(r_i)\,, \qquad v_{ij} = \frac{1}{4\pi\epsilon_0}\frac{e^2}{|\boldsymbol{r}_i - \boldsymbol{r}_j|}\,. \tag{2}$$

As the single-particle basis we use j-coupled harmonic oscillator wavefunctions $|nljm\rangle$. The oscillator parameter is fixed by minimizing the ground-state energy for each cluster separately. Thus for a selected set of single-particle states the problem is parameter free. The remaining task is to diagonalise the Hamiltonian (1) in the complete antisymmetric many-particle configuration space built up from the single-particle basis.

The best solution is provided by the complete shell-model (CSM) configuration-mixing scheme [1]. In the limit of an infinite basis the method is exact. For example, for ten electrons the single-particle basis 1s 1p 2s 1d yields such a "complete" solution with matrix dimensions of about 3000. Thus we foresee that only small clusters of up to some 20 valence electrons can be treated in this way.

For future applications we note that nuclear calculations with dimensions of 62 000 and 320 000 for the coupled and uncoupled schemes, respectively, have been reported lately. However, even with these dimensions the next shell 2p 1f could not be included completely. Restricted calculations are still possible in an even larger basis, but a safe procedure for performing the truncation is lacking. In the present calculations we use the JT-coupled scheme with the code MSH7000 [1] in the exactly solvable 1s 1p 2s 1d space. The method of calculating excitation spectra and spectroscopic quantities such as electromagnetic transitions is straightforward and sufficiently documented, so we need not comment it here in more detail.

An optimal way to truncate the many-particle configuration space is to let the dynamics of the system itself define the truncation. This can be done by using a variational principle to extract from the Hamiltonian the mean field each of the particles feels due to its interaction with all the others. The symmetries (here particle number and total angular momentum) broken by the mean field can be restored by projection either after or before the variation. A versatile family of such many-body models (MONSTER, VAMPIR) have been developed [2] and applied [11] e.g. to medium-heavy nuclei during the last decade. With only minor modifications we have now applied these models to atomic clusters represented by the Hamiltonian of eqs. (1) and (2).

For clusters with an even number of electrons we first seek the number- and angular-momentum- projected VAMPIR mean field of the lowest state. Angular momentum and parity $J^\pi = 0^+, 2^+, 4^+, \ldots$ only can be reached by projection from the axially and reflection-symmetric intrinsic mean field. The Hartree–Fock–Bogoliubov (HFB) transformation underlying the VAMPIR mean field is cabable of self-consistently attaining axially symmetric shapes and pairing properties dictated by the two-particle (here ee) interaction. Then the ground and excited states are calculated with MONSTER by diagonalising the Hamiltonian (1) in the space spanned by the projected two-quasi-electron (2qe) wavefunctions

$$|NJ^\pi M\rangle = f_0 P(JM, K=0; N)|q\rangle + \sum_{\mu\nu} f_{\mu\nu} |q_\mu q_\nu; NJ^\pi M\rangle, \qquad (3)$$

where $|q\rangle$ stands for the VAMPIR mean field, M (K) for the angular-momentum projection to the laboratory (body) z-axis, P for the angular-momentum and particle-number projection operator and f_0 and $f_{\mu\nu}$ for the amplitudes of the various wavefunction components. The exact form of the 2qe part is

$$|q_\mu q_\nu; NJ^\pi M\rangle = \sqrt{\tfrac{1}{2}}[P(JMK;N)q_\mu^\dagger q_\nu^\dagger + \pi(-)^{J-K} P(JM,-K;N)q_{\bar{\mu}}^\dagger q_{\bar{\nu}}^\dagger]|q\rangle\,, \qquad (4)$$

where the parity $\pi = \pi_\mu \pi_\nu$ and $K = K_\mu + K_\nu$. For clusters with an odd number of electrons these equations are replaced by appropriate 1qe expressions.

In the 1s 1p 2s 1d basis all 2qe configurations can be included in eq. (3). For numerical feasibility in larger bases we truncate the number of configurations by the 2qe energy $E_{\mu\nu}^{\text{qe}} = E_\mu^{\text{qe}} + E_\nu^{\text{qe}}$, by the total K value or by both so that for a given J the number of 2qe configurations is typically 150. This truncation does not invalidate the results for the excitation-energy regime under discussion. Calculations of electromagnetic transition rates with the wavefunctions (3) are straightforward and they are in progres.

Our main aim so far has been to test and compare the new techniques with each other and also with other methods such as the local spin-density approximation (LSDA) [12]. In fig. 1 we compare for small sodium clusters the binding energies per electron calculated with SCM, MONSTER and LSDA. For cluster sizes $N = 1$ and 2 the CSM and MONSTER give identical energies, since both use exactly the same (1- and 2-electron) configuration space. The result is highly nontrivial because the methods used to reach it are entirely different. This is a pleasing confirmation of the formal correctness of the two models. The large discrepancy observed for the open-shell clusters with $N = 3$ to 6 is somewhat disturbing though possible since now the projected 1qe and 2qe configuration space is smaller than the complete CSM space. The same oscillator length is used in the CSM and

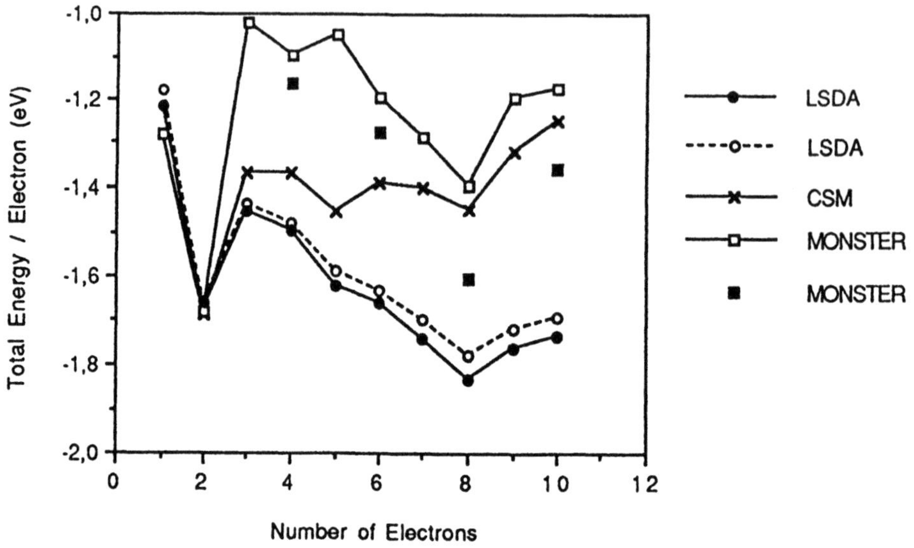

Fig. 1. Binding energies per electron for the smallest Na_n clusters calculated with various approaches using the same jellium Hamiltonian of eq. (1). Local spin-density approximation (LSDA) with self-consistent densities (solid line with bullets) and with Gaussian single-electron wavefunctions (dashed line with circles), CSM in the complete 1s 1p 2s 1d space (solid line with crosses), MONSTER in the 1s 1p 2s 1d space (solid line with open squares) and in the 1s 2s 3s 1p 2p 3p 1d 2d space (solid squares).

MONSTER. Both models fail to reproduce the trend for larger clusters as calculated with the LSDA. The LSDA results obtained with the minimal basis of pure Gaussian (HO) wavefunctions demonstrate that the detailed basis is not essential. The main reason for the discrepancy is evidently that the 1s 1p 2s 1d space is too small. This is demonstrated by the drop in the MONSTER energies when the space is enlarged to 1s 2s 3s 1p 2p 3p 1d 2d space. Varying the oscillator parameter would still bring about some extra binding.

The ionisation potentials of fig. 2 are obtained from the differences of total energies of a cluster Na_n with $N = n$ electrons and of the ionised one with $N = n - 1$ electrons. Crucial is that the oscillator parameters have to be found separately for the two clusters by minimising the total energy. The general trend and even the absolute values of the experimental ionisation potentials [13] are well reproduced by all approaches: the larger dips at the major shell closings between $N = 2$ and 3 and between $N = 8$ and 9 are clearly visible.

Importantly, a finer detail—the odd–even staggering—is not accounted for by the entirely spherical CSM and LSDA models. This is related to the fact that in (small) clusters the odd–even staggering is caused by the Jahn–Teller effect, i.e. the non-spherical shape of the cluster [8]. Thus the odd–

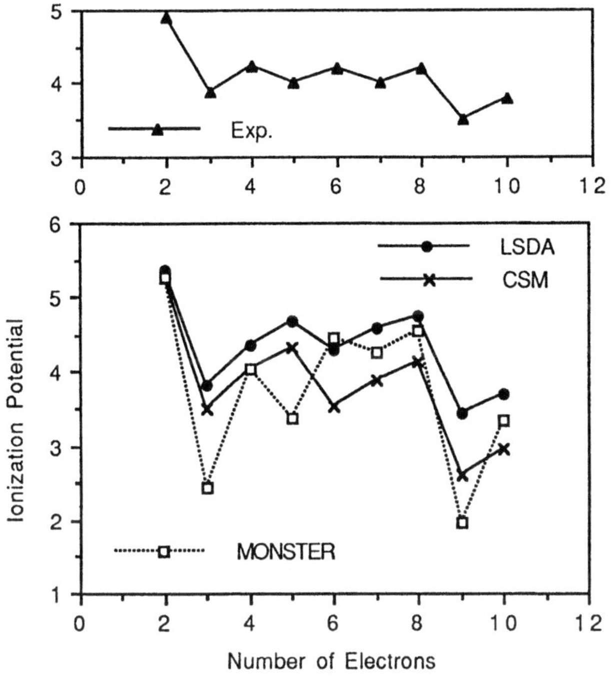

Fig. 2. Experimental ionisation potentials [13] (upper panel) for the smallest Na_n clusters compared with calculated ones (lower panel) using various approaches. LSDA (solid line with bullets), CSM (solid line with crosses), MONSTER (dotted line with open squares).

even staggering of the ionisation potentials of metallic clusters should not be mixed with the odd–even staggering present in the binding energies of atomic nuclei, which is caused by the strong pairing force characterising the nucleon–nucleon interaction. The electron–electron interaction (repulsion) does not produce any pairing: on the contrary, the self-consistent VAMPIR mean fields point to anti-pairing results for electrons. Even more clearly this is seen in the ordering of the levels with different angular momenta, which obey Hund's rules for atomic electrons.

We believe that the odd–even staggering of the MONSTER/VAMPIR ionisation potentials shown in fig. 2 is due to the fact that the projected 2qe states (3) of the clusters with even numbers of electrons contain significantly more correlations than the projected 1qe states of those with odd numbers of electrons. This difference of the two approximations always yields extra binding for clusters with even numbers of electrons.

Excitation spectra for Na_{10} in figs. 3 and 4 clearly demonstrate the repulsive nature of the ee interaction leading to a level ordering typical of an atom with two electrons in d and s orbitals outside a rigid core. Note also the exact degeneracies of excitations belonging to a given spin multiplet

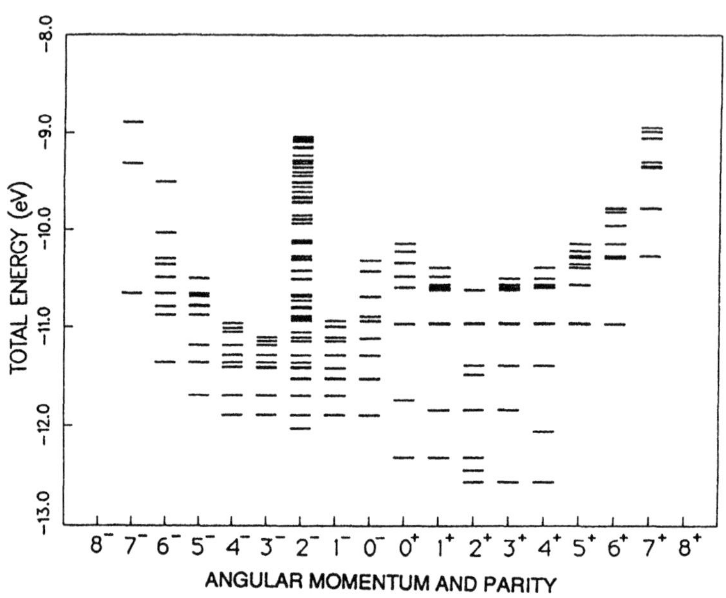

Fig. 3. Lowest excitations of Na_{10} cluster, with total angular momentum and parity indicated, calculated with CSM in the complete 1s1p2s1d space. For $J^\pi = 2^-$, 60 lowest excitations are shown.

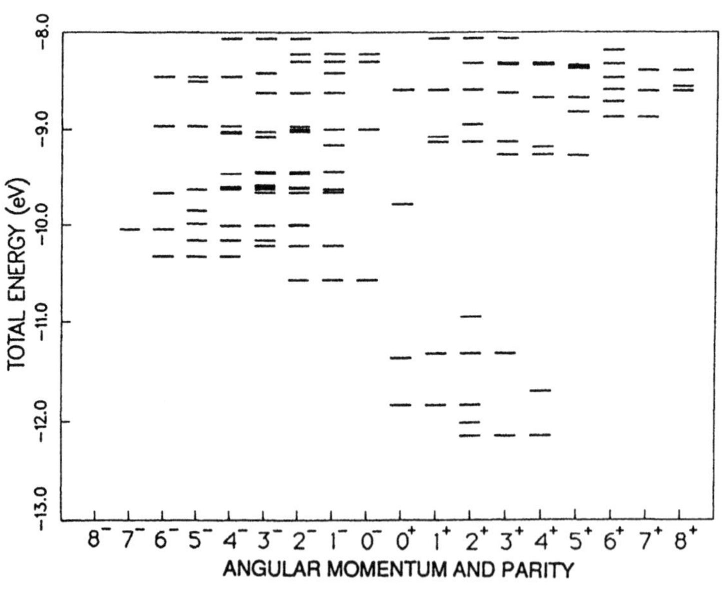

Fig. 4. Lowest projected 2qe excitations of Na_{10} cluster calculated with MONSTER in the 1s 1p 2s 1d space.

($2S+1$ members for a given L), a natural consequence of the neglected spin-orbit interaction. Although calculated within jj coupling, our eigenstates are indeed exact eigenstates of L and S.

The differences between the CSM and MONSTER spectra are caused by the projected 2qe approximation of the latter. States with seniority 3 and higher for states with $\pi = +$ (for states with $\pi = -$ the relation seems to be more complex) are not explicitly included in the MONSTER wavefunction, though the number projection brings in correlations up to 6qe. The much higher level density of the CSM spectra is best seen for the $J^\pi = 2^-$ states. However, since for one-particle, one-hole excitations (1p1h is equivalent to seniority $v = 2$ or to 2qe) the two models give similar results, the photoabsorption predictions are expected to be similar for the two models.

The excitation spectrum of Na_{20} in fig. 5 demonstrates that also heavier clusters can be treated with MONSTER. The single-electron basis includes all orbitals from the $N = 0$, 1, 2, 3 and 4 oscillator shells. The increased level density as compared to the Na_{10} case is remarkable. Here the cutoff is chosen so that the maximum number of states for a given J is of the order of 100 and 200 for negative- and positive-parity excitations, respectively. We notice that e.g. the density of the 1^- states, important for the photoabsorption properties, is actually not so different from the densities in the lighter clusters. Quantitative work on this aspect is in progress.

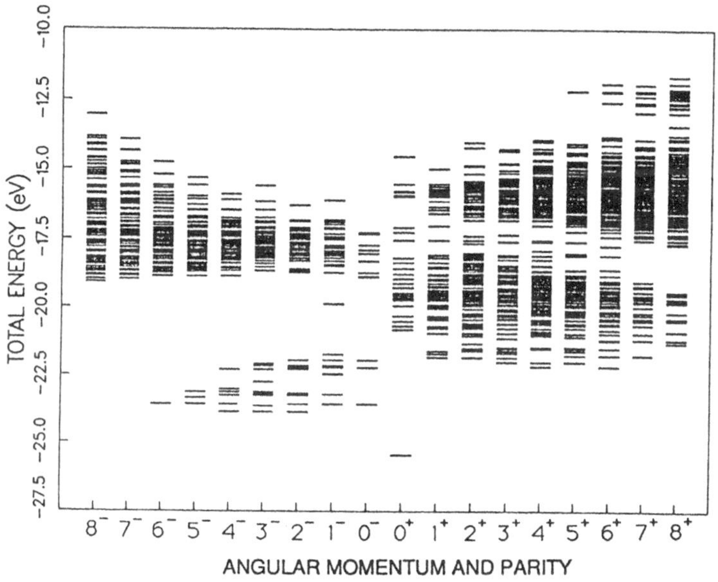

Fig. 5. Lowest projected 2qe excitations with positive parity of Na_{20} cluster calculated with MONSTER in the 1s 1p 2s 1d 2p 1f 3s 2d 1g space.

So far we have applied the CSM to photoabsorption by the Na_8, Na_9 and Na_{10} clusters. We calculate cross sections, integrated over a single excitation, by the relation

$$\int_{\Delta\lambda} \sigma(\lambda) \mathrm{d}\lambda = \frac{8\pi^3 \hbar c \alpha}{3} \frac{1}{E_{\mathrm{fi}}} \frac{(J_{\mathrm{f}} \| \boldsymbol{r} \| J_{\mathrm{i}})^2}{2J_{\mathrm{i}}+1} \,. \tag{5}$$

To obtain the curves in fig. 6 we have convoluted the cross section (5) using a Gaussian with a phenomenological width of 30 nm.

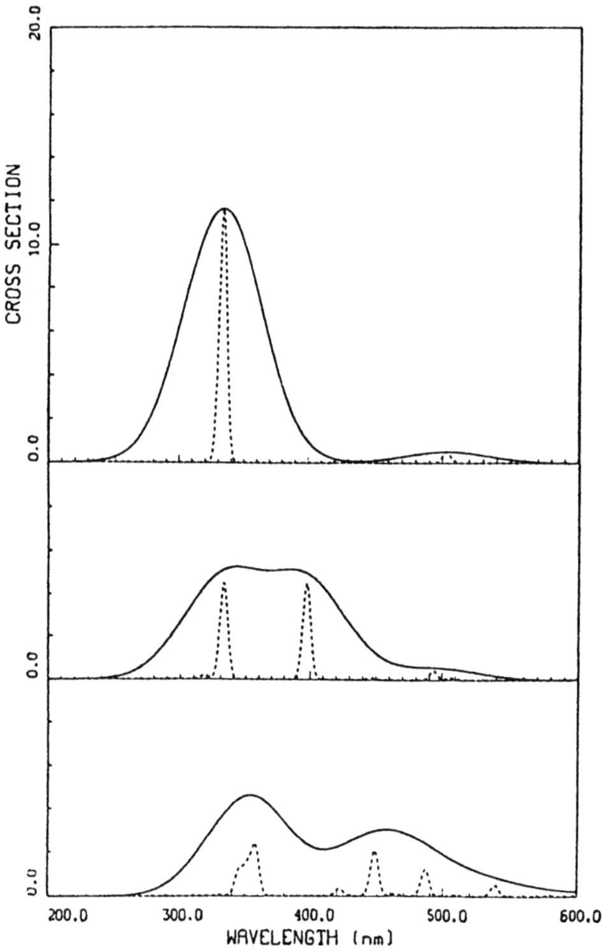

Fig. 6. Photoabsorption spectra of Na_8, Na_9 and Na_{10} clusters calculated with eq. (5) from CSM electric-dipole matrix elements. The initial state is the—generally degenerate—ground state. Cross sections are convoluted with two Gaussians using widths of 30 nm (solid line, scale in Å^2) and 3 nm (dashed line, scale in 10 Å^2).

Qualitatively, the microscopic cross sections behave similarly to those calculated for a spherical and spheroidal jellium plasma [6]. The single peak in Na_8 would correspond to a single plasma resonance of a spherical jellium and the two peaks in Na_9 and Na_{10} to the two modes possible in a spheroidal jellium. Microscopically, the $\boldsymbol{r}$ operator seems to favour single-particle excitations. This is just the opposite of the collective plasma mode. For example, in Na_8 the strong excitation at 330 nm is mainly a particle–hole state of $1d(1p)^{-1}$ character. It would be tempting to identify this excitation with the large hump seen in experiment at about 480 nm. Similarly the small transitions seen at lower energy seem to have their experimental counterparts. The calculated absolute cross sections agree very well with the experimental ones. However, the calculated and observed excitation energies differ considerably, which calls for further theoretical study with larger single-particle bases.

In summary, we wish to emphasize that the rather ambitious microsopic models developed in nuclear-structure physics over the past decades seem to have found an important new area of application in the physics of atomic clusters. Since the two-body interaction is simple and certain for clusters, which is not the case in the nuclear realm, the models can be here tested with greater confidence. This favourable feature is unfortunately somewhat offset by the apparent need for ever larger bases, both for an accurate study of small clusters and for even a summary study of large clusters.

References

[1] J. B. French, E. C. Halbert, J. B. McGrory and S. S. M. Wong, Adv. Nucl. Phys. **3** (1969) 193.

[2] K. W. Schmid, F. Grümmer and A. Faessler, Phys. Rev. **C29** (1984) 291; 308; K. W. Schmid and F. Grümmer, Rep. Prog. Phys. **50** (1987) 731.

[3] W. A. de Heer, W. D. Knight, M. Y. Chou and M. L. Cohen, Solid State Physics **40** (1987) 93.

[4] H. Nishioka, K. Hansen and B. R. Mottelson, Phys. Rev. **B42** (1990) 9377 .

[5] J. Pedersen, S. Bjørnholm, J. Borggreen, K. Hansen, T. P. Martin and H. D. Rasmussen, Nature **353** (1991) 733.

[6] K. Selby, V. Kresin, J. Masui, M. Vollmer, A. Scheidemann and W. D. Knight, Z. Phys. **D19** (1991) 43.

[7] V. Bonacic-Koutecky, P. Fantucci and J. Koutecky, Chem. Rev. **91** (1991) 1035.

[8] Z. Penzar and W. Ekardt, Z. Phys. **D17** (1990) 69.

[9] G. Bertsch and W. Ekardt, Phys. Rev. **B32** (1985) 7659; M. Brack, Phys. Rev. **B39** (1989) 3533.

[10] E. Lipparini and S. Stringari, Z. Phys. **D18** (1991) 193.

[11] E. Hammarén, K. W. Schmid, F. Grümmer, B. Fladt and A. Faessler, Nucl. Phys. **A437** (1985) 1; **A454** (1986) 301; E. Hammarén, K. W. Schmid, F. Grümmer and A. Faessler, Phys. Lett. **B171** (1986) 347.

[12] O. Gunnarsson and B. I. Lundqvist, Phys. Rev. **B13** (1976) 4274.

[13] M. M. Kappes, M. Schär, U. Röthlisberger, C. Yeretzian and E. Schumacher, Chem. Phys. Lett. **143** (1988) 251.

Dipole Excitations of Closed-Shell Alkali-Metal Clusters

C. Guet
Département de Recherche Fondamentale Matière Condensée,
Service de Physique Atomique, Centre d' Etudes Nucléaires de Grenoble
85X, 38041 Grenoble Cedex, France

W. R. Johnson
Department of Physics, University of Notre Dame
Notre Dame, IN 46556

Excitation energies and associated oscillator strengths for dipole-excited states of alkali-metal clusters – treated as jellium spheres – are calculated in the random-phase approximation(RPA). Closed-shell systems with 8, 20, 34, 40, 58, and 92 delocalized electrons are considered. The ground state is described in the Hartree-Fock(HF) approximation. The excitation spectrum is determined by solving the RPA equations. Exchange contributions are taken into account completely. The theoretical oscillator strength distribution is found to be very different for neutral and charged clusters. Static polarizabilities of the clusters are calculated and compared with experimental values and with other calculations.

I. INTRODUCTION

It is now well established [1] that gross properties of simple metallic clusters with from ten to a few hundred atoms can be understood in terms of the quantal arrangement of delocalized electrons moving in their mutual field and a smooth positive background (jellium model). To a first approximation, the dynamics associated with the moving conduction electrons govern the stability of the metallic cluster as well as its response to an external electromagnetic field. In that respect, the optical response of alkali-metal clusters has been the object of thorough experimental investigations revealing the existence of a giant dipole resonance interpreted as an collective electronic excitation (surface plasmon). [2–4] A substantial theoretical effort [5–9] has accompanied these experimental findings. In analogy with the infinite case, the finite electron many-body system is usually described in the local-density approximation(LDA). This has the advantage of taking the strong screening of the Coulomb interaction into account in a rather simple way. However, it probably fails to describe the outer part of the electronic distribution correctly, since the local exchange potential falls faster than $1/r$. This fact might explain why static polarizabilities in the LDA are smaller than the measured ones.[10] Although most of the discrepancy probably originates from the crudeness of the jellium approximation, it is worth trying to assess carefully the LDA. In this paper, we calculate the dipole oscillator strength distribution and the static polarizability of closed-shell sodium clusters by using the RPA. As in other jellium models, it is assumed that the positive ions form a homogeneous, sharp-edged spherical distribution. Our calculations differ from previous work based on the LDA in that the exchange interaction is taken into account fully. The ground state is described in the HF approximation, leading to a non-local HF central potential. A complete basis of single-particle states is constructed by confining the cluster to a cavity of large but finite radius, thereby discretizing the continuum. The spectrum of physical dipole-excited states is then obtained by solving the RPA matrix equation. These states are used to determine oscillator strengths for dipole transitions from the ground state as well as the ground state static dipole polarizability. The resulting oscillator strengths and polarizabilities are discussed in the light of LDA predictions and experimental data. The major approximation remains the assumption of a uniformly charged background. We discuss how a diffuse surface modifies the results.

II. HARTREE-FOCK APPROXIMATION FOR THE GROUND STATE

A jellium cluster consists of free electrons in a homogeneous, positively-charged background. In alkali metals, the itinerant electrons are the valence electrons, one electron per atom. The Hamiltonian describing the electron is written:

$$h_0 = \frac{1}{2}p^2 + V_{\mathrm{bkg}}(r) + U(r) \tag{1}$$

where $V_{\mathrm{bkg}}(r)$ is the positively-charged background potential, and $U(r)$ accounts in some approximation for the electron-electron interaction. Since we assume a constant-density distribution for the jellium ions, we have:

$$V_{\mathrm{bkg}}(r) = \begin{cases} -\frac{A}{2R}\left[3 - \left(\frac{r}{R}\right)^2\right] & r \leq R \\ -\frac{A}{r} & r > R \end{cases} \tag{2}$$

where A is the number of ions and $R = A^{\frac{1}{3}} r_S$. The Wigner-Seitz radius r_S is assigned its bulk value, *i.e.* $4a_0$ for sodium (a_0 being the Bohr radius). The many-body Hamiltonian describing the system made of Z delocalized electrons in the presence of A ions is then:

$$H = H_0 + V, \tag{3}$$

$$H_0 = \sum_{i=1}^{Z} h_0(r_i), \tag{4}$$

$$V = \frac{1}{2}\sum_{ij}\frac{1}{r_{ij}} - \sum_i U(r_i). \tag{5}$$

In the present work, the potential U of Eq.(1) is the non-local HF potential defined as:

$$V_{\rm HF}\, u_a(\mathbf{r}) = \sum_b \int \frac{d\mathbf{r}'}{|\mathbf{r}-\mathbf{r}'|}[u_b^\dagger(\mathbf{r}')u_b(\mathbf{r}')\, u_a(\mathbf{r}) - u_b^\dagger(\mathbf{r}')u_a(\mathbf{r}')\, u_b(\mathbf{r})] \tag{6}$$

where the u's are the single-particle wave functions and the summation extends over occupied states. Restricting to closed-shell systems, we are led to solve coupled radial HF equations for the occupied orbitals. This is a standard procedure in atomic physics; a numerical code has been written to yield single-particle orbitals u_a and eigenenergies (ϵ_a) to a very high accuracy (typically one part in 10^8). As an illustration, we show in the top panel of Figure 1 the ground-state density $\rho_{\rm HF}(r)$ for Na_{93}^+ obtained from the numerical solution to the HF equations.

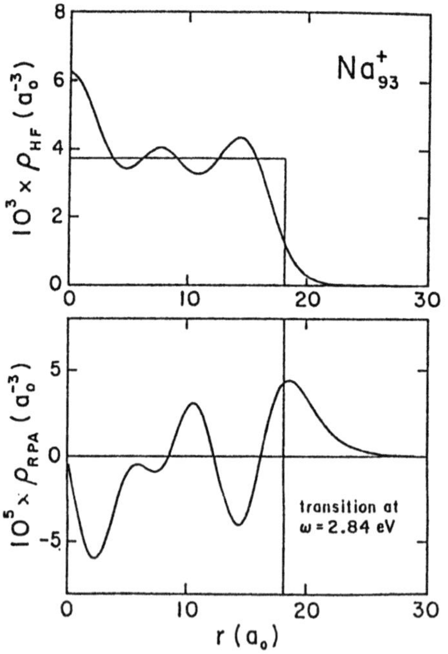

Figure 1. Top panel: radial electronic density $\rho_{\rm HF}(r)$ for the ground state of Na_{93}^+ ($r_S = 4a_0$). Bottom panel: radial transition density $\rho_{\rm RPA}(r)$ for the dipole transition at 2.84eV which carries 40.6% of the f sum rule in Na_{93}^+. The radius of the jellium background sphere is indicated by the vertical bar. (Units: a_0^{-3})

III. RANDOM-PHASE APPROXIMATION FOR DIPOLE EXCITATIONS

Previous microscopic calculations[5–7, 9] of the optical response of small alkali-metal clusters have been carried out within the framework of the RPA which allows one to take into account the strong screening of the electron-electron interaction. In all these calculations, it has been found that the average dipole resonance frequencies are systematically blue-shifted with respect to experiment.[2–4] They all start from an effective density-dependent interaction, thus avoiding the numerical complications that result from the exact treatment of exchange. Here we study the vibration modes when the ground state is a Slater determinant obtained by solving the HF equations discussed above. There are different ways of deriving the RPA equations which may be put into matrix form:

$$\begin{pmatrix} A & B \\ B^* & A^* \end{pmatrix} \begin{pmatrix} X^k \\ Y^k \end{pmatrix} = \omega_k \begin{pmatrix} X^k \\ -Y^k \end{pmatrix} \tag{7}$$

The matrix A contains matrix elements of the Coulomb interaction between particle-hole excitations, whereas the matrix B is composed of matrix elements of that interaction between the ground state and two-particle–two-hole excitations:

$$\begin{aligned} A_{ma,nb} &= (\epsilon_m - \epsilon_a)\delta_{ab}\delta_{mn} + < mb|V|an > \\ B_{ma,nb} &= < mn|V|ab > \\ < ij|V|kl > &= < ij|\frac{1}{|\mathbf{r}-\mathbf{r}'|}|kl > - < ij|\frac{1}{|\mathbf{r}-\mathbf{r}'|}|lk > . \end{aligned} \tag{8}$$

The indices a, b (n, m) refer to the hole (particle) states. The positive eigenvalues ω_k of Eq.(7) are the excitation energies of the system. The corresponding eigenvectors, representing the physical states, are expressed as linear combinations of forward-going and backward-going amplitudes X^k_{ma} and Y^k_{ma}. A numerical solution of the eigenvalue problem (7) requires a complete set of single-particle states. In order to discretize the continuum states, it is convenient to confine the cluster to a cavity of finite radius. By choosing this cavity radius sufficiently large, the low-lying bound states in the cavity can be brought arbitrarily close to the actual HF bound states. Although the cavity spectrum is discrete, it is infinite. A finite HF pseudospectrum is built by expanding the HF orbitals in terms of a finite number of B-splines. The low-lying states in this pseudospectrum can be adjusted very accurately to the HF states by an appropriate choice of the number and order of the B-splines used. For more details see Ref.[11].

An angular reduction of the RPA equation is performed with the restriction to dipole excitations. An external electric dipole field gives rise to a set of excitation channels. The radial amplitudes which are associated with these channels are expanded in terms of the pseudostates. In order to reach a very high accuracy, 50 B-splines of order 7 have been used. As an example, we have 10 channels for $Z = 40$, resulting in matrices A and B each of dimension 466×466. The FORTRAN routine RSG from the EISPACK library [12] is used to find the eigenvalues and eigenvectors of the matrix equation (7).

The dipole transition amplitude from the ground state to the kth excited state is expressed in terms of reduced matrix elements of the dipole operator d as:

$$q_k = \sum_{am} \left(X^k_{ma} - Y^k_{ma}\right) < u_m||d||u_a >, \tag{9}$$

and the corresponding oscillator strength is given by:

$$f_k = \frac{4}{3}\omega_k q_k^2. \tag{10}$$

The Thomas-Reiche-Kuhn f sum rule, $\sum_k f_k = Z$, is satisfied to better than one part in 10^5, providing a numerical test of the RPA calculation. In the lower panel of Figure 1 we show the transition density $\rho_{\rm RPA}(r)$ for the 5$^{\rm th}$ excited state of Na^+_{93} which exhausts 40.6% of the f sum rule. Generally, the transition density for the k$^{\rm th}$ excited state is given by:

$$\rho^k_{\rm RPA}(\mathbf{r}) = \sum_{ma} \left(X^k_{ma} - Y^k_{ma}\right) u^\dagger_m(\mathbf{r}) u_a(\mathbf{r}). \tag{11}$$

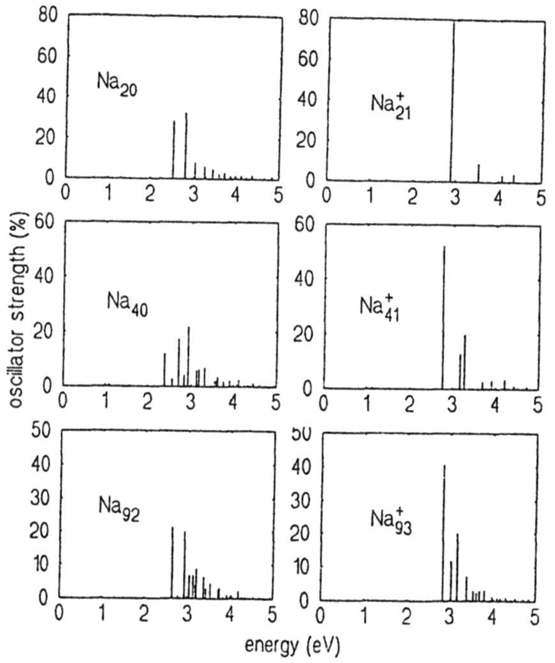

Figure 2. Oscillator-strength distribution expressed as a percentage of the f sum rule given as a function of the excitation energy for Na and Na^+ clusters with $Z = 20$, 40, and 92 delocalized electrons ($r_S = 4a_0$

The oscillator strength distributions for the lightest systems, Na_8 and Na_9^+, agree with the previous LDA results of Ref.[7]; a single line at 2.66eV (ω_{LDA} = 2.8eV) exhausts more than 70% of the f sum rule for Na_8 and a corresponding line at 2.95eV (ω_{LDA} = 3.1eV) exhausts more than 90% of the sum rule for Na_9^+. IN Figure 2 we show the oscillator-strength distributions for three larger closed-shell systems (Z = 20, 40, and 92), both neutral and singly charged. To compare the results for different cluster sizes, the percentage of the sum rule contributed by each line is plotted. A few observations ought to be made. As already shown by RPA calculations based on LDA, a few states exhaust the f sum rule. These states, which lie below the ionization threshold, around 3.4eV and 4.6eV for neutral and charged clusters, respectively, are bunched and lead to the observed giant dipole resonance. There is a marked difference between neutral and charged clusters. In the later case, which is characterized by a relatively higher ionization potential, the dipole oscillation is shared by fewer states (a single peak exhausts 80% of the sum rule in case of Na_{21}^+), but lies at about the same frequency. As the size of the cluster grows, the splitting into several states also grows. A weak dependence of the average excitation frequency on the number of atoms is observed. A calculation on Na_{58}, not shown in the figure, confirms this trend. These lines, even for the heavier systems, lie well below the classical Mie frequency 3.4eV. A straightforward comparison with experimental data on photoabsorption cannot be performed since the broad peaks which are observed are due to the coupling of the electronic vibrations to the ionic jellium density fluctuations,[13] a fact which the present model does not embrace. The observed photoabsorption cross section for Na_{20} shows two broad peaks; a stronger peak centered around 2.4eV and and a slightly weaker but broader peak at 2.75eV.[4] This is in harmony with the theoretical prediction for Na_{20} which gives two lines at 2.5eV and 2.8eV having about equal strength. With regard to RPA calculations based on LDA, a systematic – although small – red shift of the main lines is observed, but the shift is not sufficient to resolve the discrepancy with experiment. The differences between the present calculations of resonance frequencies and LDA calculations (as well as experimental data) are consistent with our prediction of the static dipole polarizabilities which are given in terms of the f-distribution through the relation

$$\alpha_{RPA} = \sum_k \frac{f_k}{\omega_k^2}. \tag{12}$$

These polarizabilities, are reported in Table 12. Our RPA calculation predicts static polarizabilities which are systematically lower than experimental values by about 15% but higher than the values obtained in the framework of the LDA by about 5%. The LDA mean field falls faster than the $1/r$ asymptotic HF potential, leading to a smaller radius and thus a smaller static polarizability. We see, however, that this effect does not improve the agreement with experiment sufficiently.

A	8	20	34	40	58	92
α_{RPA}	755	1808	2806	3529	4619	7178
α_{exp}	879 ± 17	2138 ± 43	3520 ± 17	4090 ± 77		
α_{LDA}	716	1715	2698	3328	4492	

TABLE I. Static dipole polarizabilities α_{RPA} (a_0^3) of Na clusters ($r_S = 4a_0$) for different sizes A compared to experimental values α_{exp} [10] and to LDA predictions α_{LDA}[14].

The major approximation in the jellium model lies in an arbitrary choice of the positive background distribution: a sharp-edged distribution with an assigned value for r_S. A small variation of the mean jellium background will lead to a change of the mean electronic distribution of the same order; and this in turn will lead to a much larger change of the dipole polarizabilty which scales as the volume of the electronic distribution.

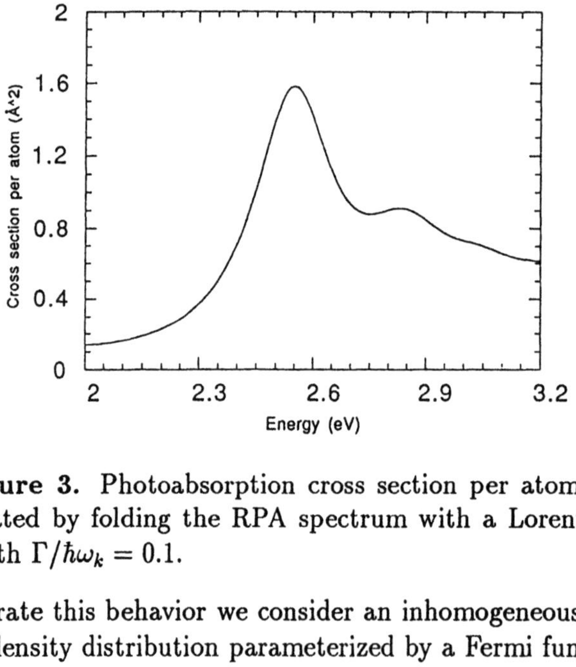

Figure 3. Photoabsorption cross section per atom for Na_{20}^+ calculated by folding the RPA spectrum with a Lorentzian shape of width $\Gamma/\hbar\omega_k = 0.1$.

To illustrate this behavior we consider an inhomogeneous but smooth background density distribution parameterized by a Fermi function:

$$\rho(r) = \frac{\rho_0}{1 + \exp\left[\frac{r-R}{\beta}\right]}. \qquad (13)$$

We assume that the central density, ρ_0 is the bulk value fixed by r_S and we adjust the surface diffuseness parameter β in order to obtain a static polarizability in agreement with the experiment.In case of Na_{20} we find a diffuseness parameter $\beta \approx 0.7a_0$. The corresponding RPA calculation leads to a strong redistribution of the oscillator strength . On one hand the mean resonance frequency is red-shifted due to a larger value of $< r^3 >$ and, on the other hand, the giant dipole oscillation is split into more states. In order to compare to experimental data we follow the prescription of Ref.[7] by folding the oscillator strength distribution with a Lorentzian shape of width $\Gamma/\hbar\omega_k = 0.1$. The resulting photoabsorption cross section is shown on Figure 3.The calculated distribution is in very good agreement with the experimental data of Ref. [4] although an overall energy shift of about 0.15 eV still remains.

In the paragraphs above, we have shown that replacing RPA calculations of the optical response of metallic clusters made using a density-dependent interaction(LDA) by RPA calculations based on the HF theory improves to a certain extent the agreement between calculations in the jellium model and experimental data. However, systematic quantitative differences between theory and experiment remain. More progress in understanding these differences will probably be made by addressing the limitations of the jellium model, rather than by improving the treatment of the finite electron system.

The authors would like to thank S. Blundell, M. Brack, C. Bréchignac, M. Hansen, H. Nishioka and C. Yannouleas for stimulating discussions.

REFERENCES

[1] For a review, see W.A. de Heer, W.D. Knight, M.Y. Chou, and M.L. Cohen, Solid State Physics **40**, 93 (1987).

[2] C. Bréchignac, P. Cahuzac, F. Carlier, and J. Leygnier, Phys. Rev. Lett. **63**, 1368 (1989).

[3] K. Selby, V. Kresin, J. Masui, M. Vollmer, W.A. de Heer, A. Scheidemann, and W.D. Knight, Phys. Rev. B **43**, 4565 (1991).

[4] S. Pollack, C.R.C. Wang, and M.M. Kappes, J. Chem. Phys. **94**, 2496 (1991).

[5] W. Ekart, Phys. Rev. B **31**, 6360 (1985).

[6] D.E. Beck, Phys. Rev. B **35**, 7325 (1987).

[7] C. Yannouleas, R.A. Broglia, M. Brack, and P.F. Bortignon, Phys. Rev. Lett. **63**, 255 (1989).

[8] M. Brack, Phys. Rev. B **39**, 3533 (1989).

[9] G.F. Bertsch, Computer Physics Comm. **60**, 247 (1990).

[10] W.D. Knight, K. Clemenger, W.A. de Heer, and W.A. Saunders, Phys. Rev. B **31**, 445 (1985).

[11] W.R. Johnson, S.A. Blundell and J. Sapirstein, Phys. Rev. A **37**, 307 (1988).

[12] B.S. Garbow, J.M. Boyle, J.J. Dongarra, and C.B. Moler, *Lecture Notes in Computer Science* **51**, (Springer-Verlag,Berlin,1977), p.250.

[13] G.F. Bertsch and D. Tomanek, Phys. Rev. B **40**, 2749 (1989).

[14] P. Stampfli and K.H. Bennemann, in *Physics and Chemistry of Small Clusters*, ed. by P. Jena, B.K. Rao and S.N. Khanna, 473 (1986), and references therein.

Some Considerations on "Isoarithmic" and Isoelectronic Clusters

Wanda Andreoni[1] and Ursula Röthlisberger[1,2]

[1]*IBM Research Division, Zurich Research Laboratory, 8803 Rüschlikon, Switzerland*
[2]*Institute for Inorganic, Physical and Analytical Chemistry, University of Bern, 3000 Bern 9, Switzerland*

We briefly summarize trends in the structural and electronic properties of small clusters of Na, Mg and Si. In particular, the validity of the spherical jellium picture is discussed and the applicability of the concepts used to classify magic and non-magic number clusters in alkali metals is examined for other systems.

Introduction

One of the ultimate goals of the theoretical investigation of small clusters is to provide a simple scheme for the classification of their structural as well as their electronic properties. One of the first steps in this search is the study of general trends, which should lead to the identification of valuable classification parameters. While in the study of solids one usually aims at (and can restrict oneself to) classifying the structural properties of isoelectronic systems [1], in the case of finite-size aggregates we are confronted with two — most of the time — distinct problems, namely the classification of the equilibrium structures of clusters with the same number of atoms (which we define as "isoarithmic") and the classification of the electronic properties of isoelectronic clusters. In either of them, the question arises, "what special characteristics do the magic-number (MN) clusters have?" Clearly, this whole project is quite formidable. A large data base is also required before one can start to build such a general and simple scheme.

In this paper, we shall discuss our findings for a still restricted data base, i.e., some microclusters of Na, Mg, Al and Si, which were obtained with the Car-Parrinello method [2,3].

Clusters with the Same Number of Atoms

In the search for trends in the physical properties of aggregates with the same number of atoms, we have considered the 13-atom clusters of Na, Mg, Al and Si in detail [4]. The low-energy patterns presented indeed no common characteristics, apart from the fact that non-crystalline exotic structures were highly favored over crystalline arrangements. While only pentagonal rings were found for the multitude of quasi-degenerate isomers of Na_{13}, hexagonal rings were also observed in Mg_{13}. The pentagonal motif had indeed been previously identified as a characteristic feature of the structures of sodium microclusters [5]. While for Na, Mg and Si the icosahedron turned out to be highly unstable, for Al_{13} a slight Jahn-Teller induced distortion was able to stabilize an icosahedron-like geometry. The reason for this can be understood in terms of the spherical jellium model, for which 40 electrons correspond to an electronic shell closing. In Al_{13} the number of valence electrons is 39 and thus corresponds almost to the shell filling. In fact from experiments the anion Al_{13}^{-} can be classified as a MN cluster [6]. Si_{13} is also observed to be magic, in the sense that it is particularly unreactive to several molecules [7]. In contrast with previous predictions [8] of an icosahedral structure that could explain this special inertness, we find that it cannot be simply explained in terms of a particularly rigid and compact atomic configuration. The low-energy structures are, however, characterized by a seed unit that is either a trigonal prism or a trigonal antiprism.

This type of trend from Na to Mg to Si also manifests itself in our results for the 10 and 20-atom clusters [5,9,10], i.e., the presence of pentagonal rings in Na, of trigonal prisms as central seeds in Si and the character of Mg, being somehow intermediate between Na and Si. The calculated structures of lowest energy [10] for Mg_{10} and Mg_{20} are illustrated in Fig. 1(a) and (b). Mg_{10} is a tetracapped trigonal prism (TTP) similar to Si_{10}, while Mg_{20} can be described as a three-layered structure with two capping atoms. These geometries are confirmed by other independent calculations that also used the Car-Parrinello method [11,12].

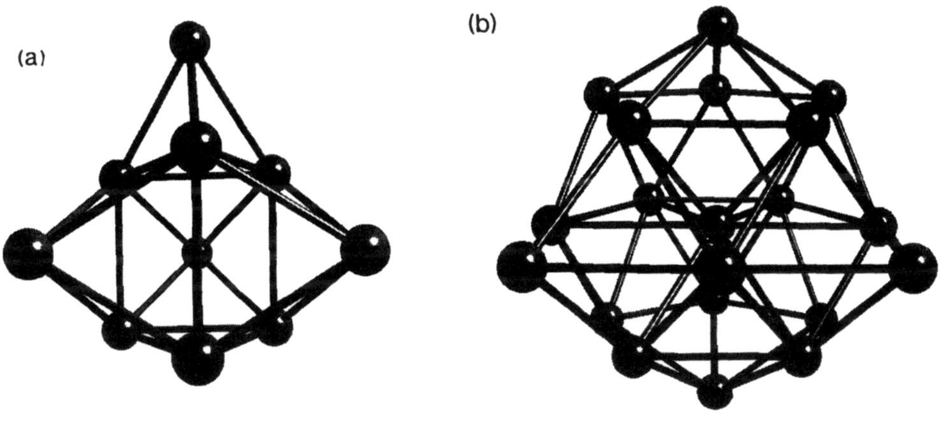

Fig. 1. Lowest-energy structures of (a) Mg_{10} and (b) Mg_{20}.

Clusters with the Same Number of Valence Electrons

In this section we shall consider isoelectronic clusters with 20 and 40 electrons, which correspond to electronic shell closing in the spherical jellium model and to MN clusters for alkali metals. Among the issues of interest, we have considered the following: Is shell closing true for magic-number clusters of elements other than the alkalis? Can we define some – either structural or electronic – simple parameter that distinguishes magic from non-magic number clusters? Is the angular momentum l also a good quantum number for the classification of one-electron states in clusters of elements other than simple metals? To what extent are spherical jellium models valid for Na and Mg clusters and which are the temperature effects?

We have chosen two parameters that give a "global" description of structural and electronic properties, respectively: the eccentricity of the structure η and the spherical-shell-closing deviation parameter E_L. The definitions are as in Ref. 5: $\eta = 1 - I_{min}/I_{av}$, where I_{min} and I_{av} are the minimum and the average moment of inertia; $E_L = 1 - (\sum_{locc} \sum_{l=0}^{l=L} w_l^i)/n_{el}$, where n_{el} is the number of electrons and w_l^i is the weight of the l-component in the expansion of the i-th orbital ψ_i in spherical harmonics centered at the center of the cluster. The sum runs only over the occupied molecular orbitals (OMOs). The calculated values are reported in Table 1.

Table 1. 20-e and 40-e systems. Al_{13} is also included for comparison.

System	η	$E_{L=2}$	$E_{L=3}$
Na_{20}	0.03	0.056	
Mg_{10}	0.05	0.049	
Si_5	0.01	0.088	
P_4	0	0.094	
Mg_{20}	0.051		0.06
Si_{10}	0.066		0.03
Al_{13}	0.027		0.02

As representatives of the 20-e systems we have considered Na_{20} which is an MN cluster, Si_5 which is definitely a non-MN, and Mg_{10}. For the small Mg clusters, abundance spectra are not available so that MNs are not yet experimentally defined. Also, the theoretical situation is still unclear, owing to the interplay of van-der-Waals and chemical bonding in the cohesion of Mg microclusters which the currently available calculations are not able to describe on the same footing. The results we give here are within the LDA scheme. Very recent calculations [12] including gradient corrections to the LDA exchange-correlation functional predict no change in the bonding picture at least in the range of 10 to 20 atoms we consider here. For the 40-e systems, the representatives chosen are Mg_{20} and Si_{10}, the latter being a MN in the abundance spectra.

We recall that the sequence of levels in a spherical jellium model is 1s, 1p, 1d, 2s, 1f, 2p etc., so that in the ideal case $E_{L=2} = 0$ for a 20-e system and $E_{L=3} = 0$ for a 40-e system. Having defined the reference model, small values can be expected only for systems with rather delocalized electron states. As an example, we note that a value $E_{L=2} = 0.05$ for a 20-e system (such as Na_{20}) must be considered small, since it says that only 5% of the spectrum of the electronic states deviates from the "ideal" composition. Looking at the trend, we see that the structural parameter η does not distinguish between MNs and non-MNs, and that the electronic parameter E_L is more appropriate. However, when we extend this analysis to an isoelectronic and especially stable molecule with strongly directional bonds, such as P_4, we find that

$\eta = 0$ (as it is a tetrahedron) and that the HOMO-LUMO gap is a few eV large, but the value of $E_{L=2}$ denotes a relatively high degree of hybridization with orbitals of higher angular momentum (Table 1). Although such an hybridization is generally allowed in the T_d symmetry, we recall for comparison for Na_8 in the stellated tetrahedron structure the OMOs are fully describable as 1s and 1p. Looking at Mg_{20} and Si_{10} in Table 1, we see that — as expected — both parameters are larger than in Al_{13}, which has 39 electrons and an especially symmetric structure. We note in particular that $E_{L=3}$ is amazingly small for Si_{10}, which reflects the fact that the structure is rather compact and the bonding is not as directional as in bulk silicon.

Investigating the electronic structure in greater detail means explicitly inspecting the hybridization of the cluster orbitals ψ_i and the nature of the electronic potential. Our results for the occupied states of the systems mentioned above can be summarized as follows: significant hybridization is present only in the HOMOs for Na_{20}, Mg_{10} and Al_{13}, while it starts in the middle of the OMOs for Si_{10} and Mg_{20}, thus covering a few eV down from the HOMO. For sake of comparison, we find that hybridization is present throughout the spectrum for non-MN clusters, such as Na_{10} and Na_{13} [5] and also for all Si clusters other than Si_{10} — including Si_{13}. Also, we found that in Na_8 this hybridization is essentially absent for the OMOs but is present in the empty states. Na_8 is particularly interesting, since this property remains true for all the low-energy isomers and is only weakly affected at finite temperature [5]. Regarding the structural dependence of the above considerations, we have verified that for Mg_{10} and Na_{20} these are largely independent of the specific low-energy isomer considered. In contrast, for Si_{10} the only other isomer we have considered, the tetracapped octahedron (TO), presents a large hybridization for most of the OMOs. Indeed, in our calculations [9] the TO is by 0.65 eV higher than the TTP, while for Mg_{10} and Na_{20} we find topologically similar and quasi-degenerate isomers.

Regarding the nature of the electronic potential, we have considered it in detail for sodium clusters in Ref. 5. In particular, in the case of Na_{20} we have found that the spherical component is dominant and that the deviation from spherical symmetry is spatially localized. Indeed, this is particularly high in those regions in which capping atoms are located and both the atomic and the electronic densities are relatively high and anisotropic. This is reflected, for instance, in the fact that the two highest OMOs present significant hybridization. Looking at the radial variation of the spherical component [here reported in Fig. 2(a)] and decomposing it into the pseudopotential, electrostatic and exchange-correlation contributions, we learn that a strong cancellation occurs between the first two terms, which are more strongly and directly dependent on the atomic structure. The same analysis made for Na_{10} [Fig. 2 (b)] and Mg_{10} (Fig. 3) shows that such a cancellation is much less effective. In Figs. 2 and 3 we also compare the Wood-Saxon (WS) potential with our LDA potential. The parameters for Na are taken from Ref. 13 and for Mg we follow the same prescriptions. The WS potential is a very good approximation for Na_{20} beyond 4 a.u.; it corresponds to somehow averaging out the oscillations in Na_{10}, but is

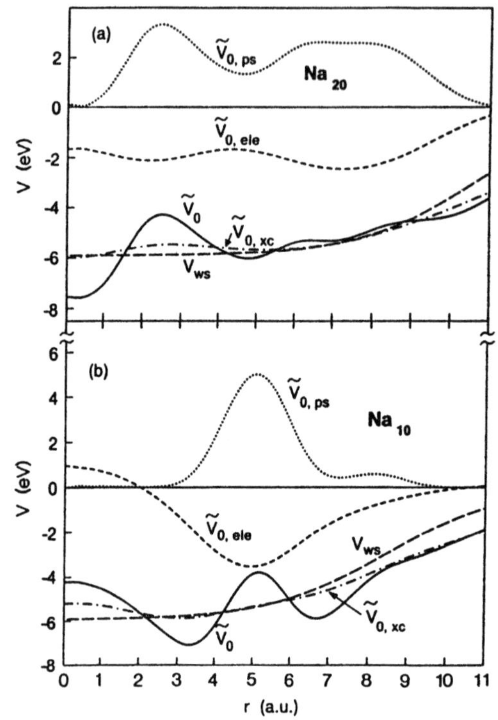

Fig. 2. Spherical component of the local electronic potential and its contributions (see text). V_{ws} is the Wood-Saxon model. From [5].

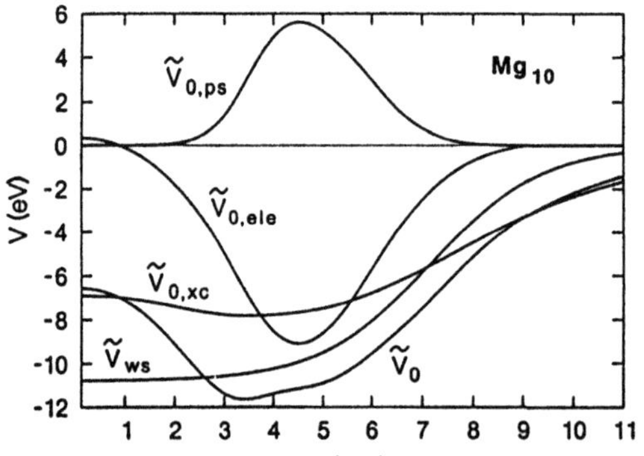

Fig. 3. Same as Fig. 2 for Mg_{10}.

definitely too high and too short-ranged for Mg_{10}. In Mg_{10}, one can indeed obtain a better agreement beyond 3 a.u. by increasing the Wigner-Seitz radius parameter. We note, however, that the transferability of the Wood-Saxon potential with fixed parameters is good from Na_8 to Na_{20}, but rather poor from Mg_{10} to Mg_{20}.

Finally we remark that temperature effects for the small Na clusters tend towards increasing the hybridization of the OMOs and, correspondingly, towards decreasing the HOMO-LUMO gaps. However, this does not imply that the magic-number clusters defined by the spherical-shell closing lose meaning. As mentioned above, Na_8 keeps a very small value of $E_{L=1}$ at least up to 500 K, and non-magic number clusters such as Na_{10} and Na_{13} are found to be even more affected by temperature, both in the electronic spectrum and in the structural properties. In particular, they easily transform to non-rigid systems. For a detailed study, the reader is referred to Ref. 5.

Conclusions

The considerations made above must be regarded only as first steps of a program aimed at finding a classification scheme for either "isoarithmic" or isoelectronic clusters. Using the spherical jellium model for the classification of the electronic states of small clusters is analogous to using the free electron model as basis paradigm for the energy bands in solids. As such, its application is wider than for alkali metals but still limited. In considering the criteria leading to MNs, we see that in silicon the key criterion has not yet been identified, in spite of the fact that a correlation exists between the calculated fragmentation energies, the HOMO-LUMO gaps and the appearance of peaks in the abundance spectra. Inspection of the electronic structure and the preferred geometry indicate that Si_{10} is somewhat special among the silicon microclusters. We also notice that in Na, Mg and Si there is no preference for highly symmetric structures, even in the case of MNs.

An example of a totally different system is C_{60}, the magic number *par excellence.* Here the structure is the key factor for the stability [14], the symmetry is very high and the system also corresponds to a closing of electronic shells. The electronic potential is dominantly spherical, as witnessed by the fact that most of the OMOs keep a dominant *l*-character. The electron density is localized on the atomic shell and the ordering of the energy levels is of the type 1s, 1p, 1d, 1f, 1g, 1h etc. This hints at a new paradigm for the classification of the electron states in fullerenes.

References

1. See, e.g., W. Andreoni, G. Galli: Phys. Chem. Minerals **14**, 389 (1987) and references therein.
2. R. Car, M. Parrinello: Phys. Rev. Lett. **55**, 2471 (1985)
3. For recent reviews on the application to clusters, see R. O. Jones: Angew. Chem. **103**, 647 (1991); W. Andreoni, Z. Phys. D **19**, 31 (1991); V. Kumar: *Proc. 8th Natl. Workshop on Atomic and Molecular Physics*, Hyderabad, Dec. 1990, ed. A.P. Pathak (Nova Publication)
4. U. Röthlisberger, W. Andreoni, P. Giannozzi: J. Chem. Phys. (in press)
5. U. Röthlisberger, W. Andreoni: J. Chem. Phys. **94**, 8129 (1991)
6. R.I. Leuchtner, A.C. Harms, A.W. Castelman Jr.: J. Chem. Phys. **94**, 1093 (1991); A. Nakajima *et al.*: Chem. Phys. Lett. **177**, 297 (1991)
7. M.F. Jarrold, J.E. Bower, K.M Creegan: J. Chem. Phys. **90**, 3615 (1989); M.F. Jarrold, U. Ray, K.M. Creegan: *ibid.* **93**, 224 (1990); U. Ray, M.F. Jarrold: *ibid.* **94**, 2631 (1991)
8. J.R. Chelikowsky, J.C. Phillips: Phys. Rev. Lett. **63**, 1653 (1989)
9. W. Andreoni, G. Pastore: Phys. Rev. B **41**, 10243 (1990)
10. U. Röthlisberger, W. Andreoni: To be published
11. V. Kumar, R. Car: Phys. Rev. B **44**, 8243 (1991); V. De Coulon *et al.*, Z. Phys. D **19**, 173 (1991)
12. V. de Coulon, P. Ballone, J. Buttet: Preprint
13. W.D. Knight *et al.*: Phys. Rev. Lett. **52**, 2141 (1984)
14. H. Kroto: Science **242**, 1139 (1988)

MAGNETIC PROPERTIES OF TRANSITION- AND RARE-EARTH METAL CLUSTERS

P. Jensen[1], *G. Pastor*[2], *and K.H. Bennemann*[1]

[1] Institute for Theoretical Physics, Freie Universität Berlin,
W-1000 Berlin 33, Germany

[2] Universität zu Köln, Theoretische Festkörperphysik, Zülpicher Str. 77,
W-5000 Köln 41, Germany

Abstract: The magnetic properties ($\mu_i(n), T_c(n)$, spin-order) of transition metal clusters like Fe_n, Co_n, Cr_n, and of rare-earth metal clusters like Gd_n, Tb_n, are discussed, including the behaviour of an ensemble of clusters in an external magnetic field.

Recently, Callaway et al.[1] and in more detail Pastor et al.[2] studied the ground state magnetic properties like local magnetic moment $\mu_i(n)$ at site i in a cluster of n-atoms, the spin-order in such clusters, and also the size-dependence of the Curie temperature[3], $T_c(n)$. This was motivated by the following interesting physical questions: what is the interdependence of lattice structure and magnetism, s. f.c.c. vs. b.c.c. Ni_n, etc.; how changes local spin behaviour to itinerant behaviour in Ni_n, Fe_n etc., as a function of cluster size; how is magnetic frustration handled in small Cr_n, Mn_n clusters (via distortion?); how important is magnetic anisotropy in small clusters? Using an electronic theory, Hubbard tight-binding-like Hamiltonian, one obtains[2] for Fe_n, Cr_n, Ni_n, etc., that $\mu_i(n)$ depends typically on cluster site i and size n, but

$$\mu_i(n) \gtrsim \mu_b, \tag{1}$$

where μ_b refers to the bulk magnetic moment. One obtains furthermore, that in Ni_n and Fe_n magnetism depends sensitively on the atomic structure, that spin order occurs already in small clusters and that $T_c(n)$ is smaller in clusters[3]. Corresponding to the situation at surfaces, the magnetic anisotropy $E_{\text{anis}} \sim K_2 \cos^2\vartheta + \cdots$, is expected to be larger in clusters than in bulk. Results on the spin-dependent electronic density of states, $N_{i\sigma}(\varepsilon)$, indicated many cluster specific interesting features, reflecting many body effects and in particular magnetism, s. for example, $N_{i\sigma}(\varepsilon, n)$ for Cr_n.[2].

In view of these results, the very recently observed Stern-Gerlach deflection of a cluster ensemble in an inhomogeneous magnetic field B by de Heer et al.[4] for Fe_n and by Bucher et al.[5] for Co_n, Tb_n, etc., seemed very puzzling. Note that an asymmetric deflection profile and a low magnetization $\langle \mu_z(B) \rangle$ is observed. For Fe_n and Co_n this low magnetization can be explained by a superparamagnetic model[3,5] (Langevin behaviour), whereas for Gd_n (and presumably also for Tb_n) a strong deviation from the Langevin behaviour is obtained. For the latter clusters, at first sight, the experimental results seem to suggest $\mu_i(n) < \mu_b$ as being responsible for the low magnetization[3,4,5], which is not expected physically.

In the following we use our results obtained for $\mu_{i(n)}$ and $T_c(n)$ to explain in particular the experimental results observed for cluster ensembles. The control parameters of our theory, details can be found in ref. 3, are (a) T - the cluster temperature, (b) T_c - the spin order critical temperature, (c) K_2 - the magnetic anisotropy, and (d) the blocking temperature T_{bl}, resulting, for example, from an energy barrier due to magnetic anisotropy[6], which hinders the combined total cluster magnetic moment $\vec{\mu}_n$ to align in the inhomogeneous magnetic field B. For $T < T_c$ the clusters behave like superparamagnets. It is assumed[4,5] that before entering the Stern-Gerlach magnet the clusters rotate rapidly and that $\vec{\mu}_n$ is tied to a crystal axis due to magnetic anisotropy, thus $\vec{\mu}_n$ rotates with the cluster.

Results of our physical model for the magnetic behaviour of an ensemble of transition-metal and rare-earth clusters, which in general have a much stronger magnetic anisotropy, are the following: First, due to the possibility of transferring angular momentum to other internal degrees of freedom,

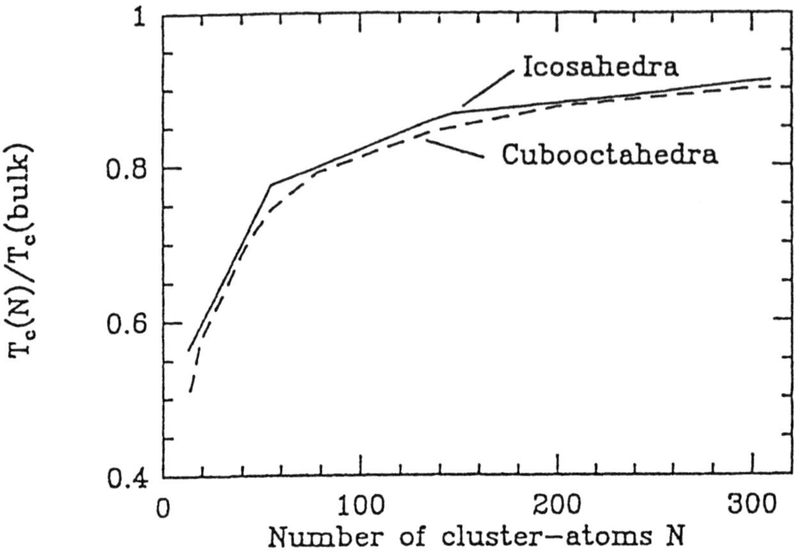

Fig. 1: Calculated Curie-temperature $T_c(n)$ as a function of the number n of atoms for icosahedral and cubooctahedral clusters.

asymmetric deflection occurs for clusters larger than a critical size n_c ($n_c \sim 3 \rightarrow 4, \cdots$), since energy gain by aligning $\vec{\mu}_n$ along $\vec{B}$ is possible and not prevented by angular momentum conservation like in atoms. Secondly, to assure that $T < T_c$, $T \sim 200 \div 300$K typically[4,5], we show molecular field type results for $T_c(n)$ in Fig. 1. Note that $T_c \approx \frac{2}{3}\, T_c^{bulk}$ for $n \sim 50$. Results for $\langle \mu_z(B) \rangle$ are shown in Figs. 2 and 3. In these two figures we assumed that the cluster rotation does not prevent the cluster magnetization to reach its equilibrium value. Note the significance of T vs. T_{bl} and of K_2: $\langle \mu_z(B) \rangle$ decreases for increasing K_2 as physically expected.

Thus, $\langle \mu_z \rangle$ is expected to be much smaller for Tb_n than for Co_n in the case $T < T_{bl}$, for example. $\langle \mu_z \rangle$ should increase with n, since T_c increases, and since possibly the atomic structure changes (s. $Ni_n \xrightarrow[n]{} f.c.c.$, etc). The comparison with experimental results[4,5] in Fig. 3 indicates discrepancies. These are presently unclear, but might have an interesting physical origin (resulting from the cluster rotation with frequency ω_{rot}). Note, if $\omega_{rot} > \omega_{rel}$, $\omega_{rel} \approx \omega_\nu \cdot \exp(-nK_2/T)$, where ω_ν denotes the gyromagnetic

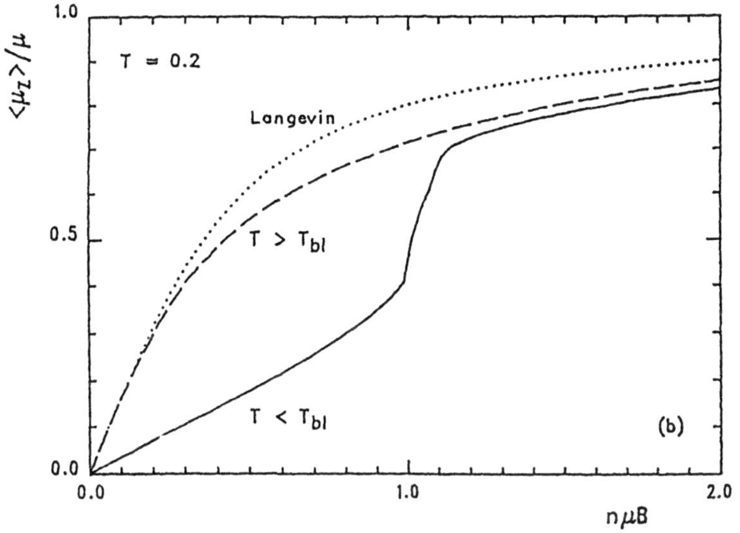

Fig. 2: Comparison of $\langle \mu_z \rangle / \mu$ for the two limiting cases $T < T_{bl}$ and $T > T_{bl}$ at $nK_2 = 1.0$ and $T = 0.2$. The dotted curve refers to the Langevin function obtained for $T > T_{bl}$ and $K_2 = 0$.

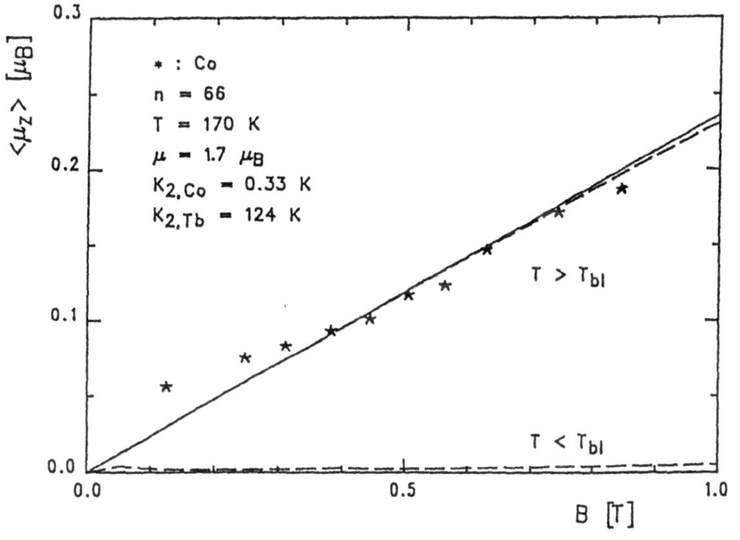

Fig. 3: $\langle \mu_z \rangle$ vs. applied magnetic field B for Co_n- (full lines) and Tb_n-clusters (dashed lines) with n=66 and the cluster temperature $T = 170$K. For the anisotropy energy K_2 we take the bulk values (Co: 0.33 K; Tb: 124 K), and the cluster Curie temperature $T_c = \frac{2}{3} T_c^{bulk}$. The two limiting cases $T > T_{bl}$ (superparamagnetic case in comparison with $T_c \to \infty$) and $T < T_{bl}$ are depicted for Tb. For a better comparison of the theoretical results we choose for the atomic magnetic moment of Tb the Co- moment: $\mu_{Co} = 1.7\mu_B$. The parameters n and K_2 indicate that the case $T > T_{bl}$ is appropriate for Co and the case $T < T_{bl}$ for Tb. The symbols depict the experimental results for Co_{66}- clusters[5].

precession frequency, one expects $\langle\mu_z\rangle \to 0$, since the moments have no time to relax in the direction of $\vec{B}$. From the standpoint of the rotating cluster, its magnetic moment experiences a fluctuating magnetic field $\vec{B}(\omega_{rot})$.

Under certain conditions the situation of a rotating magnetic cluster in a static magnetic field B corresponds to the one of a non-rotating cluster suspended, for example, in a liquid, and experiencing a fluctuating magnetic field[7] $B(\omega)$. In Fig. 4 we show results for this latter case.

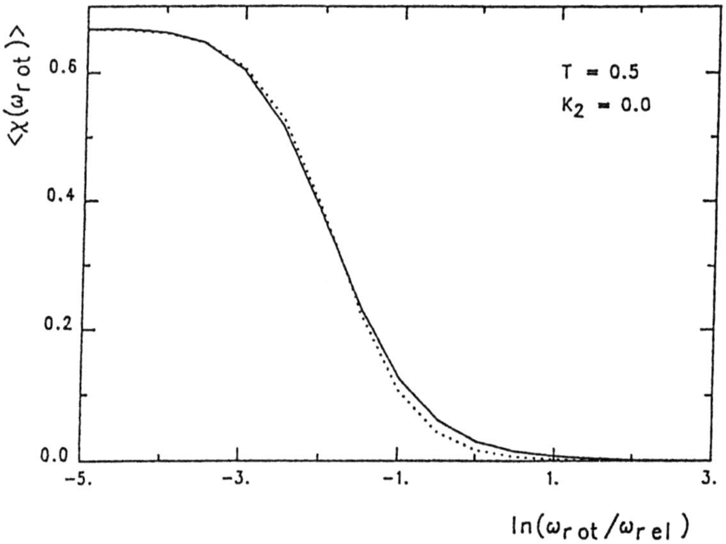

Fig. 4: Frequency dependent magnetic susceptibility $\chi(\omega_{rot})$ vs. $\omega_{rot}/\omega_{rel}$ for $T = 0.5\ \mu B$ ($B = 1$ T) and $K_2 = 0$ for superparamagnetic clusters in an oscillating external magnetic field $\vec{B}(\omega_{rot})$ as obtained from a linear relaxation model: $\dot{\mu}_z(t) \propto \mu_z(t) - \langle\mu_z\rangle$. $\tau_{rel} = 2\pi/\omega_{rel}$ is the characteristic relaxation time of the clusters (full line). For comparison the Debye-type susceptibility $\chi(\omega_{rot}) = \chi_o(1 + (\omega_{rot} \cdot \tau_{rel})^2)^{-1}$ is shown (dotted line).

In summary, we have studied the magnetic behaviour of clusters. The magnitude of the magnetic moments $\mu_i(n)$ depends on its position inside the cluster, the cluster size n, and its atomic structure, s. f.c.c. vs. b.c.c. like Ni_n, Fe_n, etc. We obtained $\mu_i(n) \gtrsim \mu_b$. Short range spin order occurs already in small clusters, ferromagnetic like in Ni, Fe, Co and antiferromagnetic like in Cr, Mn. Itinerancy increases with size in transition metal clusters, and magnetic anisotropy is larger for rare-earth metal clusters than for transition metal clusters. For a cluster ensemble the magnetization $\langle\mu_z(B,T,n)\rangle$ exhibits the following properties: The cluster ensemble behaves as a superparamagnet (Langevin system), if $T < T_C$, and $T > T_{bl}$, and $\omega_{rot} < \omega_{rel}$. In the other cases $\langle\mu_z\rangle$ can be much lower due to e.g. an increasing anisotropy energy K_2 (Fe $\rightarrow$ Co $\rightarrow$ Tb), causing $T < T_{bl}$. Furthermore, $\langle\mu_z\rangle$ decreases with temperature and increases with n, if $T < T_C$.

References

1. K. Lee, J. Callaway, K. Kwong, R. Tang, and A. Ziegler, Phys. Rev. B 31, 1796 (1985)
2. G.M. Pastor, J. Dorantes-Dávila, and K.H. Bennemann, Sol. St. Comm. 60, 465 (1986); Phys. Rev. B 40, 7642 (1989)
3. P.J. Jensen, S. Mukherjee, and K.H. Bennemann, Z. Phys. D 21, 349 (1991) and Proceedings Intern. Symposium on the Physics and Chemistry of Finite Systems: From Clusters to Crystals, Oct. 1991, Richmond/USA
4. W.A. de Heer, P. Milani, and A. Châtelain, Phys. Rev. Lett. 65, 488 (1990)
5. J.P. Bucher, D.C. Douglass, and L.A. Bloomfield, Phys. Rev. Lett. 66, 3052 (1991); S.N. Khanna and S. Linderoth, Phys. Rev. Lett. 67, 742 (1991)
6. I.S. Jacobs and C.P. Bean, in: *Magnetism*, edited by G.T. Rado and H. Suhl (Academic Press, NY., 1963), Vol. III
7. M. Hanson, J. Magn. Magn. Mater., 96, 105 (1991), and private communication (1991)

Lecture Notes in Physics

For information about Vols. 1–374
please contact your bookseller or Springer-Verlag

Vol. 375: C. Bartocci, U. Bruzzo, R. Cianci (Eds.), Differential Geometric Methods in Theoretical Physics. Proceedings, 1990. XIX, 401 pages. 1991.

Vol. 376: D. Berényi, G. Hock (Eds.), High-Energy Ion-Atom Collisions. Proceedings, 1990. IX, 364 pages. 1991.

Vol. 377: W. J. Duschl, S.J. Wagner, M. Camenzind (Eds.), Variability of Active Galaxies. Proceedings, 1990. XII, 312 pages. 1991.

Vol. 378: C. Bendjaballah, 0. Hirota, S. Reynaud (Eds.), Quantum Aspects of Optical Communications. Proceedings 1990. VII, 389 pages. 1991.

Vol. 379: J. D. Hennig, W. Lücke, J. Tolar (Eds.), Differential Geometry, Group Representations, and Quantization. XI, 280 pages. 1991.

Vol. 380: I. Tuominen, D. Moss, G. Rüdiger (Eds.), The Sun and Cool Stars: activity, magnetism, dynamos. Proceedings, 1990. X, 530 pages. 1991.

Vol. 381: J. Casas-Vazquez, D. Jou (Eds.), Rheological Modelling: Thermodynamical and Statistical Approaches. Proceedings, 1990. VII, 378 pages. 1991.

Vol. 382: V.V. Dodonov, V. I. Man'ko (Eds.), Group Theoretical Methods in Physics. Proceedings, 1990. XVII, 601 pages. 1991.

Vol. 384: M. D. Smooke (Ed.), Reduced Kinetic Mechanisms and Asymptotic Approximations for Methane-Air Flames. V, 245 pages. 1991.

Vol. 385: A. Treves, G. C. Perola, L. Stella (Eds.), Iron Line Diagnostics in X-Ray Sources. Proceedings, Como, Italy 1990. IX, 312 pages. 1991.

Vol. 386: G. Pétré, A. Sanfeld (Eds.), Capillarity Today. Proceedings, Belgium 1990. XI, 384 pages. 1991.

Vol. 387: Y. Uchida, R. C. Canfield, T. Watanabe, E. Hiei (Eds.), Flare Physics in Solar Activity Maximum 22. Proceedings, 1990. X, 360 pages. 1991.

Vol. 388: D. Gough, J. Toomre (Eds.), Challenges to Theories of the Structure of Moderate-Mass Stars. Proceedings, 1990. VII, 414 pages. 1991.

Vol. 389: J. C. Miller, R. F. Haglund (Eds.), Laser Ablation-Mechanisms and Applications. Proceedings. IX, 362 pages, 1991.

Vol. 390: J. Heidmann, M. J. Klein (Eds.), Bioastronomy - The Search for Extraterrestrial Life. Proceedings, 1990. XVII, 413 pages. 1991.

Vol. 391: A. Zdziarski, M. Sikora (Eds.), Ralativistic Hadrons in Cosmic Compact Objects. Proceedings, 1990. XII, 182 pages. 1991.

Vol. 392: J.-D. Fournier, P.-L. Sulem (Eds.), Large-Scale Structures in Nonlinear Physics. Proceedings. VIII, 353 pages. 1991.

Vol. 393: M. Remoissenet, M.Peyrard (Eds.), Nonlinear Coherent Structures in Physics and Biology. Proceedings. XII, 398 pages. 1991.

Vol. 394: M. R. J. Hoch, R. H. Lemmer (Eds.), Low Temperature Physics. Proceedings. XXX,XXX pages. 1991.

Vol. 395: H. E. Trease, M. J. Fritts, W. P. Crowley (Eds.), Advances in the Free-Lagrange Method. Proceedings, 1990. XI, 327 pages. 1991.

Vol. 396: H. Mitter, H. Gausterer (Eds.), Recent Aspects of Quantum Fields. Proceedings. XIII, 332 pages. 1991.

Vol. 398: T. M. M. Verheggen (Ed.), Numerical Methods for the Simulation of Multi-Phase and Complex Flow. Proceedings, 1990. VI, 153 pages. 1992.

Vol. 399: Z. Švestka, B. V. Jackson, M. E. Machedo (Eds.), Eruptive Solar Flares. Proceedings, 1991. XIV, 409 pages. 1992.

Vol. 400: M. Dienes, M. Month, S. Turner (Eds.), Frontiers of Particle Beams: Intensity Limitations. Proceedings, 1990. IX, 610 pages. 1992.

Vol. 401: U. Heber, C. S. Jeffery (Eds.), The Atmospheres of Early-Type Stars. Proceedings, 1991. XIX, 450 pages. 1992.

Vol. 402: L. Boi, D. Flament, J.-M. Salanskis (Eds.), 1830-1930: A Century of Geometry. VIII, 304 pages. 1992.

Vol. 403: E. Balslev (Ed.), Schrödinger Operators. Proceedings, 1991. VIII, 264 pages. 1992.

Vol. 404: R. Schmidt, H. O. Lutz, R. Dreizler (Eds.), Nuclear Physics Concepts in the Study of Atomic Cluster Physics. Proceedings, 1991. XVIII, 363 pages. 1992.

New Series m: Monographs

Vol. m 1: H. Hora, Plasmas at High Temperature and Density. VIII, 442 pages. 1991.

Vol. m 2: P. Busch, P. J. Lahti, P. Mittelstaedt, The Quantum Theory of Measurement. XIII, 165 pages. 1991.

Vol. m 3: A. Heck, J. M. Perdang (Eds.), Applying Fractals in Astronomy. IX, 210 pages. 1991.

Vol. m 4: R. K. Zeytounian, Mécanique des fluides fondamentale. XV, 615 pages, 1991.

Vol. m 5: R. K. Zeytounian, Meteorological Fluid Dynamics. XI, 346 pages. 1991.

Vol. m 6: N. M. J. Woodhouse, Special Relativity. VIII, 86 pages. 1992.

Vol. m 7: G. Morandi, The Role of Topology in Classical and Quantum Physics. XIII, 239 pages. 1992.

Vol. m 8: D. Funaro, Polynomial Approximation of Differential Equations. X, 305 pages. 1992.

Vol. m 9: M. Namiki, Stochastic Quantization. X, 217 pages. 1992.

Vol. m 10: J. Hoppe, Lectures on Integrable Systems. VII, 111 pages. 1992.

GPSR Compliance

The European Union's (EU) General Product Safety Regulation (GPSR) is a set of rules that requires consumer products to be safe and our obligations to ensure this.

If you have any concerns about our products, you can contact us on ProductSafety@springernature.com

In case Publisher is established outside the EU, the EU authorized representative is:

Springer Nature Customer Service Center GmbH
Europaplatz 3
69115 Heidelberg, Germany

Batch number: 10838062

Printed by Printforce, the Netherlands